PORTS AND NETWORKS

Written by leading experts in the field, this book offers an introduction to recent developments in port and hinterland strategies, operations and related specializations. The book begins with a broad overview of port definitions, concepts and the role of ports in global supply chains, and an examination of strategic topics such as port management, governance, performance, hinterlands and the port-city relationship. The second part of the book examines operational aspects of ports, and maritime and land networks. A range of topics are explored, such as liner networks, finance and business models, port-industrial clusters, container terminals, intermodality/synchromodality, and handling. The final section of the book provides insights into key issues of port development and management, from security, sustainability, innovation strategies, transition management and labour issues.

Drawing on a variety of global case studies, theoretical insights are supplemented with real world and best practice examples. This book will be of interest to advanced undergraduates, postgraduates, scholars and professionals interested in maritime studies, transport studies, economics and geography.

Harry Geerlings is Professor in Governance of Sustainable Mobility at the Erasmus School of Social and Behavioural Sciences (ESSB) of the Erasmus University Rotterdam, The Netherlands.

Bart Kuipers is Senior Researcher at the Erasmus School of Economics (ESE) at the Urban, Port and Transport Economics Department of the Erasmus University Rotterdam, The Netherlands.

Rob Zuidwijk is Professor of Ports in Global Networks at the Rotterdam School of Management (RSM), Erasmus University Rotterdam, The Netherlands.

PORTS AND NETWORKS

Strategies, Operations and Perspectives

Edited by Harry Geerlings, Bart Kuipers and Rob Zuidwijk

Routledge
Taylor & Francis Group

LONDON AND NEW YORK

First published 2018
by Routledge
2 Park Square, Milton Park, Abingdon, Oxon OX14 4RN

and by Routledge
605 Third Avenue, New York, NY 10017

Routledge is an imprint of the Taylor & Francis Group, an informa business

British Library Cataloguing-in-Publication Data
A catalogue record for this book is available from the British Library

Library of Congress Cataloging-in-Publication Data
Names: Geerlings, H., editor. | Kuipers, Bartholdt, 1959– editor. |
    Zuidwijk, Rob, editor.
Title: Ports and networks : strategies, operations and
    perspectives / edited by Harry Geerlings, Bart Kuipers and Rob Zuidwijk.
Description: Abingdon, Oxon ; New York, NY : Routledge, 2018. |
    Includes bibliographical references and index.
Identifiers: LCCN 2017011733 | ISBN 9781472485007 (hbk) |
    ISBN 9781472485038 (pbk) | ISBN 9781315601540 (ebk)
Subjects: LCSH: Harbors. | Harbors—Management. | Shipping. |
    Sustainable development. | Globalization.
Classification: LCC HE551 .P7587 2018 | DDC 387.1068—dc23
LC record available at https://lccn.loc.gov/2017011733

ISBN: 978-1-4724-8500-7 (hbk)
ISBN: 978-1-4724-8503-8 (pbk)
ISBN: 978-1-315-60154-0 (ebk)

Typeset in Bembo
by Apex CoVantage, LLC

English editing and formatting by Sharon N. Welsh

Cover photos © Mr Siebe Swart; use courtesy of The Port of Rotterdam Authority

The editing of this book has been financially supported by the interuniversity Leiden Delft Erasmus (LDE) – Center of Metropolis and Mainport.

CONTENTS

PART 3
Ports and networks: perspectives 283

FIGURES

TABLES

BOXES

FORMULAE

CONTRIBUTORS

Behzad Behdani is Assistant Professor in Logistics and Operations Research at Wageningen University and Research Centre, The Netherlands.

Frans A. J. van den Bosch is Professor of Management Interfaces between Firms and Business Environment at the Department of Strategic Management and Entrepreneurship, Rotterdam School of Management (RSM), Erasmus University Rotterdam, The Netherlands.

Héctor J. Carlo is Associate Professor in the Industrial Engineering Department at University of Puerto Rico–Mayagüez (UPRM), Puerto Rico.

Francesco Corman is Assistant Professor at the Institute for Transport Planning and Systems of ETH in Zürich, Switzerland.

Rommert Dekker is Professor of Quantitative Logistics and Operations Research at the Erasmus School of Economics (ESE) at the Erasmus University Rotterdam, The Netherlands.

Harry Geerlings is (Port) Professor in Governance of Sustainable Mobility at the Erasmus School of Social and Behavioural Sciences (ESSB) of the Erasmus University Rotterdam, The Netherlands.

Arie Gelderblom is Senior Researcher at SEOR, a research institute associated with the Erasmus University Rotterdam, The Netherlands.

Changqian Guan is Professor in Intermodal Freight Transportation at the Department of Marine Transportation of the US Merchant Marine Academy, New York, USA.

Hercules Haralambides is Professor of Maritime Economics and Logistics at the Erasmus University Rotterdam, The Netherlands.

Leonard Heilig is a PhD candidate at the Institute of Information Systems of the University of Hamburg, Germany.

Rick M. A. Hollen is Managing Partner at Erasmus Centre for Business Innovation, and Research Associate at the Department of Strategic Management & Entrepreneurship of the Rotterdam School of Management (RSM), Erasmus University Rotterdam, The Netherlands.

Martijn van der Horst is Senior Researcher and Lecturer in Port Economics at the Urban, Port and Transport Economics Department of the Erasmus University Rotterdam, The Netherlands.

Hein A. M. Klemann is a Professor in Economic History at the Erasmus University Rotterdam, The Netherlands.

Jaap de Koning is a Professor of Labour Market Policy at the Erasmus School of Economics (ESE) at the Erasmus University Rotterdam, The Netherlands.

Rob Konings is Senior Researcher at the Faculty of Architecture of the Delft University of Technology, The Netherlands.

Bart Kuipers is Senior Researcher at the Erasmus School of Economics (ESE) at the Urban, Port and Transport Economics Department of the Erasmus University Rotterdam, The Netherlands.

Peter de Langen is Owner and Principal Consultant of Ports & Logistics Advisory based in Spain and Visiting Professor at the Copenhagen Business School, Denmark.

Derk Loorbach is Director of DRIFT and Professor of Socio-economic Transitions at the Faculty of Social Science, both at the Erasmus University Rotterdam, The Netherlands.

Larissa van der Lugt is Senior Researcher and Project Coordinator of the Urban, Port and Transport Economics Department at the Erasmus University Rotterdam, The Netherlands.

Olaf Merk is Administrator of Ports and Shipping at the International Transport Forum (ITF) of the Organisation for Economic Co-operation and Development (OECD) in Paris, France.

Judith Mulder is a Postdoctoral Researcher at the Erasmus School of Economics (ESE) of the Erasmus University Rotterdam, The Netherlands.

Rudy R. Negenborn is Associate Professor at the Department of Maritime and Transport Technology of the Delft University of Technology, The Netherlands.

Michiel Nijdam is Corporate Strategist at the Port of Rotterdam Authority, The Netherlands.

Kees Jan Roodbergen is Professor of Quantitative Logistics at the University of Groningen, The Netherlands.

Christa Sys is Professor of Transport, Logistics and Ports at the Department of Transport and Regional Economic at the Centre for Maritime and Air Transport (C-MAT) of the University of Antwerp, the University of Ghent and at the University College Ghent, Belgium.

Lóránt Tavasszy is Professor in Freight Transport and Logistics at the Delft University of Technology, The Netherlands.

Thierry Vanelslander is tenure track Professor at the Department of Transport and Regional Economics of the University of Antwerp, Belgium.

Albert Veenstra is Professor of International Trade Facilitation and Logistics at the Technical University of Eindhoven and Scientific Director of the Dutch Institute for Advanced Logistics (Dinalog), The Netherlands.

Tiedo Vellinga is Professor of Ports and Waterways at the Faculty of Civil Engineering and Geosciences of the Delft University of Technology, The Netherlands.

Iris F. A. Vis is Professor of Industrial Engineering at the Faculty of Economics and Business of the University of Groningen, The Netherlands.

Henk W. Volberda is Professor of Strategic Management and Business Policy and Director Knowledge Transfer at the Rotterdam School of Management (RSM), Erasmus University Rotterdam, The Netherlands.

Stefan Voß is Professor and Director of the Institute of Information Systems at the University of Hamburg, Germany.

Bart Wiegmans is Senior Researcher at the Faculty of Civil Engineering of the Delft University of Technology, The Netherlands.

Shmuel (Sam) Yahalom is Distinguished Service Professor at The State University of New York and Professor of Economics and Transportation in the Graduate Program of International Transportation Management in SUNY Maritime College, USA.

Kees Zandvliet is Senior Researcher at SEOR at the Erasmus University Rotterdam, The Netherlands.

Rob Zuidwijk is Professor of Ports in Global Networks at Rotterdam School of Management (RSM), Erasmus University Rotterdam, The Netherlands.

ABBREVIATIONS

ABP2020	Algeciras BrainPort 2020
AEO	Authorised Economic Operator
AGV	Automated guided vehicle
AIS	Automatic identification systems
ALV	Automated lifting vehicle
API	Application programming interface
APS	Advanced planning and scheduling
ASC	Automated stacking crane
ATC	Automated transfer crane
AVI	Automatic vehicle identification
BAP	Berth allocation problem
BAU	Business-as-usual
BIMCO	Baltic and International Maritime Council
BOO/BOT	Builds, (owns) and operates
CAAP	Clean Air Action Plan
CARB	California Air Resources Board
CCS	Carbon capture and storage
CFC	Chlorofluorocarbon
CH_4	Methane
CITOS	Computer Integrated Terminal Operations System
C-MAT	Centre for Maritime and Air Transport
CO_2	Carbon dioxide
COP21	Conference of Parties #21
CSR	Corporate social responsibility
DGPS	Differential GPS
DPM	Diesel particulate matter
ECA	Emission control area

ECT	Europe Container Terminals
EDI	Electronic data interchange
EEDI	Energy Efficiency Design Index
EIS	Efficiency Incentive Scheme
EMAS	European Union's Eco-Management and Audit Scheme
EMS	Environmental management systems
EPA	Environmental Protection Agency
ERP	Enterprise resource planning
ESI	Environmental Ship Index
ESPO	European Sea Ports Organisation
EU	European Union
EUR	Erasmus University Rotterdam
EURECA	Effective use of reefer containers for conditioned products through the port of Rotterdam
FCD	Floating car data
FCL	Full container load
FDCA	2,5-Furandicarboxylic acid
FDE	Foreign direct investment
FOB	Free on board
GCI	Global Competitiveness Index
GDP	Gross domestic product
GHG	Greenhouse gas
GNP	Gross national product
GPS	Global positioning system
GRI	Global Reporting Initiative
GTO	Global terminal operators
HC	Hydrocarbon
HFC	Hydrofluorocarbon
HPA	Hamburg Port Authority
HPH	Hutchinson Port Holdings
IAME	International Association of Maritime Economists
IAPH	International Association of Ports and Harbours
ILO	International Labour Organization
IMO	International Maritime Organization
IoT	Internet of Things
IPCSA	International Port Community Systems Association
ISO	International Organization for Standardization
IT	Information technology
ITF	International Transport Forum
ITS	Intelligent transport systems
ITT	Inter-terminal transportation
ITTRP	Inter-terminal truck routing problem
IUCN	International Union for Conservation of Nature and National Resources

KPI	Key performance indicators
LBS	Location-based services
LCL	Less than container load
LIVRA	Logistical Chain Information Waterways Rotterdam-Antwerp
MARPOL	International Convention for the Prevention of Pollution from Ships
MEL	Maritime economics and logistics
MEPC	Marine Environment Protection Committee
MIDA	Maritime industrial development area
NCMS	National Center for Manufacturing Sciences
NDRC	National Development and Reform Commission
NGO	Non-governmental organisation
NIOD	Netherlands Institute of War Documentation
NO_x	Nitrogen oxides
NWO	Dutch Science Foundation
OCR	Optical character recognition
OECD	Organisation for Economic Co-operation and Development
PA	Port authority
PCS	Port community systems
PDC	Port development companies
PEF	Polyethylene furanoate
PET	Polyethylene terephthalate
PFC	Perfluorocarbon
PI	Performance indicators
PIANC	World Association for Waterborne Transport Infrastructure
PM	Particulate matter
PPI	Port performance indicator
PPP	Public private partnership
PRISE	Port River Information System Elbe
PSBR	Public sector's borrowing requirements
PVE	Preparatory Vocational Education
QC	Quay crane
QCAP	Quay crane assignment problem
QCSP	Quay crane scheduling problem
QoL	Quality of life
RFID	Radio frequency identification
RMG	Rail mounted gantry
RMGC	Rail mounted gantry crane
RoI	Return on investment
RoRo	roll-on/roll-off
RSC	Rail Service Centre
RSM	Rotterdam School of Management
RTG	Rubber-tyred gantry
RTGC	Rubber-tyred gantry crane

RTLS	Real-time location system
RU	Railway undertakings
SaaS	Software-as-a-service
SBA	Social Benefit Analysis Framework
SC	Straddle carrier
SD	Sustainable development
SEEMP	Ship Energy Management Efficiency Plan
SEPA	State Environmental Protection Administration
SME	Small and medium-sized enterprises
SO_x	Sulphur oxides
SOE	State-owned enterprises
SPL	Smart Port logistics
SSTL	Smart and Secure Trade Lanes
STS	Ship-to-shore
SVE	Secondary Vocational Education
TBL	Triple Bottom Line
TEN-T	Trans-European Transport Network
TEU	Twenty-foot equivalent unit
TIR	Third Industrial Revolution
TOS	Terminal operating systems
TRAIL	Transport, infrastructure and logistics
TU Delft	Delft University of Technology
UNFCCC	United Nations Framework Convention on Climate Change
UPRM	University of Puerto Rico–Mayagüez
VAL	Value-added logistics
VANET	Vehicular ad hoc network
VMRS	Vessel movement reporting systems
VOT	Values of time
VTIS	Vessel traffic information system
VTS	Vessel traffic service
WCED	World Commission for Environment and Development
WEF	World Economic Forum
WIRA	Waterfront Industry Reform Authority
WPCI	World Ports Climate Initiative

INTRODUCTION

Harry Geerlings, Bart Kuipers and Rob Zuidwijk

Much of the world's welfare today has been produced, or is facilitated, by sea ports and their related activities. Ports are the locations where trade, logistics and production converge. Ports and their network connections have experienced unprecedented growth over the last decades: many ports in the world have benefited from the increase in international trade. This growth can be primarily explained by the flourishing economies of the Asian countries and to the related process of globalisation; with the integration of the world market, economic growth and higher levels of income, transport has become a major economic activity. In this context, an efficient transport system is a crucial precondition for port development and an asset in local, regional and international mobility.

This book starts with presenting the development of ports over the last decades, where we see that ports grew, together with the emerging global economy, into global hubs for large-scale efficient trade and shipping. From this perspective, ports play an important role in modern societies and make a substantial contribution to the GDP of cities and hinterland regions. For many products, production and consumption are scattered worldwide, and ports play an important role in connecting these points of production and consumption and establishing global supply chains. As such, ports can be considered as nodes in global logistics networks, where maritime transport and hinterland transport meet. The accessibility of ports is an important indicator of economic performance. To connect with the hinterland, ports make use of different modalities, such as trucks, trains, vessels and pipelines. Therefore, both excellent infrastructures and logistic systems are required to serve businesses and consumers, but also to support the competitive position of the port.

At the same time, all these activities generate negative effects, such as emissions and noise, which need to be addressed as well. This challenge is best described as the 'need for a sustainable development' in ports and their related networks. For many ports, growth has gone hand in hand with the emergence of large-scale fossil-based industries in port areas, which has made these ports dependent on fossil fuel–based trade and efficient bulk logistics. However, the fossil fuel–based production, trade and logistics have started to erode. This requires a more fundamental change than can be achieved only through technological innovation, optimisation or planning: a transition towards sustainable port activities is required. This will be addressed in the book as well.

While there is a rich literature on ports, port management, logistics and sustainable transport, this is the first book that provides a multidisciplinary introduction to these domains in an integrated way. The idea behind this textbook is to present an introduction to ports and their hinterland related networks, but also to present the related side effects. The case studies and illustrations within the book have a slight bias towards Western European ports (Rotterdam, Antwerp, Hamburg), but the theories are of use for ports in general. Therefore, the cases can be understood as an inspiration for port development in emerging economies and also in economies in transition. It contributes to a port environment that is fit for the new challenges that ports are facing today.

The book contains 22 chapters and is structured around three themes:

Part 1:   Ports and networks: strategies
Part 2:   Ports and networks: operations
Part 3:   Ports and networks: perspectives

The chapters in these three parts are briefly discussed next.

Part 1 – Ports and networks: strategies

The first part of the book not only provides an introduction to the different fields, but also forms the basis for Parts 2 and 3. In a way the chapters of Part 1 'set the scene' and deal with the 'rules of the game'.

In Chapter 1, Nijdam and Van der Horst provide the basic knowledge of port definitions, main actors, functions and concepts and the role of ports in global supply chains in their overview contribution.

Ports are increasingly seen as nodes in global supply chains. Zuidwijk deals in Chapter 2 with the dynamics in supply chains that are strongly experienced in the port business. Understanding the functioning and priorities in supply chains offers a first perspective in the operations of ports. Zuidwijk sees ports as enablers of green and secure shippers' global supply chains. The 'Environmental Ship Index' and 'Customs Data Pipeline' are presented as examples of those enablers.

Ports cannot be seen as stand-alone phenomena, but are part of a network and have a function for forelands and hinterlands. This hinterland is to a large extent related to the national economies and the regional economic (production- and services-based) logistic systems; ports have a 'strategic connectivity' with other ports and nodes. This implies that seaports have a strategic importance and a wider impact on the regional, national and European economy. Van den Bosch, Hollen and Volberda provide in Chapter 3 insight into the strategic value creation of ports for national economies, illustrated by the port of Rotterdam and the Dutch economy.

The management of ports is an essential function. Because of the changing (power) relationships between important stakeholders in ports, the role of port authorities is challenged. Examples are the ongoing formation of new forms of cooperation in container liner networks, such as new alliances (2M, THE Alliance, Ocean Alliance), and the equal important development of global terminal operators. At present there are different institutional arrangements: state owned, privatised and so forth. In addition, national and regional governments become shareholders in port authorities. In Chapter 4 Van der Lugt explains the different institutional arrangements and illustrates what these dynamics mean for the functioning of port management and port authorities. She presents a new strategic scope for port authorities, like becoming a cluster or chain manager or (international) entrepreneur.

Haralambides continues in Chapter 5 with the subject of port management and relates it to institutional reform. After explaining the driving forces behind port reform, he presents different forms of public involvement in the port industry and explains some major issues related to government retrenchment in ports. The chapter presents a broad and rich overview of 30 years of port reform and includes relevant theories and measures. The examples presented of his own experience as president of the Italian port of Brindisi make this contribution even more valuable.

Ports are often located in cities, close to consumers and producers. This proximity of port and city can bring conflicts, but can also provide unique development opportunities. Merk presents in Chapter 6 an overview of the port-city interface and of the effective management of this interface. He pays attention to issues like how to manage port-city conflicts, how to arrange for 'peaceful co-existence' between ports and cities, and how to create synergies between ports and cities by coupling their respective strengths. How to mitigate the environmental impacts of ports? How to mobilise the port as a driver for the urban economy? As the lead author of the OECD publication 'The Competitiveness of Global Port-Cities', Merk presents insights from this influential research program.

The assessment of a port is increasingly presented in term of port performance indicators. Port performance indicators are important statistical tools to measure the performance of different port activities and to enable a clear assessment of the performance of port activities within a port and between different ports. Port performance indicators are usually used in focus areas like economics and finance, operations and development needs. In Chapter 7 Yahalom and Guan present a broad overview of port performance indicators for a wide range of port activities. They present general indicators for ports, operational indicators especially aimed at terminal operations, and in addition analysis, financial and economic indicators. They also pay attention to social-economic, environmental and government issues.

As ports are increasingly confronted with congestion (on the road as well as on rail and on the water), accessibility has become a key port performance indicator. Policies related to guaranteed accessibility of ports are increasingly focused on management issues (orgware and software), instead of on the building of new infrastructure (hardware). Policies focusing on modal shift and variabalisation of costs for the users of infrastructure are receiving increased attention. The accessibility of the port by means of dynamic traffic management and the measures is an important issue.

In Chapter 8, Corman and Negenborn present an overview of different approaches to increase accessibility. Per hinterland modality they present a number of approaches to improve accessibility by means of hardware, software and orgware solutions. They illustrate these solutions with examples from the port of Rotterdam.

Public authorities in particular invest in port infrastructure and hinterland connections. The connection of the port with the hinterland is vital for the functioning of a port. An important question with respect to investment of most ports in hinterland connections is, who benefits from the return on investment: the port region or the hinterland? Often a social cost-benefit analysis is executed to answer the question with respect to the return on investment. In Chapter 9, Sys and Vanelslander present such a social cost-benefit analysis framework for a specific port hinterland project, a road project improving the hinterland links of the port of Zeebruges.

In their research on coordination issues between port stakeholders, Van der Horst and De Langen illustrate in Chapter 10 the importance of bottlenecks in coordination issues between parties responsible for the functioning of maritime transport chains. Increasingly coordination issues are seen as the key for solving bottlenecks in hinterland accessibility. In their chapter they present an overview of the most important coordination issues and some of the initiatives in maritime transport chains to overcome bottlenecks in coordination; examples of horizontal and vertical integration are illustrated. Special attention is also paid to the issue of extended gates for the port of Rotterdam.

Part 2 – Ports and networks: operations

In Part 2, material is provided on the operational aspects of ports and networks. The main subsystems are maritime networks, port networks, and land networks. Maritime networks involve the worldwide transport services offered by shipping liners.

The main players and their services and different types of networks are considered in Chapter 11 by Mulder and Dekker on liner networks. A variety of planning problems are reviewed, and the development of maritime networks is illustrated by means of a case study of Indonesia.

Particular attention is given by Veenstra to the business model of the shipping liners in Chapter 12 by considering the revenues and costs of maritime shipping.

Maritime networks interface with land networks via ports. Maritime logistics not only considers transportation activities, but also transshipment, handling and storage of freight. An increasing portion of freight is handled in a standard loading unit, the container. More specifically, in Chapter 13, Vis, Carlo and Roodbergen focus on design, planning and operations in container terminals.

In Chapter 14, information management in ports is considered by Heilig and Voß. They provide a framework to categorise the various systems for port-centric information management.

Land side operations are studied with a focus on the new concept of synchro-modality in Chapter 15 by Tavasszy, Behdani and Konings. This concept brings the use of various transport modes to the next level by combining the use of road, rail and inland waterway networks in a dynamic and integral way. Ports do not only act as global hubs, but also as industrial clusters.

The operational aspects of the industrial seaport are considered in Chapter 16 by Kuipers. He presents the different forces underlying the location of industrial com-plexes in seaports and also presents different types of seaport industrialisation, with most attention given to the 'modern' port industry and the chemical industry. The green-ing of port business and the potential the biobased and circular economy are offering, together with current practices like industrial ecology, co-siting and the realisation of crossovers between port and city, are issues addressed in this contribution. The port region of Teesside (UK) is presented as a case study for the general issues addressed.

Part 3 – Ports and networks: perspectives

Part 3 offers an overview of different perspectives relevant for port studies. This part starts with a contribution by Klemann, who presents in Chapter 17 a historical perspective on ports and shipping. This perspective is very important for the under-standing of the current functioning of seaports. Why did the port of Rotterdam become the largest port of Europe, and why was it the largest port in the world for nearly four decades? What are the underlying dynamics responsible for important changes in the position of ports amid their forelands and hinterlands?

Sustainability might be the perspective on ports that has received most atten-tion during the past few years because of the heavy impact of port and port-related issues on the climate and on the local environment. Ports, and especially industrial ports, are heavy producers of CO_2, fine particles and other emissions and noise. Geerlings and Vellinga present in Chapter 18 an overview of sustainability issues in seaports and current policies as practiced by the port of Rotterdam and some other important ports in Western Europe to increase the sustainability perfor-mance of ports.

The attacks on the World Trade Center and on other targets in the USA on 9/11 resulted in maximum attention on safety and security in intercontinental flows of goods and persons. The tragic events provoked a long list of safety and security measures, increasing the visibility and transparency of international logistics opera-tions, especially in the container business. These measures had a big impact on the underlying supply chain management practices, related to the increased transpar-ency of chains. Guan and Yahalom give an overview of the safety and security perspective in Chapter 19, combining this perspective with the most important supply chain issues. Trade and transport are not able to function optimally without a proper information infrastructure. The complexity and scale of modern container operations is dependent on information systems, and the safety and security perspective depends on the information infrastructure.

Geerlings and Wiegmans present an overview of recent innovations shaping the port business in Chapter 20. They pay attention to recent trends and actual developments in port and hinterland innovations and the nascent phenomena of 'the Internet of Things'.

In Chapter 21, De Koning, Zandvliet and Gelderblom give an overview of current issues from the labour perspective. The impact of major historical transitions is very visible in the seaport environment. From an innovation perspective, the impact of containers and of oil and related petrochemical industries is mentioned.

These innovations resulted in a system change in the global economic organisations. This type of system change is often called a 'transition'. It is clear that these transitions have a big impact on ports. The current transition that is changing the port landscape is the transition to sustainable economic activities related to renewable sources of energy and feed stocks, like biomass. But how to realise and manage this much-needed transition? Loorbach and Geerlings present in Chapter 22 the latest insights in transition management aimed at the port business. This transition perspective means a strategic move for the total port business, having implications for ports and networks as well as port operations. The transition perspective therefore places the total of the presented port perspectives 'into perspective'.

The book is written primarily for educational purposes, for use either in courses at universities or in other education programs or self-study. Each chapter contains a general introduction on the topic and the structure of the sections, an introduction to the discipline, a case study/illustration and an inside perspective about the expected future developments in the specific study domain. We purposely limited the number or references in each contribution and added suggestions for further readings.

We think the book can also be beneficial to researchers, practitioners (from the private and public sectors) who are engaged in ports, logistics and related research and practice, because the main aim is to provide insights into the integrated approach that is needed in this domain.

If you have any suggestions related to this book in general or parts of it or have questions, please contact us.

Harry Geerlings, Bart Kuipers
and Rob Zuidwijk (editors)
Rotterdam, 2017

PART 1

Ports and networks

Strategies

1

PORT DEFINITION, CONCEPTS AND THE ROLE OF PORTS IN SUPPLY CHAINS

Setting the scene

Michiel Nijdam and Martijn van der Horst

1.1 Introduction

This first section starts from a broad definition of a seaport. Primarily, ports serve as transfer points in a transport chain. Besides this transport function, they also function as a location for (petrochemical) industry and logistics activities. The section will discuss these three port functions. For each of the functions it will become clear that they are highly different. Each port function has different firms that provide the services, different geographical scales of competition and different ways to measure port performance. Given a port's many functions, it is difficult to measure port performance. Worldwide, throughput is most used in the industry. However, this does not provide information on the (regional) economic benefits of the port. Therefore, port-related employment and value added are also used as port performance indicators. The three port functions and the performance of the port will be illustrated for the case of Rotterdam, being Europe's largest port. Also the performance of the port will be discussed. Several port performance indicators are discussed for the Rotterdam situation. The section illustrates some measures that have been taken in the past to increase the value added of the Rotterdam port region and to maintain its position in worldwide supply chains.

Section 1.2 will introduce the primary function of a port and the most important companies active in ports, such as deep-sea terminal operators, shipping lines, port-related transport companies, transport intermediaries (forwarders and brokers) and the port authority. Section 1.3 will discuss the three port functions (transport node, location for the [petro]chemical industry and location for logistics activities). Section 1.4 deals with measuring the economic performance of ports. In Section 1.5 the focus will be on how the port of Rotterdam could increase its value added. Section 1.6 will give some summarising conclusions.

1.2 The primary function of a port and its main actors

Many different definitions exist about ports. An often used textbook definition can be found in Stopford (2009: 81), who defines a port as 'a geographical area where ships are brought alongside land to load and discharge cargo – usually a sheltered deep-water area such as a bay or river mouth'. The Port Working Group of the European Commission (1975: 6) defines a port as

> an area of land and water [. . .] to permit, principally, the reception of ships, their loading and unloading, the storage of goods, the receipt and delivery of these goods by inland transport and can also include the activities of businesses linked to sea transport.

From both definitions the first and most important function of a port becomes clear: ports are nodes in transport chains. Looking at the second part of the definition of the European Commission (1975), it makes clear that a port is also a location for economic activities related to the handling of ships and cargo in a port. In this section we will discuss the primary function of a port, namely its role as a node in transport chains. In the next section we will extend the function of ports to other activities.

The transport node function of a port is similar to other nodes in transport chains, such as airports or stations for public transport. But ports are to a large extent focused on cargo. Cargo includes bulk cargo, general cargo, containers, roll-on/roll-off and project cargo.

- Bulk cargo can be defined as cargo that is transported unpackaged in large quantities. Often a distinction is made between dry and liquid bulk cargo. The main dry bulk cargoes are coal, iron ore and grain.
- Liquid bulk products are usually split up in two groups: crude oil and refined oil products. Of course, the latter category consists of a variety of products, such as naphtha, gasoline, diesel fuel, kerosene and liquefied petroleum gas.
- General cargo, sometimes also referred to as 'break bulk', includes steel, non-ferrous metals; and paper, wood and fruit if transported on pallets and in rolls, bales, sacks, 'big bags', packages and bundles. In contrast with dry and liquid bulk, general cargo can be counted apiece if the goods are loaded and unloaded.
- Container cargo is cargo stored in standardised boxes, generally 20 or 40 feet in length. Containers are units that are used for a large variety of goods, such as electronics, textiles, chemicals and machinery. Because of the standardised measures, containers can be handled in the same way in container ports and in port-related transport, resulting in low transport costs.
- Roll-on/roll-off cargo is 'wheeled cargo'. It encompasses all transport flows where vehicles like cars, trucks, semi-trailer trucks and trailers drive on and off the ship.
- A final commodity group is project cargo, consisting of all kinds of special cargoes that need to be transported overseas and require a special transport

solution. Examples include bulky and heavy objects such as yachts, generators, transformers, ships' engines and wind turbines.

Ports are nodes in transport chains because there is a need for the exchange of cargo. De Wit and Van Gent (2001) distinguish three reasons why these nodes are necessary. First, ports are needed to connect different modes of transports. Each mode of transport is characterised by different technical–operational and economic features. For example, seaborne transport vessels are equipped for long-distance, intercontinental transport, with large quantities of cargo, while truck transport is suitable for short-distance, continental transport, bringing the cargo to its final destination.

A second reason why a port exists is that the demand for transport is spatially diverse. That is, production and consumption of goods do not take place at the same location (country, region, city). For example, large quantities of grain are produced in the United States, but are processed and consumed in other places worldwide. In general, no country in the world is self-sufficient. Each country sells what it produces and tries to acquire what it needs. In this respect, it is important to note that ports are facilitators of international trade. The demand for port services is derived from the demand for transport services, and the demand for transport services depends on international trade flows. Changes in international trade patterns and transport networks have a huge impact on ports (De Langen et al., 2012). Because there are trade flows worldwide with different origins and destinations and with a different 'thickness' (small/large quantities of cargo flows), there is a need for consolidation and deconsolidation. In ports, goods could be combined in case of consolidation with the purpose to realise economies of scale and make transport more efficient.

Third, ports could facilitate temporary storage (buffer). A node is a buffer where goods are temporarily stored. Temporary storage could help to match the difference in scale between transport modes (De Langen et al., 2012). For example, a ship carries up to 300,000 tons of cargo, while the connecting transport modes, such as truck, barge and train, can only transport much smaller quantities. As a consequence, temporary storage is needed. Temporary storage could also be needed for logistics or marketing reasons. For example, creating a storage place in the port that is close to the market could help to shorten the 'time-to-market' of the cargo for its end user.

Important actors in the port are the deep-sea terminal operators, the shipping lines and port-related transport companies for railway transport, inland waterway transport and truck transport (see Figure 1.1). Besides these actors, two intermediaries play a role, namely, the forwarder and the shipbroker. In the remainder of this section we discuss the major roles and the clients of these actors, including the role of providers of nautical services (pilots, towage companies and mooring companies), and the port administration or port authority.

- **Deep-sea terminal operator** (also called stevedore or terminal operating company). The main activity of a deep-sea terminal operator is loading and

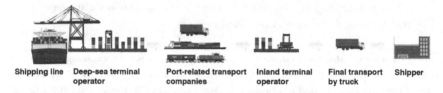

Shipping line   Deep-sea terminal   Port-related transport   Inland terminal   Final transport   Shipper
                 operator   companies   operator   by truck

FIGURE 1.1   Main actors in the maritime transport chain of containers

unloading ships and the temporary storage of the goods. A deep-sea terminal operator focuses mainly on the handling of one cargo type, because the technologies with regard to handling and storage of the different commodities vary. In general, deep-sea terminal operators have two different customers: shipping lines and the importing or exporting companies (shippers). In some transport chains, where cheap and reliable access to port services is important, shippers choose to vertically integrate. In this case, they have their own terminal operating company, or they have a majority share in a terminal. Some terminal operators exploit terminals in different ports all over the world, while others are only active in one port.

- **Shipping line**. The core business of shipping lines is to operate ships and provide shipping services. In maritime transport, a distinction can be made between liner shipping and tramp shipping. Liner shipping offers transport with a fixed route and fixed schedule. This can be compared with a bus line service on a route with a fixed timetable and stops. Shipping lines in tramp shipping do not have a fixed route. Ships sail from one port to another depending on the demand for cargo. Liner shipping is offered for containers, cars and RoRo-cargo (roll-on/roll-off–cargo), while tramp shipping is used for commodities such as oil and iron ore. Shipping companies do not necessarily own ships. In many cases, specialised ship-finance and ship-owning companies play a role and charter ships for a relatively long period to the shipping companies. The main business of shipping companies is to sell the capacity of the ship to their customers; this is mainly the shipper. In many cases, the shipper does not directly negotiate with the shipping line but it has outsourced the management and purchasing of transport flows, including the selection of shipping companies, to intermediary companies like forwarders or shipping agents (De Langen et al., 2012).

- **Forwarder**. Because of market knowledge, a forwarder has an overview of transport possibilities and tariffs, and the forwarder's role is to provide a door-to-door transport solution for a shipper. Based on his expertise and experience of the market, the forwarder works for the shippers as an independent intermediary in transport organisation. The forwarder has many functions (Ducruet and Van der Horst, 2009). Besides the selection of the transport company and negotiation about the freight rates, the forwarder can be seen as a 'consolidator'. By doing so he collects small shipments from shippers, consolidates these shipments into large loads and presents the consolidated shipment to transport carriers. Besides that, he takes care of the transport documents, including notification of the required customs documents.

- **Shipbroker.** Another intermediary in the port is the shipbroker. He acts as the agent for the shipowner and is responsible for the business of the ship. This includes obtaining cargo from shippers, arranging port activities of the ship like (un)loading of cargo, customs clearance and insurance.

- **Port-related transport companies for transport by rail, waterway, or road.** Different transport modes can be used to bring the goods from the port to their destination in the hinterland. Railway transport could be competitive for serving the hinterland on a long distance (from 300 km). Transport by waterway also has a competitive advantage for long-distance transport. More than railway transport, waterway transport is appropriate for large cargo volumes of non-time-critical cargoes. Inland water transport has a strong position for bulk cargo. Road transport is the most used transport mode for hinterland transport. This is mainly because of its flexibility and the fact that it is able to reach all final destinations. Due to the fact that truck transport is often used, many port regions are congested.

- **Providers of nautical services: pilots, towage companies and mooring companies.** In ports, ships need various nautical services. Pilots manoeuvre ships through the port and berths and assist with (un)berthing. Pilotage organisations are traditionally public organisations, because of their crucial role in securing the nautical safety in the port. Towage companies provide towage services with tugboats that assist the ship while manoeuvring in the port. Mooring companies are specialised companies to tie the ships to the quay. The providers of the nautical services get paid by shipping lines for their services.

- **Port authority.** Ports have a governing body called a port authority or port administration. The port authority is a public or private institution that is responsible for the planning, development and safety of the port. Port authorities can be classified as landlords, tool ports or service ports (World Bank, 2007). Unlike service ports or tool ports, landlord ports are not involved in terminal operations and do not invest in the port superstructure such as cranes and warehouses. Landlord port authorities provide the land and basic infrastructure to the private sector and leave all other activities to private firms (World Bank, 2007). The landlord port authority has two revenue drivers. It collects port dues from the shipping companies and land rents from the tenants in the port area like terminal operators, manufacturers or other port-related industries.

1.3   Three port functions

Influenced by external developments in the economy, society or technology, many ports worldwide have changed and expanded their role.[1] Besides the link between maritime and continental transport flows, industrial complexes have been developed in ports, containing a large number of different, sometimes related industrial functions. Moreover, ports offer a variety of different logistics services. The functions

TABLE 1.1 Different functions of a port

Characteristics	Port as a transport node	Port as a location for industrial activities	Port as a location for logistics activities
Function of port	Cargo handling Storage Trade	Cargo handling Storage Trade Industry	Cargo handling Storage Trade Industry Distribution and value-added activities
Relevant firms in the port in the provision	Deep-sea terminal operators, providers of nautical services	Port authority (landlord), utility providers (water, heat, energy)	Forwarders and transport firms
Most important port users	Shipping lines	Port-related manufacturing firms	Shippers
Geographical level of competition	Other ports in proximity	Other 'sites' for industrial activities worldwide	Other logistics zones, either in neighbouring ports or at inland locations
Important determinants in port competition	Nautical accessibility Infrastructure and transport to hinterland Service to deep-sea vessels (port turnaround time)	Nautical accessibility Availability/price of land Proximity to sales market Investment climate Quality labour force	Infrastructure and transport to hinterland Availability/price of land Quality and flexibility of labour force
Underlying developments	Industrialisation, structural economic prosperity		Globalisation, introduction of container, environmental protection

Source: De Langen et al. (2007); Van Klink (1995); adapted by authors.

of a port as a transport node and as a location for industrial and logistics activities are summarised in Table 1.1. The table shows that every port function has relevant firms on the demand and supply side. Also the geographical scope of competition and the most important determinants in port competition clearly differ.

The (un)loading of ships and the storage of goods still are the cornerstone activities of ports. The main actors in the provision of these services are the deep-sea terminal operator and the providers of nautical services. The competitive position of a port as a transport node is primarily determined by its geographical location – how is the port located with respect to the hinterland? Often a variety of ports in one region – a port range – are locked in competition to attract cargo moving to or from the common hinterland. One of the most analysed port ranges is the Hamburg–Le Havre range. Major European ports like Hamburg, Bremen, Rotterdam, Antwerp and Le Havre compete to serve Northern, Central and Eastern Europe. In order to attract shipping lines, ports have to secure their nautical accessibility. Also having good physical infrastructure to access the hinterland and the availability of transport services are required to be successful in port competition. The performance of nautical service providers and deep-sea terminal operators is another important determinant in port competition. Their efficiency could be expressed in the 'port turnaround time', including the time spent entering the port, loading and unloading, and departing the ports.

During the mid-twentieth century, industrial activities arose in the port. An important underlying development was the industrialisation that already started in the eighteenth century. Industrial activities became dominant and were based on the availability of raw materials. After the Second World War, economic prosperity increased along with the need for oil and petrochemical products. Oil and petrochemical companies were looking worldwide for preferable places for their refineries, installations and storage tanks. Van Klink (1995) showed that a first essential location factor for those installations was the maritime access, because growth in oil transport resulted in large-scale transport by supertankers. For example, with its draught of 24 meters, the Port of Rotterdam was a favourable location in Europe. The Rotterdam port authority also invested in the development of new maritime industrial areas, anticipating demands of new port users. Besides nautical accessibility and the availability of land, the presence of a qualified labour force gained importance for industries settled in port regions.

Although the container was introduced in international trade in 1966, it can still be seen as a major development that developed ports into places for logistics activities. The container was introduced with the aim to make transport more efficient. The basic idea was that efficiency in the transport chain could be improved through a system of 'intermodalism', where container cargo could be transported with minimum interruption via different transport modes. Next to the development of the introduction of the container, two other developments emerged. The energy crises and the emerging attention for the environment also restrained the demand for energy and brought the growth in oil products to an end (Van Klink, 1995). Container terminal operating companies appeared as new firms in the port, together with activities in groupage and warehousing where value-added activities for shippers could take place. Value-added activities include packaging, repackaging, filling, labelling, weighing, conditioning, palletising and so forth. For its logistics function, ports compete with neighbouring container ports with logistics zones, but also with inland locations. Because containers can be transported easily among transport modes, places in the hinterland – linked regions around inland terminals – developed into logistics centres. So, warehousing and related value-added activities became relatively footloose from the port.

With the introduction of the container and the development of intermodal transport networks, attention increased for the importance of the quality of the access to the hinterland. Together with the availability of space, and the quality and flexibility of the labour force, this can be seen as one of the major determinants in competition among container ports. Chapter 10 of this book deals with the importance of efficient port-related transport chains.

1.4 Measuring port performance

To measure the performance of ports is a complex task. Since the port is a combination of functions, activities and companies, one needs to take into account many variables originating from many companies and organisations. The performance of a single port terminal might be measured in terms of efficiency (moves per hour), while a warehouse might measure the occupancy rate and a production facility the output produced. For all of these separate functions, a company's detailed performance can be measured. It becomes somewhat difficult when one wants to measure the performance of the port as a complete system. The performance measures for the individual companies cannot be added together to get a total that says something about the whole port. The challenge is to find a measure that describes the overall port performance without reducing the explanatory power too much.

1.4.1 Port as a transport node

The throughput in tons is the most used measure for port performance and also the main performance indicator used to evaluate a port's performance. The cargo throughput reflects the competitive position of a port vis-à-vis other ports. In the case of Rotterdam, the relevant ports to compare it with are the ports in the Hamburg–Le Havre range (Figure 1.2). These ports largely compete for the same cargo and vessels because they serve the same hinterland.

The total throughput reflects to some extent the size of a port, but as every type of cargo has a different way of handling and is linked with different industries, it does not give insight into the actual performance or competitiveness of the port. The main advantage is that if one compares the throughput per cargo category and compares it to the competing ports, at least the competitive position of a port can be determined and monitored.

As an illustration, the portfolio development of the port of Rotterdam is given in Figure 1.3. The *y*-axis represents the relative percentage of growth of the cargo segment. The *x*-axis represents the competitive performance of the port of Rotterdam compared to the other ports in the Hamburg–Le Havre range. The sizes of the bubbles represent the tons throughout per cargo segment. For detailed description of the calculations, see Haezendonck et al. (2006).

This graph gives an instant indication of the competitive performance of the port, the attractiveness of the market segments the port is specialised in and the importance of these segments for the port. The graph shows that containers, crude

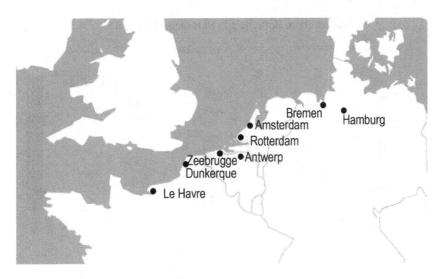

FIGURE 1.2 Main ports in the Hamburg–Le Havre range

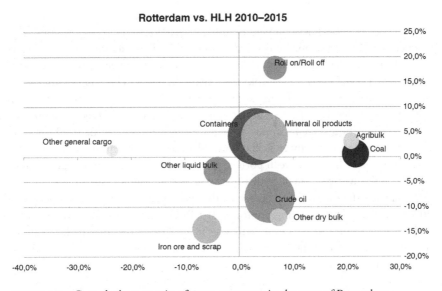

FIGURE 1.3 Growth share matrix of cargo segments in the port of Rotterdam

oil and mineral oil products are the largest cargo segments in the port. The performance of the port varies per segments. Mineral oil products showed growth in the whole Hamburg–Le Havre range, and Rotterdam gained market share in the period 2010–2015. Also, containers are a segment where both the total market grew and Rotterdam won market share of about 3%. For crude oil, on the other hand, the total market declined by 8%, but Rotterdam gained 5.6% market share in this segment.

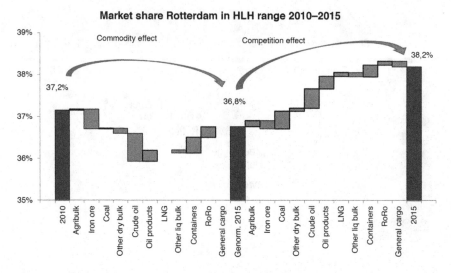

Market share Rotterdam in HLH range 2010–2015

FIGURE 1.4 Market share development split in commodity and competition effects

To relate the market development of individual cargo segments and the competitive performance of the port of Rotterdam to the overall market share of Rotterdam in the Hamburg–Le Havre range, a waterfall graph is instrumental. It can be used to separate a commodity effect from a competition effect. The commodity effect shows the development of the market share as a result of the cargo mix in a port; when a port is already important in the supply chain of a certain type of cargo and it has infrastructure geared towards it, then a general increase in demand for that cargo will create extra activity in the port. The competition effect shows to what extent cargo has shifted between ports; when a port grows in a cargo segment at a rate above the average growth of that market segment, it wins market share.

Figure 1.4 shows the development of the market share of Rotterdam between 2010 and 2015. At the starting point in 2010, Rotterdam had 37.2% market share, and the resulting market share in 2015 was 38.2%. The growth in market share is the result of −0.4% commodity effect and +1.4% competition effect. These effects are compound effects in 11 cargo segments.

1.4.2 Ports as a location for industry or logistics activities

When the port is viewed as a location, whether it is for industry or logistics activities such as warehousing, the throughput measured in tons is no longer a suitable measurement. One wants to know how much activity takes place rather than how many goods are processed. The performance indicator that best reflects the economic activities of a port is the value added, and the socio-economic goals are most related to employment in a port. These two are linked together because salaries paid are part of the total value added, which is a sum of the wages, profits and depreciations.

Next to a broader view than only cargo, one also has to expand the view in terms of geography when analysing the (socio)economic performance of a port. Port activities not only take place along the key walls, but also at some distance from the port, where companies that are an integral part of the port business community are located. For an exploration of port clusters including the functional and geographical borders of ports from an economic perspective, see De Langen (2003). Most employment is located directly along the port basins, but also a considerable number of people are working in port jobs in the areas surrounding the port basins.

In total 94,000 people work in the port, of which almost two-thirds work in the transport node function and handle cargo (Table 1.2). The other third finds

TABLE 1.2 Employed persons and added value in the port of Rotterdam (2014)

Sector and subsector	Employed people		Added value	
	EP	%	million euro	%
Transport node	59,957	63.9%	6,385	51.1%
Transport	37,214	39.7%	2,522	20.2%
Navigation	1,475	1.6%	181	1.5%
Inland navigation	7,089	7.6%	508	4.1%
Road transport	27,264	29.1%	1,618	12.9%
Rail transport	1,333	1.4%	89	0.7%
Pipeline transport	53	0.1%	125	1.0%
Services for transport	13,871	14.8%	1,922	15.4%
Handling and storage	8,872	9.5%	1,941	15.5%
Business location	33,802	36.1%	6,113	48.9%
Industry	20,422	21.8%	4,749	38.0%
Foodstuffs	2,417	2.6%	318	2.5%
Petroleum	3,363	3.6%	1,379	11.0%
Chemicals	4,740	5.1%	1,835	14.7%
Metals	3,317	3.5%	303	2.4%
Vehicles	1,963	2.1%	123	1.0%
Production of electricity	1,874	2.0%	531	4.2%
Other	2,748	2.9%	259	2.1%
Wholesale	7,982	8.5%	744	6.0%
Public and private services	5,398	5.8%	620	5.0%
Total	93,759	100.0%	12,498	100.0%

Source: Erasmus University Rotterdam (2015).

employment in production, trade or professional services. The added value represents the amount of value that is created by the companies in the port; it reflects the part of GDP that is directly created in the port. The value added is more evenly distributed between transport and other activities. The industry in the port creates the most economic value, at 38% of total added value (Table 1.2).

1.4.3   Private investments

Private investments partly represent the trust companies have in the vitality of their business in the port. Industrial companies might invest in new installations that have an economic lifetime of more than 20 years. The same is true for port equipment. When there is a growing investment in fixed assets, it reflects the future opportunities of the port. Declining investment can give an indication that the portfolio of activities in a port is less fit for the future. In the port of Rotterdam, the investment levels have grown in the period from 2010 onwards (Figure 1.5). An important driver for that is the realisation of the Second Maasvlakte, a 1,000-hectare expansion of the port that induced new private sector investments in container terminals. In addition, several industrial sites, mainly related to the oil and chemical industry, made upgrade investments in their installations.

1.4.4   Other performance indicators

The multifaceted complex that a port is made of proves to be hard to measure. Several scholars and port professionals have paid thought to developing new performance indicators for ports. Ideas are of economic, social and environmental nature and can be backward- or forward-looking.

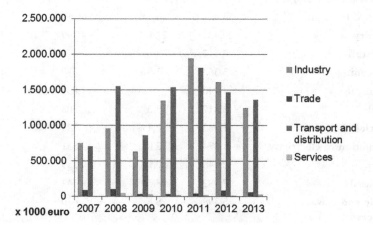

FIGURE 1.5   Port investments in greater Rotterdam per sector

Source: Authors, based on data CBS (2015).

For social issues one can think of measures related to employment, training and social involvement of port companies. Employment in the port companies can be expressed in quantities, and as described in the previous section, it can also be measured in education levels. Rising education levels in the port industries are an indication of growing automation and/or growing high-level activities in the port, such as research and development or strategic decision centres.

Ports are typically a source of noise and pollution. Controlling emissions and reducing them where possible is essential for ports to maintain their license to operate. In many cases it also limits further growth, and ports need to reduce their emission levels before they can accommodate new activities. The environmental performance of a port is indicated by all measure of nuisance and pollution; one can think of CO_2, NO_x, sulphur emissions and noise.

Innovation is another important issue for economic development. Locations where more innovation takes place are the locations that have better prospects. These innovations can increase productivity and thus enhance the competitive position, or they can even create whole new industries in a port. A measure for innovativeness can be the increased productivity levels, the number of port-related patents or the number of new products and services introduced.

1.5 How can ports increase their value added? The case of Rotterdam

Increasing the performance of a seaport is as multifaceted as the functions of the port and the goals that one wants to achieve with a seaport. If one takes into account that the purpose of a port is broader than only transport flows, performance improvement should also be linked to these goals.

Port authorities try to influence the total performance of their port, not only in efficiency but also in terms of economic value for the region. Traditionally it has been the task of a port to facilitate trade in the region and create economic prosperity. Also, in a time when port authorities are increasingly liberalised, they often still have government as their shareholders and have public tasks. Port authorities that strive for profit maximisation are scarce. Strategic options for port authorities typically include:

1   Attracting more port-related business to the port
2   Creating cluster synergies
3   Increasing the logistic value of the port.

1.5.1 Attracting port-related business

Port authorities have the port land area under management and try to fill this area with tenants that facilitate transport, such as stevedoring companies, or tenants that use maritime transport for their sourcing, such as the chemical industry. To increase the value added in the port, the port authority might target specific industries, such as companies that can add to the existing business or that are a stepping-stone for the development of a new business segment in the port. In Rotterdam, an example is Neste's refinery of biomaterials to develop a broader biobased cluster in the port.

1.5.2 Creating cluster synergies

Ports are economic clusters of related activities. By concentrating business in a specific area, companies profit from the availability of resources, the presence of supporting industries, easier input-output relations, and collective infrastructure. An example of collective infrastructure in Rotterdam is the multicore pipeline. This pipeline system was initiated by the port authority in joint venture with Vopak, a liquid bulk storage company. The pipeline connects several industrial sites in the port of Rotterdam, and capacity can be used in a flexible way by all industries.

1.5.3 Increasing logistic value

The core function of a port is linking regions through maritime transport. The value a port derives from that is related to the cargo it handles and to what extend this cargo is transshipped, further manufactured and/or managed. The transship-ment function of a port creates value in itself because by moving cargo to the right place, the cargo increases in value. A product has more value in a location close to the customer. Part of this value is captured at the port because port dues, cargo han-dling and transport is paid and employment is created as a result. If a port authority wants to increase the value that is created in a port, it can build facilities for and try to attract value added logistics companies. One way to do that is creating distribu-tion parks, as did Rotterdam. At three locations in the port area there are dedicated parks for distribution centres, so-called Distriparks; Eemhaven in the east, Botlek in the middle and Maasvlakte in the west.

These locations are specifically tailored to the warehouses that handle maritime cargo that is shipped close to the distribution parks. The Eemhaven park is filled with third-party logistics companies that handle cargo from containers. The Botlek park is in the midst of the chemical industry and hosts a set of chemical distributors. The Maasvlakte park is home to several large-scale dedicated warehouses, reflecting the size of the container operations on the terminals located close by (Figure 1.6).

1.6 Conclusion

This chapter defined the role of a port. The primary function of a port is being in node in a transport chain to exchange cargo including bulk cargo, general cargo, containers, roll-on/roll-off, and project cargo. Ports can connect different modes of transport and facilitate storage of the goods. Another reason why ports exist is because the demand of transport is spatially diverse. Ports can also include other business activities linked to sea transport. Many ports worldwide have developed as locations for industrial activities and/or logistics activities. Ports have many

FIGURE 1.6 Location of Distriparks in the port of Rotterdam: Eemhaven, Botlek and Maasvlakte

Source: Port of Rotterdam (2017).

different functions and are locations for a diverse set of companies. That makes measuring the performance of ports complex. Throughput in tons is the most common indicator for measuring port performance. However, the total throughput does not give insight into the actual performance or competitiveness of the port. The cargo throughput does reflect the competitive position of a port in comparison with other ports. In this case, it is relevant to compare the throughput with ports in proximity. Two other often used port performance indicators are value added and employment. In the last part of the chapter we concluded that port authorities try to influence the total performance of the port. Because port authorities often have governments as their shareholders and have public tasks, the number of port authorities that strive for profit maximisation are limited. Strategic options for port authorities are attracting more port-related business to the port area, creating cluster synergies and increasing the logistic value of the port.

Note

1 This is not to say that every port developed into a 'complete' port, including a large industrial complex and logistics activities.

References

CBS (2015) *Central Bureau of Statistics Netherlands*. Retrieved from www.cbs.nl

De Langen, P. W. (2003). *The Performance of Seaport Clusters*. Erim, Erasmus University Rotterdam, Netherlands.

De Langen, P. W., Nijdam, M. H., and Van der Horst, M. R. (2007) New indicators to measure port performance. *Journal of Maritime Research*, 4(1), 23–36.

De Langen, P. W., Nijdam, M. H., and Van der Lugt, L. M. (2012) *Port Economics, Policy and Management*. Rotterdam, Erasmus University Rotterdam.

De Wit, J. G., and Van Gent, H. A. (2001) *Economie en transport*. Utrecht: Lemma.

Ducruet, C. B., and Van der Horst, M. R. (2009) Transport integration at European ports: Measuring the role and position of intermediaries. *European Journal of Transport and Infrastructure Research*, 9(2), 121–142.

Erasmus University Rotterdam (2015) *Port Monitor 2013: The Economic Value of the Dutch Seaports*. Regional, Port and Transport Economics, Erasmus University Rotterdam.

Haezendonck, E., Verbeke, A., and Coack, C. (2006) Strategic positioning analysis for seaports. *Research in Transportation Economics*, 6, 141–169.

Port of Rotterdam (2017) *Graphical Maps Port of Rotterdam*. Retrieved from www.portofrotterdam.com

Stopford, M. (2009) *Maritime Economics* (3rd ed.). Abingdon, Routledge.

Van Klink, H. A. (1995) *Towards the Borderless Mainport Rotterdam: An Analysis of Functional, Spatial and Administrative Dynamics in Port Systems*. Thesis publishers.

World Bank (2007) *Port Reform Toolkit, Module 2: Alternative Port Management Structures and Ownership Models* (2nd ed.). Washington, DC: World Bank.

Suggestions for further reading

Besides the papers provided as references in the text, we recommend the following texts as suggestions for further reading.

Bird, J. H. (1971). *Seaports and seaport terminals* (Vol. 158). Hutchinson.

Brooks, M. R., & Cullinane, K. (Eds.). (2006) *Devolution, Port Governance and Port Performance.* Vol. 17, Salt Lake City, Elsevier.

Cullinane, K. (2011) *The International Handbook of Maritime Economics.* Cheltenham, Edward Elgar.

Merk, O. (2013) *The Competitiveness of Global Port-Cities: Synthesis Report.* OECD Regional Development Working Papers, 2013/13, Paris, OECD.

Talley, W. K. (2009) *Port Economics.* London, Routledge.

2

PORTS AND GLOBAL SUPPLY CHAINS

Rob Zuidwijk

2.1 Introduction[1]

This chapter describes the seaport as an enabling node in supply chains, which connects organisational, logistics, and information networks to support the seamless flow of goods. The role of the port as a *Global Hub* in logistics networks is probably the best known, and is indeed described in, for example, the port vision document of the port of Rotterdam.[2] A second role of the port as described in this particular document is as an *industrial cluster*. This refers to the location of multiple industrial activities in the port area, where preferably, synergies are achieved. For example, synergies are attained in so-called industrial ecosystems that feature the reuse of by-products of one facility by another facility as a resource (see Chapter 16). A third role that the port plays is the *Information Hub*, which facilitates the exchange of information in support of international trade and logistics (see Chapter 14). Here, international trade is supported by information and communication technologies, which progressively set the standards for the global logistics systems.[3] Presently, port community systems provide information services that facilitate the exchange of information between actors involved in trade and logistics processes in and around the port.[4] A fourth role of the port is *World Port City*, where the development of the port and the urban area around the port area go hand in hand, and where commodity trade and business services such as legal and financial services, and maritime services add value[5] (see Chapter 6).

The various roles that a port can play all create opportunities for the port to be an enabling node in global supply chains. The Global Hub aims to establish excellent port operations that allow the seamless transfer of goods between the maritime and hinterland network.[6] In principle, despite its vital role, the value added is usually limited as it is created by logistics services only. However, in the case of containerised goods, as soon as services are performed on the product, for example, in distribution centres, then the value that is added may be more substantial. Also in the case of bulk product, industrial activities add value. The location of value-adding activities in the port area or in its hinterland may bind the flow of goods to the port, which enhances its competitive position. As such, the development of the industrial cluster in the port area also drives the throughput of the port. Value can also be added by the provision of business services in the supply chain, which allows for the synergetic development of port-cities and helps establish the World Port City. The coordination of a supply chain, that is, the concerted action of the various actors in the supply chain to achieve a better performance, is driven by the exchange of information. The synchronisation of activities, for example, requires organisations to share information about planning and execution. This can be enabled by inter-organisational information systems, and the port may provide such information services in its role as an Information Hub. An upcoming challenge is how port communities may provide information services that add value to goods flows and that enforce the competitive position of the port, even if the port is not involved in the physical handling of the goods.

To further clarify how ports are enabling nodes in global supply chains, we address a number of topics. In Section 2.2, we discuss the role of ports in global supply chains, where we also distinguish between the port supply chain and the supply chain of an individual shipper. In Section 2.3 we discuss in more detail the port as an enabler in the port supply chain. In Section 2.4, we discuss the potential role of ports in establishing green and secure shippers' global supply chains. In Section 2.5, we conclude and provide some future research topics.

2.2   Ports and global supply chains

Both in the academic community and in practice, the role of ports in global supply chains has been emphasised. Robinson (2002) observes that the role of ports develops from a logistics hub, where goods are transshipped efficiently, to provide services that add value in global supply chains. He argues that ports have a significant role to play in these value chains, but he does not provide concrete examples of how this could be done. Carbone and de Martino (2003) provide some concrete examples and discuss how ports add value in the global supply chain of the car manufacturer Renault. The establishment of value-adding activities in the port area provide examples, such as logistics activities that are in line with international trade agreements, and consolidation of freight in transport.

Notteboom and Rodrigue (2005) further elaborate on the role of ports in global supply chains. They emphasise the development of ports as nodes in maritime and hinterland networks. In addition, they argue that the competitive position of ports is not only determined by the efficiency of port operations, but also by the connectivity of the port to its maritime and hinterland networks. Connectivity here relates to the extent the port connects to other ports by means of maritime services and to inland locations by means of land transport services. The quality of these services is obviously also relevant. Here the use of various modes of transport other than trucking for land transport is relevant. The use of multiple modes of transport in sequence or in parallel requires a certain level of coordination. Therefore, it can be argued that the port acts in a multimodal chain of logistics activities.

It is relevant to note that the global supply chains of shippers are in general distinct from the port supply chains, as the latter can be seen as an aggregate of the former. This is most clear in the case of containerised goods, and less obvious in the case of bulk goods. The scale growth in the maritime transport of containers is a good example of how the container chain is quite different from the supply chain of an individual shipper. Modern container vessels can carry up to 20,000 TEU,[7] which holds the cargo of thousands of shippers' supply chains. Alternatively, we can consider the total volume of containerised cargo that goes through the port, and compare this with the volume of individual shippers. For example, the port of Rotterdam has a yearly throughput of about 12,000,000 TEU.[8] In contrast, most shippers either order goods in less than container load shipments, or they ship a limited number of containers on a yearly basis. There are also companies that ship a considerable amount of containers, such as the larger retail organisations and manufacturers. For example, beer manufacturer Heineken ships 90,000 TEU on a yearly basis.[9] As a consequence, the share of an individual shipper of the total throughput of the port is below 1%, and usually much less (see Figure 2.1).

Admittedly, many smaller shippers make use of freight forwarders to ship their cargo, either Full Container Loads (FCL) or Less than Container Loads (LCL), where in the latter case the freight forwarder acts as a consolidator. Even if the

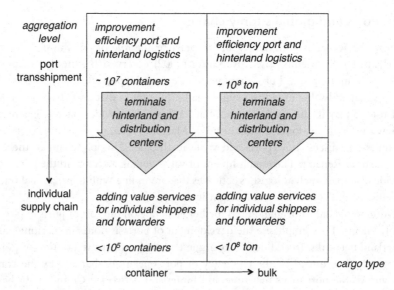

FIGURE 2.1 Distinct aggregation levels between shippers' supply chains and port chain

Source: Zuidwijk and Veenstra (2015).

freight forwarders are considered as shippers, the difference in aggregation level between shippers' supply chains and port supply chain is significant.

For liquid and dry bulk products, the situation is quite different. We illustrate this by considering the throughput of crude oil in the port of Rotterdam. In 2014, the import of crude oil to the port of Rotterdam was equal to 95 million metric tons, half of which was used by refineries in the port area. Of this volume, Shell consumed 30%, BP consumed 30% and about 20% was consumed by Esso;[10] see Figure 2.1.

Based on some statistics, we may conclude that port supply chains and shippers' supply chains differ. In the next two sections, we will argue that both types of chains provide opportunities for ports to add value.

2.3 The port as an enabler in the port supply chain

The port supply chain encompasses the maritime transportation, transshipment at the port, and land transportation. Important examples are the transportation of maritime containers and bulk cargo. As discussed in the previous section, the port supply chain of maritime containers is quite different from the shippers' supply chain, and the individual shippers' preferences are usually not incorporated into the design of the services, which are usually based on a 'one-size-fits-all' approach. However, larger shippers may have some influence, and there is a trend towards a more customised approach. In this section, we will discuss the port supply chain in more detail, and we will discuss some developments that depart from the 'one-size-fits-all' container logistics service.

Figure 2.2 provides a not necessarily comprehensive overview of the port supply chain, which actually involves a network of organisations as displayed. The logistics layer involves organisations that are engaged in the container logistics operations, such as shippers (both consigner and consignee), inland and sea carriers, and inland and sea terminal operators. In the transaction layer, organisations such as freight forwarders and shipping agents are not only involved in the orchestration of the logistics chain but are also involved in the provision of port services. Banks and insurance companies, and other business services such as legal services and commodity traders, are also involved in this transaction layer. Also involved are authorities such as port authorities, police, inspection authorities and customs, who are involved in the governance layer.

Ports aim to achieve operational excellence to enable a smooth flow of goods through the port, and they face the challenge of sustainable development and use of their logistics capabilities. The accessibility of the port can be measured by the extent to which a port is able to handle sea vessels and land transport modes both effectively and in a timely fashion. The Rotterdam Port Authority, for instance, has put a lot of effort into improving the accessibility of the port both on the land side and on the sea side. The capacity of the sea side has been enlarged by the construction of a number of deep sea terminals by competing port operators on a recent land extension, called the 'Maasvlakte 2'.[11] To enhance

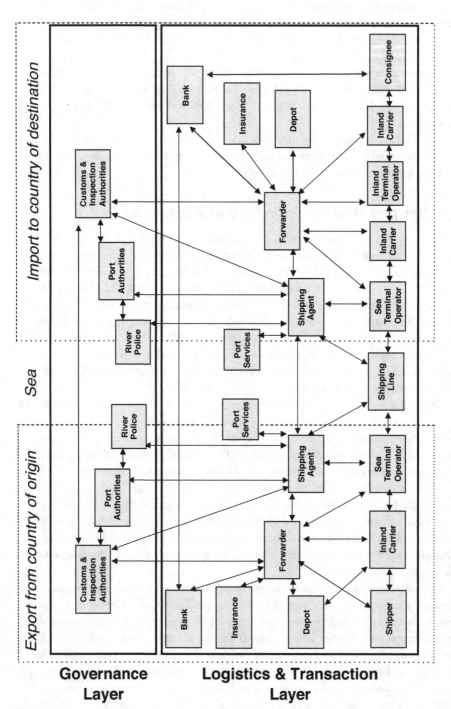

FIGURE 2.2 Layered model of port supply chain

Source: Van Baalen et al. (2009).

the accessibility on the land side, the port authority has deployed a number of instruments. Examples are the organisations introduced to improve accessibility on the road network (Traffic Management Company), the inland waterway network (NextLogic) and the rail network (Keyrail).[12] Other ports in the world, such as the ports of Los Angeles and Long Beach, have also made investments to improve their accessibility.[13]

Private companies also contribute to the accessibility or connectivity of ports. Again with respect to the case of the port of Rotterdam, Europe Combined Terminals (ECT) is developing new services to more effectively use alternative transport modes within various infrastructures, that is, other than trucking, according to the extended gate concept.[14] An extended gate is

> an inland intermodal terminal directly connected to sea port terminal(s) through high capacity transport means such as river vessel and train, where customers can leave or pick up their standardised loading units as if directly with a seaport, and where the sea port terminal can choose to control the flow of containers to and from the inland terminal.
>
> *(Veenstra et al., 2012, p. 21)*

In multimodal transport, multiple transport modes are deployed in sequence or as alternatives. Multimodal transport typically involves high-capacity transport modes such as barges and trains. These transport modes are usually more efficient and more benign to the environment and the society than transport by truck. The allocation of truck capacity to container shipments is usually one-to-one. The dispatch of trucks is demand driven and the fleet of trucks will be routed in such a way that the truck fleet is utilised as much as possible, that is, in such a way that the number of empty kilometres, for example, is minimal.

The allocation of train and river vessel capacity to demand is less straightforward, as multiple shipments need to be transported. In most cases, this coordination problem is solved by offering in advance a schedule with fixed departure times along fixed routes. Even in such a case, providing smooth connections between transport services is far from trivial. In practice, the use of a transport mode is planned in advance. Uncertainties both on the demand side and the supply side deteriorate the performance of the multimodal transportation system. The use of multiple modes of transportation in sequence results in, for example, cascaded delays driven by operational inter-dependencies.

For the aforementioned reasons, multimodal transport requires coordinative efforts and information exchange. When higher levels of information are available for the planning of network services, the high-capacity transport modes are used more often and the performance of the transportation system improves.

Exploiting this principle, 'synchromodal' transport systems are put in place, which allows for the deployment of any available transport mode in a flexible way. For example, a synchromodal transportation system will use a truck when it is idling at the right place, and it will put a container on a barge when the departure time and expected transit time of the barge allows for a timely arrival of the container at the final destination. The opportunistic allocation of containers to transport modes requires the possibility to allocate a container to transport capacity on a specific mode at the last moment. The booking of transportation without specifying the mode of transport in advance supports such flexible allocation. More details about synchromodal transportation systems are provided in Chapter 15.

On the sea side there are other challenges. One of them is the timing of deep-sea vessel arrivals. Container terminals need to plan the use of their berths in advance, and delays in the arrivals of deep-sea vessels disrupt these plans. Based on (historical) information of the navigation of deep-sea vessels, better forecasts can be produced. The next section presents a simple model to estimate the value of such better forecasts (Exposition 1).

Exposition 1: optimisation of resource availability time interval

Given a probability distribution that describes vessel arrival uncertainty, we aim to optimise the time interval in which handling resources, such as quay cranes, are available. The parameters that we optimise are the centre and the width of the interval. During the time interval, the handling of the vessel can commence. The resources have a cost rate, so setting the time interval wider will result in higher resource costs. On the other hand, given the time interval in which the resources are available, the vessel may arrive too early or too late. An early arrival results in vessel idle time *or* replanning costs. A late arrival results in resource idle time *and* replanning costs. In any case, late-arrival costs are usually higher than early-arrival costs. Given the probability distribution and the time interval, the probabilities of early and late arrival can be computed (see Figure 2.3).

We now formalise. Let τ be the centre of the interval, and Δ be the width of the interval. Then the probability that the container arrives early is equal to $F(\tau - \Delta)$, and the probability that the container arrives late is equal to $1 - F(\tau + \Delta)$. Here $F(t)$ is the cumulative probability distribution of the probability density function $f(t)$. We

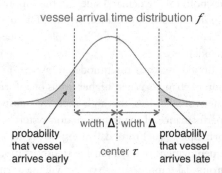

FIGURE 2.3 Optimisation of resource availability time interval

also define the cost parameters: C_- equals the cost of early arrival, C_+ the cost of late arrival, and we assume $C_- < C_+$. The resource cost rate is equal to c_r.

The expected costs of early arrival are equal to $C_-(\tau - \Delta)$, and the expected costs of late arrival are equal to $C_+(1 - F(\tau + \Delta))$. The resource costs equal $c_r \times 2 \times \Delta$. The total costs are the sum of these three cost terms. For optimal τ and Δ, we get $C_- \times f(\tau - \Delta) = C_+ \times f(\tau + \Delta)$. This basically states that the optimal interval will be shifted in such a way that the marginal earliness and lateness costs are the same. We also get $C_- \times f(\tau - \Delta) + C_+ \times f(\tau + \Delta) = 2c_r$, so the optimal interval is widened to the extent where the marginal earliness and lateness costs are equal to the marginal resource costs.

FORMULA 2.1   Expected costs of early arrival

$$C_- \times F(\tau - \Delta)$$

FORMULA 2.2   Expected costs of late arrival

$$C_+ \times (1 - F(\tau + \Delta))$$

FORMULA 2.3   Resource costs

$$c_r 2\Delta$$

FORMULA 2.4   Optimal τ and Δ

$$C_- \times f(\tau - \Delta) = C_+ \times f(\tau + \Delta)$$

FORMULA 2.5   Marginal earliness and lateness costs are equal to the marginal resource costs

$$C_- \times f(\tau - \Delta) + C_+ \times f(\tau + \Delta) = 2c_r$$

2.4   The port as an enabler of green and secure shippers' global supply chains

Supply chain coordination and information exchange is usually motivated by economic performance improvements through revenue enhancement or cost reduction. The coordination of activities in supply chains to enhance environmental or societal performance has received much less attention. There is a growing interest in the (joint) involvement of global supply chains and seaports in voluntary or regulated programs that aim to reduce the negative externalities of their logistics activities.

The example of carbon footprint reduction has received particular attention. While some programs are regulated, such as the Emission Trading Scheme for large direct emitters in Europe, other programs are voluntary, such as the supply chain–wide emission reduction programs initiated by some original equipment manufacturers (OEMs; Mattel, SCA) and retailers (Walmart, Tesco). In supply chains, the reduction of negative externalities needs to be coordinated, as the associated investments and returns are not distributed evenly. Financial incentives can be used as a means to coordinate the supply chain in such a case. Also, port authorities want to play an important role in creating incentives to reduce carbon emissions in global supply chains. The Environmental Ship Index (ESI) aims to measure greenhouse gas emissions throughout the global maritime transport chain, and the ESI is used in voluntary programs by port authorities. As such, port authorities take the initiative to incentivise other partners in the multitude of supply chains that make use of the ports to reduce emissions on a voluntary basis. Although this initiative can be seen as creating incentives in the port supply chain, it opens the door to initiatives that affect the shippers' supply chains directly.

Exposition 2: Environmental Ship Index[15]

The World Ports Climate Initiative (WPCI)[16] developed an index that rates the environmental performance of seagoing vessels. As the ESI project announced, the index can be used by port authorities to reward and promote green vessels, while carriers and ship owners can use the index as a sign of recognition of their green efforts. The index incorporates estimated emission levels of NO_x, SO_x and CO_2, while the installation of an on-board power supply is taken into account as well. The emission performance of the vessel is not based on measured emission output, but based on certified information about the design of the vessel (engines) in the case of NO_x, the type of fuel used in the case of SO_x and energy efficiency in the case of CO_2. The index also incorporates the baseline requirements of the International Maritime Organisation (IMO).[17] The scores of the individual vessels are published via the ESI website. Quite a few ports, such as the ports of Rotterdam, Hamburg, Antwerp and Los Angeles, provide a financial incentive in the form of a discount on port dues or a reward based on the ESI score of the calling vessel.

Another topic of interest in organisational networks in global supply chains is the coordinated management of risks. Because, for example, one organisation may experience the impacts of a particular risk, while another organisation may be in a better position to deal with the risk at hand, coordination may be required to manage supply chain risks effectively. The management of risks in global supply chains has received progressive attention in recent years, in particular in the management of the supplier base of a focal firm. However, the interdependencies between organisations in international trade and logistics create another need for coordination of risk management efforts.

Exposition 3: customs data pipeline

Customs not only seeks to engage global supply chains in regulated programs, but also in voluntary programs, such as Customs-Trade Partnership Against Terrorism

(C-TPAT) and Authorised Economic Operator (AEO). For the international transport of containers, Dutch and UK customs aim to make further steps. It is worthwhile to consider the collateral benefits of new business models where customs organisations have access to supply chain operational data via a 'data pipeline' to better assess the believability of declarations, or even further, to assess the risks of global supply chains based on system audits instead of inspections of individual transactions (Hesketh, 2009; Heijmann and Hesketh, 2014).

Customs that operate in seaports are engaged in the port supply chains and need to supervise the import and export of millions of containers on a yearly basis. On the other hand, they need to scrutinise the trade in each of the individual shippers' supply chains. The 'data pipeline' as discussed in Exposition 3 enables the sharing of cargo information between partners in shippers' supply chains and governmental agencies. One may investigate the improvement of container logistics performance through controlling the cargo level. The main rationale is that common container logistics practice is focused on the handling and transportation of the standardised loading units, and does not fully reap the potential of optimising the handling and transportation of the cargo inside the containers. Some of these benefits have already been recognised by the business world. The practice of trans-loading cargo from one type of container to another, for example, is justified by better utilisation of land transport means.

Organisations may also collaborate and share information to reduce the environmental footprint of their supply chain. In order to do so, the environmental footprint of a product needs to be measured and accounted for. Even port authorities can be held responsible for the footprint of the supply chains that use the port as a node in their logistics network. The governance of the corresponding 'chain of custody' may vary. For sustainable hardwood, the certification is very strict; the customer must be assured that the specific item has been harvested responsibly. For sustainable energy, the certification is based on accounting principles: the amount of energy sourced in an environmental responsible way must exceed the amount of green energy sold to the markets. Information systems play a key role in the deployment of these various chains of custody.

The aforementioned topics highlight the potential of using cargo information for new control concepts in container logistics. They also allow for the development of new business models that incorporate the foreseen benefits in a value proposition.

Of particular interest are information systems that support port communities, the so-called port community systems. In order to reduce market risks, port community systems may develop into platforms which are the equivalent of public infrastructures. The development of port IT services on these platforms could be done by (commercial) third parties. The pricing of these IT services on such a platform can be tuned to appreciate that users are both producers of data and consumers of services that use the data.

Global supply chains are initiated by global trade. The adding value of information systems that support the management of global trade processes can be measured in terms of enhanced operational performance of the supply chain. Moreover, information systems help make logistics execution in supply chains more reliable and thereby help reduce the risks associated with financial settlement. This creates better conditions, for example, a letter of credit and insurance in a financial settlement of an international trade transaction. The interactions between trade processes and logistics execution in global supply chains is receiving progressive attention, both for containerised and bulk products.

2.5 Conclusions

In this chapter, we have reflected on the role of seaports in global supply chains. We have distinguished between port supply chains and shippers' supply chains, where the two supply chains act on different levels of aggregation. The port plays an important role in the port supply chain, and modern ports are progressively able to act as value-adding hubs within these supply chains. However, ports are more and more challenged to act in individual shippers' supply chains to accommodate value-adding services that are more specific. Customs authorities, for example, are already facing this challenge. The upcoming role of electronic commerce in global supply chains may trigger a more active role of ports in these supply chains.

Notes

1 Some parts of this chapter have also been presented in a Dutch publication: Rob Zuidwijk and Albert Veenstra (2015). De haven in wereldwijde netwerken en verlader-ketens. *Tijdschrift Vervoerswetenschap* 51(3): 15–25, and in the inaugural address of the author. The booklet of the inaugural address, which explains how ports are connected, and connect organisational, logistics, and information networks, is available via http://repub.eur.nl/pub/79091.
2 The port vision document is accessible via www.havenvisie2030.nl.
3 For example, electronic commerce has a progressive impact on international logistics; see for example http://fortune.com/2016/01/14/amazon-china-earns-its-ocean-shipping-license/.
4 Port community systems are discussed in Baalen et al. (2008).
5 The OECD report "The Competitiveness of Global Port-Cities: Synthesis Report," available via www.oecd.org, addresses that seaports do not always reap the full potential of synergies between ports and cities.
6 On the maritime network, deep-sea shipping services connect seaports globally, while the hinterland network connects the seaport with inland locations.
7 A TEU or "Twenty Foot Equivalent Unit" is a standardised measure of container volume. Most containers used in maritime transport are either one or two TEU.
8 Container throughput in 2015 of the port of Rotterdam was equal to 12,234,535 TEU (port of Rotterdam).
9 This estimate is based on the Dutch news item 'Heineken verscheept exportbier via Alpherium naar Rotterdam.' www.logistiek.nl, October 4, 2010.
10 These figures have been derived from the Facts & Figures Rotterdam Energy Port and Petrochemical Cluster (www.portofrotterdam.com). The other half of the import volume was forwarded to the port of Antwerp (25%–30%), Germany (15%–17%), and to the port of Flushing (5%–7%).

11 In 2015, both APM Terminals and Rotterdam World Gateway, an international consortium consisting of four global shipping lines and terminal operator DP World, opened deep deep-sea terminals on this port extension.

12 See the section on accessibility in the vision document of the port of Rotterdam.

13 The Alameda corridor is a fast rail connection between the ports of Los Angeles and Long Beach, and the national rail system near downtown Los Angeles; see www.acta.org.

14 ECT has developed European Gateway Services; see www.europeangatewayservices.com.

15 For details, see www.environmentalshipindex.org.

16 See http://wpci.iaphworldports.org.

17 See www.imo.org.

References

Carbone, V., and Martino, M. D. (2003) The changing role of ports in supply-chain management: An empirical analysis. *Maritime Policy & Management*, 30(4), 305–320.

Heijmann, F., and Hesketh, D. (2014) *The Pipeline Interface*. Retrieved from www.cassandra-project.eu

Hesketh, D. (2009) Seamless electronic data and logistics pipelines shift focus from import declarations to start of commercial transaction. *World Customs Journal*, 3(1), 27–32.

Notteboom, T., and Rodrigue, J.-P. (2005) Port regionalization: Towards a new phase in port development. *Maritime Policy and Management*, 32(3), 297–313.

Robinson, R. (2002) Ports as elements in value-driven chain systems: The new paradigm. *Maritime Policy & Management*, 29(3), 241–255.

Veenstra, A., Zuidwijk, R., and Van Asperen, E. (2012) The extended gate concept for container terminals: Expanding the notion of dry ports. *Maritime Economics and Logistics*, 14, 14–32.

Zuidwijk, R., and Veenstra, A. (2015) The value of information in container transport. *Transportation Science*, 49(2), 675–685.

Suggestions for further reading

Besides the papers provided as references in the text, we recommend the following text as a suggestion for further reading on the development of information and communication systems in a seaport environment.

Van Baalen, P., Zuidwijk, R., and Van Nunen, J. (2009) Port inter-organizational information systems: Capabilities to service global supply chains. *Foundations and Trends in Technology, Information and Operations Management*, 2(2–3), 81–241.

3

HOW PORTS CREATE STRATEGIC VALUE FOR THEIR COUNTRY

Frans A. J. van den Bosch, Rick M. A. Hollen and Henk W. Volberda

3.1 Introduction

Authorities of leading ports face two interrelated strategic challenges. The first is to contribute to the international competitive position of the port. This competitive position is usually indicated by a port's market share with respect to its captive region in terms of volumes of containers and/or bulk goods handled. The second challenge, which is the focus of this chapter, is to contribute to the international competitive position of the region or country where the port is situated. This is particularly important for port authorities that are public organisations or, when corporatised, whose shares are (partly) held by a local, regional and/or national government. These port authorities are expected to improve the possibilities of organisations situated within and outside the port area to benefit, from an internationally competitive stance, from the port.

The value of a port for its region and country is typically expressed in terms of quantitative indicators such as direct and indirect value added and employment. From a strategic point of view, however, these indicators of economic value need to be complemented with qualitative indicators associated with innovative and competitive dynamics, which underlie international competitiveness. By so doing, one arrives at the strategic value of ports for firms in their region or country. The more this strategic value will be assessed, understood and communicated to the various stakeholders involved, the stronger the governmental and societal support for existing and new activities in the port, and the larger also the expected willingness of these stakeholders to help further increase this value – for instance, by adapting regulations or providing financial resources. This, in turn, will also contribute to a port's own international competitiveness.

The following key question is addressed in this chapter: how do ports create strategic value for their country? It starts with an introduction of the concept of strategic value before going on to distinguish and elaborate different ways in which a port can create strategic value, which are then interlinked in a conceptual framework. The chapter continues with illustrative examples of how Europe's largest port, the port of Rotterdam, creates more strategic value for its home country, The Netherlands. Finally, implications for port authorities are discussed regarding their role in increasing a port's strategic value, followed by a conclusion of the chapter.

3.2 Strategic value

For a long time the economic importance of ports has dominated the policy debate about their significance for the countries in which these ports are located. This economic notion of value indicates what has been realised in a certain period because of a port's presence, such as the amount of value added and the number of people working in the port. From a strategic point of view, however, it is highly important to also look at how firms in a port's region/country benefit from the port in terms of the port's contribution to the sustainable international competitiveness of these firms. The latter is captured in the notion of strategic value. The strategic value of a port for its country is made up of the port's economic value for the country (quantitative part) and the difficult-to-copy contribution to the sustainable international competitiveness of firms in the country where the port is located (qualitative part); see also Box 3.1. This contribution is determined by several factors, as will be elaborated upon later in this chapter.

BOX 3.1 ECONOMIC VALUE AND STRATEGIC VALUE OF PORTS FOR THEIR COUNTRY

- A port's economic value consists of the (direct and indirect) value added, (direct and indirect) employment, size of investments and other quantified indicators of what has been economically realised in the country in which the port is located in a particular period because of the port's presence.
- A port's strategic value consists of its economic value plus the port's difficult-to-copy (typically qualitative) contribution to the sustainable international competitiveness of firms in its country.

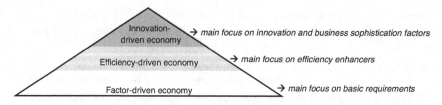

FIGURE 3.1 WEF pyramid of economic development phases and corresponding competitive focus

Source: Van den Bosch et al. (2011), based on the World Economic Forum (WEF) Global Competitiveness Report series.

In order to improve understanding of the concept of strategic value, the notion of international competitiveness requires a closer look. International competitiveness is a relative concept that is best measured according to an external benchmark. The renowned Geneva-based World Economic Forum (WEF) discerns between three economic development stages of countries, each with an increasing focus on innovation and business sophistication factors. As depicted in the pyramid (see Figure 3.1), these stages correspond to (1) factor-driven economies (the pyramid's lowest level), (2) efficiency-driven economies, and (3) innovation-driven economies (the highest level; Schwab, 2015). A country can also be in a transition from, for example, the second to the third phase of economic development.

The higher a country is positioned in this pyramid, the higher its Global Competitiveness Index (GCI). Hence, countries with an innovation-driven economy are considered more competitive than countries with a mainly efficiency- or factor-driven economy. Also, there are differences in the international competitiveness of countries that find themselves in a similar development phase. For instance, in 2015 the efficiency-driven economy of China (GCI ranking #28) was more internationally competitive than the efficiency-driven economy of Indonesia (ranking #37). In 2015, the ten most competitive economies were respectively Switzerland (GCI ranking #1), Singapore, the United States, Germany, The Netherlands, Japan, Hong Kong SAR, Finland, Sweden and the United Kingdom. All of these ten economies are primarily innovation-driven. The fact that the highest degree of international competitiveness is innovation-driven means that in order for leading ports in high-income countries, such as the ports of Antwerp, Busan, Felixstowe, Hamburg, Hong Kong, Le Havre, Los Angeles, Rotterdam, Singapore, Tokyo and Valencia, to increase their strategic value, these ports have to contribute in particular to the international innovation-driven competitiveness of (firms in) their country.

3.2.1 Determinants of international competitiveness

In his influential book, *The Competitive Advantage of Nations* (1990), Michael Porter introduced an important strategic framework for analysing international

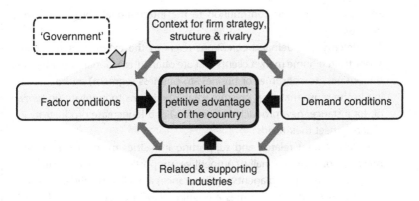

FIGURE 3.2 Diamond Framework: determinants of a country's international competitiveness

Source: Adapted from Figure 3.5 in Porter (1990).

competitiveness of (firms in) an industry, a cluster of industries (such as a port), region or country. Due to its shape, it has been labelled the Diamond Framework. This framework, depicted in Figure 3.2 with regard to the international competitiveness of a country, is used to analyse how four interacting determinants stimulate firms to innovate, increase their productivity and, as a result, improve the international competitiveness of their country.

As described in Box 3.2, these determinants are (1) factor conditions, (2) demand conditions, (3) related and supporting industries, and (4) the context for firm strategy, structure and rivalry. In addition, the framework takes account of the role of the 'government' in influencing these determinants and their interactions. Informal and formal relationships between organisations play an important role in the dynamic interactions (see the small arrows in Figure 3.2) between these determinants. For instance, the interactions between 'demand conditions' and 'related and supporting industries' will become stronger when the relations between demanding clients and suppliers are less characterised by mere market transactions and, instead, more by recurrent and relational interactions.

BOX 3.2 THE DETERMINANTS OF INTERNATIONAL COMPETITIVENESS OF THE DIAMOND FRAMEWORK

- The determinant factor conditions refers to production factors (such as natural resources, quality of work force and capital resources), physical infrastructure (such as berths, roads and railways) and knowledge infrastructure (including information and communication technology (ICT) platforms and knowledge institutions). As these become more advanced and specialised, their contribution to the international competitiveness increases.

- The determinant demand conditions refers to the extent that there are sophisticated home market clients that are often internationally oriented (in the business-to-consumer or business-to-business segment) and demanding in the sense that they pressure established firms to innovate (in terms of, for instance, new products and services) and to increase productivity in order to meet their needs.
- The determinant related and supporting industries mainly refers to the presence of internationally competitive supporting industries (such as legal services, shipping agents and insurance) as well as related industries. When firms in these industries are embedded in leading networks and operate internationally, they catalyse high-quality or low-cost services, contributing to competiveness.
- The context for firm strategy, structure and rivalry particularly emphasises two aspects: (1) how firms are established, organised and deal with their stakeholders, and (2) the intensity of rivalry in firms' home bases, pressuring these firms to be flexible and engage in improvement and renewal.
- Governmental and quasi-governmental organisations on different levels can influence these four determinants and their interactions. For instance, their regulations, policies and/or investments can stimulate or inhibit firms to reach higher levels of renewal and international competitiveness.

Source: Porter (1990). For further readings, also see Huybrechts et al. (2002) and Van Den Bosch et al. (2011).

The four determinants and their interactions can be applied to ports as well as to countries (see Van Den Bosch et al., 2011). The so-called port diamond and national diamond reflect the international competitive position at respectively the level of a port and a country. Importantly, these 'diamonds' are related to each other. The larger the contribution of the determinants of competitiveness on a port level to a country's competitiveness, the higher the port's strategic value for this country. That contribution should be unique (that is, difficult to substitute) in the sense that other ports are unable to provide such a contribution in the nearby future. Illustrations of the determinants as regards the port of Rotterdam are provided in this chapter's case study.

3.2.2 Strategic connectivity

Over time, scholars such as Moon, Rugman and Verbeke (1995) suggested to expand Porter's Diamond Framework by incorporating the fact that strategic connections between countries enable these countries to partly access and utilise one another's determinants of competitiveness. The same applies to ports: by developing strategic connections with other ports (or comparable logistic hubs), a port becomes able to access and utilise part of the determinants of competitiveness that are present in these ports, thereby strengthening determinants of competitiveness present in its own port. The more connectivity between ports contributes to the international competitiveness of their country, the more these ports create strategic value for their country.

Strategic connectivity has been defined as the set of interorganisational relationships within and between ports, and between ports and inland logistic and/or industrial centres, that enable the organisations involved to strengthen their sustainable international competitiveness (see, for instance, Hollen, 2015, and Van Den Bosch and Hollen, 2015). In order for this competitiveness to be truly sustainable, all the organisations involved should benefit from these relationships by appropriating part of the accrued value. These organisations may include firms, port authorities, knowledge institutions, governments and other types of organisations, and comprise different kinds of cooperation (strategic alliances, joint ventures, etc.) focused on strategic renewal and specialisation. The more ports are externally connected in these ways, the harder it is to be 'replaced' by an alternative port and, as such, the higher the strategic value of these ports for their country will be(come). Strategic connectivity may, for example, enable unique access and utilisation of high-quality knowledge, supplier or customer networks in another port.

Strategic connectivity is distinct from structural connectivity. The latter form of connectivity relates to a port's structural network position – including its hub and broker functions – in global value-driven chain systems (Huybrechts et al., 2002; Robinson, 2002), commonly enabled by physical infrastructure. In particular, it refers to the number, frequency and capacity of good flows between ports (and other logistic/industrial hubs) in transport networks (including ship, barge, truck, rail and pipeline movements), which is quantifiable. Strategic connectivity, in contrast, is commonly assessed qualitatively and involves difficult-to-acquire and difficult-to-copy resources. Whereas structural connectivity is mainly about good flows, strategic connectivity is centred in particular around strategic information and knowledge flows between organisations. These information and knowledge flows are increasingly important for orchestrating and optimising physical good flows. This can be witnessed from the growing utilisation of big data in logistics.

A distinction can be made between strategic connectivity between ports (or port-related hubs in general, such as inland terminals) within the same country (national strategic connectivity) and strategic connectivity between ports (port-related hubs) in different countries (international strategic connectivity). The degree of strategic connectivity of a port is partly determined by the (inter)national networks of the firms located there. The more geographically extended these networks are, stretching out over multiple ports, the larger a port's strategic connectivity. Hence, one way to enhance strategic connectivity is by attracting firms that operate internationally and which use both their internal network (that is, within the firm itself) and external network (with their stakeholders in various parts of the world) to become more innovative and productive.

The conceptual framework (see Figure 3.3) depicts the triple strategic value creation of ports in terms of their contribution to a country's international

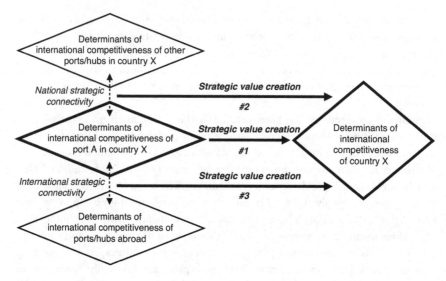

FIGURE 3.3 Triple Strategic Value Creation Framework: how port A can create strategic value for the international competitiveness of country X

Source: Adapted from Van den Bosch et al. (2011).

competitiveness. It suggests that there are at least three ways in which a port creates strategic value for its country:

- Strategic value creation through the determinants of a port's international competitiveness and the interactions between these determinants, which collectively constitute the port's 'diamond' (see also arrow #1 in Figure 3.3);
- Strategic value creation through strategic connectivity of the port with other ports/logistic hubs in the same country (national strategic connectivity; see also arrow #2);
- Strategic value creation through strategic connectivity of the port with ports/ logistic hubs abroad (international strategic connectivity; see also arrow #3).

Strategic connectivity at the national and international level can include both downstream and upstream integral chains of logistics/production networks. Downstream strategic connectivity is about a port's relationships with logistic/industrial hubs in its hinterland and overseas ports that are supplied (for instance, after transshipment) from the port. In contrast, upstream strategic connectivity pertains to a port's relationships with feeder ports in its foreland (which are often abroad), from which goods flow to the port before they are, in turn, transported to its hinterland.

   In sum, ports can contribute to the international competitiveness of their country in three ways (see also Box 3.3), which are illustrated with regard to the port of Rotterdam in the next section.

BOX 3.3 THREE WAYS IN WHICH A PORT CAN CONTRIBUTE TO STRATEGIC VALUE CREATION FOR ITS COUNTRY

- Gradually improve the port's various determinants (see Box 3.2) of international competitiveness and reinforce the interactions between these determinants;
- Enlarge the degree and quality of a port's downstream and upstream national strategic connectivity;
- Enlarge the degree and quality of a port's downstream and upstream international strategic connectivity.

3.3 Strategic value creation by ports for their country: the port of Rotterdam case

The port of Rotterdam, Europe's largest seaport and one of the world's most important junctions of good flows, contributes to the international innovation-driven competitiveness of its home country, The Netherlands, in several ways. Next, illustrations are provided of how this strategic value is created. In doing so, first we show how various determinants of international competitiveness of the port contribute to the international competitiveness of (firms in) The Netherlands. Then, examples are provided of how its national strategic connectivity contributes to the country's innovation-driven international competitiveness, before going on to show how this is the case also for the international strategic connectivity of the port of Rotterdam.

3.3.1 Strategic value creation through determinants of international competitiveness of the port of Rotterdam: illustrative examples

The factor conditions of the port of Rotterdam yield a unique and irreplaceable basis for the international competitiveness of The Netherlands, providing firms and consumers in this country optimal access to world markets. It is adjacent to deep seawater and located at the intersection of two large rivers, forming an important gateway to over 500 million European inhabitants. It allows the world's largest ships to moor 24 hours a day and, as illustrated in Box 3.4, it offers a strong and advanced infrastructure. Companies involved in import, export, transit and re-export activities benefit from this, as it has a positive effect on their value added and results in a better competitive position. Large international companies in The Netherlands that are not located in the port of Rotterdam area benefit from, for instance, the advanced transport, handling, storage and distribution options of the port of Rotterdam to sell their products worldwide.

BOX 3.4 ILLUSTRATIVE EXAMPLES OF ADVANCED FACTOR CONDITIONS IN THE FORM OF INFRASTRUCTURE

- Multimodal infrastructure: A multimodal physical infrastructure that unites various dedicated modes of transport, including deep-sea, short sea and inland shipping, pipeline transport (including the 'Multicore' pipeline bundle), road and rail transport.
- Energy infrastructure: Advanced energy infrastructure that includes the transport, generation and storage of energy (oil, coal, electricity, liquified natural gas [LNG], biofuels, biomass, wind power) and which enables the port to be a leading energy port in Europe; high-quality steam grid, heat and CO_2 connections with the surrounding area; plug-and-play energy infrastructure at the Maasvlakte 2 area.
- Knowledge infrastructure: Extensive knowledge network in proximity of the port area, including centres of excellence and other knowledge institutes such as the Erasmus University Rotterdam, Delft University of Technology and the Shipping and Transport College; world-class process operator and nautical training facilities; innovation facilities such as RDM's Centre of Expertise and Innovation Dock; 'smart grid' infrastructure for the 'modality' ICT to enable large and advanced data streams.

The newly constructed Maasvlakte 2 area, which has extended the port with around 1,000 hectares in space for commercial use, offers sufficient possibilities to realise the growth needed to maintain the status of the port of Rotterdam as the 'Gateway to Europe' in the future, which is important also to firms in The Netherlands located outside the port area. By having created this land extension alongside the Dutch coast, the port could decrease a previously existing 'factor disadvantage' – that is, the lack of space for firms in the port itself and in the country in general – in an innovative way. This is beneficial for The Netherlands as, for instance, innovative firms are no longer deemed to find another (foreign) port to locate or extend their business operations.

Many leading firms are located in the port of Rotterdam, such as AkzoNobel, APM Terminals, Boskalis, Broekman Logistics, Europe Container Terminals (ECT), Huisman Itrec, Huntsman, IHC Merwede, Maersk, Mammoet, Neste, SBM Offshore, Shell, Van Oord and Vopak. These and other firms, many being active internationally and/or highly specialised, are catalysts of competition and innovations in infrastructure, clusters and knowledge development in the port and The Netherlands in general. They stimulate competition and innovation in particular because these firms are highly demanding in terms of their needs of products or services. Their presence in the port of Rotterdam stimulates and pressures firms in the port and elsewhere in the country to innovate and increase their productivity in order to meet these lead users' buying needs. Various examples of how these demand conditions contribute to specialisation, innovation and internationalisation can be seen in maritime sectors – for example, when looking at the business ecosystems of firms like Huisman Itrec and SBM Offshore. As regards design and production of drilling, lift and pipelay equipment, The Netherlands is one of the leading countries, and these highly demanding lead users in the port of Rotterdam have contributed to this achievement.

In addition, there are a large number of firms in the port of Rotterdam with a strong international competitive position that form part of related and supporting industries. These supplying firms are encouraged by various leader firms in their international network to boost their innovative performance, of which also customers in the port of Rotterdam and elsewhere in the country benefit. Examples of suppliers to the leader firms are ABB, Croon, Heinen & Hopman, Keppel Verolme, Nemag, RH Marine Group and Wärtsilä. These and other suppliers in the port, many of them with a presence in multiple countries, are keen to be productive and innovative in order to stay internationally competitive. Their presence in the port of Rotterdam is beneficial for the sustainable international competitiveness of The Netherlands.

Moreover, the port of Rotterdam offers a business context that stimulates renewal, innovation, productivity enhancement and strategic collaboration as well as competition between firms. In this way, the port contributes to the international competitiveness of The Netherlands. The large number of industrial and logistic firms in the port ensures a high level of competitive rivalry in various market segments. Besides competition, there are several collaborative networks in the port, such as chlorine- and steam-related collaborations, which improve production efficiency and reliability of supply, from which clients outside the port benefit. Another example is that an innovative payment system and structure enhancing procedures initiated largely in this port have contributed to the country's strong position in the Western European inland shipping sector.

3.3.2 Strategic value creation through national strategic connectivity: illustrative examples

The port of Rotterdam is strategically connected to other ports in The Netherlands. For example, it cooperates with the port of Amsterdam in areas such as safety, environmental sustainability, ICT and railway transport to parts of their captive hinterland. The collaboration between these ports does not necessarily imply less competition between the two. It does lead, however, to an enhanced international competitiveness of firms in and beyond both ports by providing these firms with the opportunity to benefit from the strengthening of the 'Dutch Mainport Network' (which is also stimulated by the national government). Box 3.5 provides an illustrative example of strategic connectivity between the ports of Rotterdam and Amsterdam. Clients of both ports benefit from enhanced strategic connectivity as innovative services are developed based on, for instance, the integration of logistic knowledge. It benefits also related and supporting industries within both ports, resulting in increased growth and specialisation.

BOX 3.5 ILLUSTRATIVE EXAMPLE OF STRATEGIC CONNECTIVITY WITH THE PORT OF AMSTERDAM: PORTBASE

The Port of Rotterdam Authority is increasingly taking on the role of supply chain coordinator and facilitator by, among others, investing in Portbase, a joint logistics communication system with the Port of Amsterdam Authority (as a second investing partner) established in 2009. Portbase offers a broad range of intelligent services for the efficient mutual exchange of port-transcending information between the ports' customers (such as shippers and carriers) and between other firms involved and governments (customs). All of the information exchanges are conducted via a central point. Hence, the organisations that use Portbase no longer need to maintain a multiplicity of bilateral connections, and are enabled to further optimise their operational processes and cooperation within the chain system. Overall, Portbase reinforces the integral Dutch logistics system and, as such, the country's international competitiveness.

Another example of national strategic connectivity of the port of Rotterdam is the case of inland terminal Alpherium in the city of Alphen aan den Rijn, as further elaborated in Box 3.6. In order to accommodate the anticipated growth in container flows, it becomes increasingly important for (firms in) ports to invest in strategic interlinkages with this type of terminals. The realisation of Alpherium has strengthened the logistics chain in The Netherlands by, for instance, fostering the use of cargo bundling, a relatively innovative concept in logistics. Besides, it has increased transport reliability and, by enlarging the possibilities for intermodal cargo flows to the port and hinterland, it has enhanced accessibility of both Alphen aan den Rijn and the port of Rotterdam.

BOX 3.6 ILLUSTRATIVE EXAMPLE OF STRATEGIC CONNECTIVITY WITH AN INLAND PORT: THE ALPHERIUM CASE

Alpherium is the largest inland shipping terminal for containerised transshipment in The Netherlands. The terrain is property of the Port of Rotterdam Authority and the principal initiators of the terminal were Van Uden Group and Heineken. Van Uden Group invested €15 million in the construction and is transport user and operator of the terminal. Heineken, which had been looking for an alternative for transporting beer containers from its brewery in Zoeterwoude to the ports of Rotterdam and Antwerp by truck, acted as launching customer. With the terminal's opening in 2010, new possibilities for effective cargo bundling arose. That is, after barges with full beer containers from Alpherium have been unloaded at the Port of Rotterdam, large importing firms including Blokker, Intertoys and Zeeman Textiles make use of these same barges to import full containers through the Alpherium terminal.

3.3.3 Strategic value creation by international strategic connectivity: illustrative examples

The international structural connectivity of the port of Rotterdam is large. For instance, it has the most liner shipping services in Europe and is the most important port of call at the Europe–Far East services. All the large shipping companies call at the port. But what about its strategic connectivity? First of all, the internationalisation policy of the Port of Rotterdam Authority (see Dooms et al., 2013) through joint ventures and strategic alliances has led to a growing network of global strategic partnerships. For instance, it established two joint ventures with the Omani government to develop a large port-industrial complex and freezone in Sohar (see Box 3.7).

BOX 3.7   ILLUSTRATIVE EXAMPLE OF AN INTERNATIONAL PORT PARTICIPATION: SOHAR PORT

In 2002, the Port of Rotterdam Authority established two joint ventures – (1) the Sohar Industrial Port Company (SIPC) and (2) Sohar International Development Company (SIDC) – with the Sultanate of Oman to develop and manage Sohar Port & Freezone, a deep-sea port-industrial complex and a freezone in Oman in the proximity of the Strait of Hormuz. SIPC has focused mainly on fostering industrial-scale investments in mining, quarrying and mineral processing activities across the Batinah region, using the Port of Sohar as an important international hub for trades in bulk commodities. The port's major clusters are petrochemicals, metals and logistics. SIDC has focused primarily on developing the freezone.

Firms in the Rotterdam area and elsewhere in The Netherlands have benefitted from the participations. The increased exposure of Dutch know-how contributed to demand in Sohar for the expertise and related products and services of firms such as Arcadis, BAM Group, Royal Haskoning DHV, Tebodin and Van Oord. Apart from these new orders for Dutch firms in areas such as port design and construction, the participations enabled the Port of Rotterdam Authority to acquire a broader basis of knowledge of port management, which is useful for new port participations and for its 'home base', the port of Rotterdam. Also, the ties with internationally leading firms located in both the ports of Sohar and Rotterdam, such as C. Steinweg Handelsveem, could be strengthened. In addition, the port developments in Sohar meant a new stepping-stone for Dutch firms for doing business in the Middle East.

The Port of Rotterdam Authority also participated in, for instance, the development of Porto Central, an industrial port in Brazil, and made cooperation agreements with ports in Nangang (China) and Constanta (Romania) as well as with Qatar Petroleum-RLC. Closer to home, the Port Authority encouraged strategic connectivity between the chemical and refinery complexes in the port of Rotterdam and comparable complexes in the port of Antwerp in order for firms in Rotterdam and elsewhere to profit from a more integrated industrial complex.

International strategic connectivity of the port of Rotterdam is not only driven by the Port of Rotterdam Authority. For instance, research indicates that also organisations such as the Dutch Tax and Customs Authority and firms play an important role (see Box 3.8), thereby improving the international competitiveness of The Netherlands (Van Den Bosch and Hollen, 2015).

BOX 3.8 ILLUSTRATIVE EXAMPLE OF HOW THE TAX AND CUSTOMS AUTHORITY AND FIRMS CREATE STRATEGIC VALUE

Traditionally, the Dutch Tax and Customs Authority controlled 100% of port-related transactions. In 2007, it started with providing Authorised Economic Operator (AEO) certificates to firms considered to be in self-control and trustworthy. Firms with an AEO certificate are controlled less strictly, which saves them significant time. In addition, the Tax and Customs Authority initiated the implementation of 'Smart and Secure Trade Lanes' (SSTLs) between the port of Rotterdam and, for instance, the ports of Shenzhen and Shanghai, meaning that entire end-to-end supply chains (rather than single firms) are being certified.

This shift from a transaction-level to a chain-level customs control enhances the strategic connectivity of the port Rotterdam with the ports of Shenzhen and Shanghai, thereby contributing to the international competitiveness of both The Netherlands and China.

The development of SSTLs is increasingly driven by private firms instead of by customs authorities. Firm-driven SSTLs are perceived by firms as being less bureaucratic and, hence, as more valuable.

Besides, the strategic value of the port of Rotterdam for The Netherlands increases as the number of internationally operating firms that are located in the port gets larger. Examples of such firms are Broekman Logistics, Huntsman, Shell and Vopak, which belong to the top of the world in their industry. Such firms in fact benefit from several 'diamonds' (that is, the determinants of international competitiveness and their interactions; see Figure 3.2): from the port of Rotterdam 'diamond' but also from the 'diamonds' of hubs elsewhere to which these firms are strategically connected. For instance, the R&D efforts of Shell in the United States become available to the Shell Pernis complex in the port of Rotterdam as well, and innovations introduced by Hutchison Port Holdings are available to its subsidiary ECT in the port of Rotterdam (and vice versa).

3.4 Implications for port authorities

This chapter started with the notion that port authorities of leading ports which are considered as being of national interest face a challenge to contribute to the creation of strategic value for the country, besides strengthening the international competitive position of the port. Doing so is particularly important because most port authorities either are (quasi-)governmental entities or have governmental stakeholders and, hence, are expected by a country's society to improve the possibilities of firms – also outside the port area – to benefit from the port in terms of their international competitiveness. Port authorities ought to assess this strategic value of the port in addition to the typically reported economic value, and to communicate this value to the port's stakeholders; see also Box 3.9. Indeed, better insights into a port's strategic value will make these stakeholders (such as the national government) more willing to proactively contribute to further increase this value over time, which in turn contributes to a port's international competitiveness.

BOX 3.9   KEY MANAGERIAL IMPLICATIONS FOR PORT AUTHORITIES

- Investigate and periodically assess and report the strategic value created by the port for the country.
- Invest – partly through co-creation with port-related firms – in the creation of strategic value in line with the three levels of analysis of the Triple Strategic Value Creation Framework (see Figure 3.3).
- In order to do so, (re)consider to extend or change the port authority's existing business model.

The dominant port governance model in large and medium-sized ports is the landlord model, which implies that a port authority is primarily responsible for, and focused on, the economic exploitation, infrastructure maintenance and long-term development of its port area. In this chapter it was explained that in order for a port authority to contribute to a port's strategic value creation for its country, it needs to focus partly on improving the determinants of international competitiveness of the port, such as the factor conditions and the context in which firms in the port compete and cooperate with one another. In addition, as explained, a port authority has to foster strategic connectivity of (firms in) the port with (firms in) other ports, or logistic hubs in general, within the country as well as abroad. As suggested in prior studies (see Hollen et al., 2013; Van Der Lugt et al., 2013; Verhoeven, 2010), a port authority may have to strategise beyond its traditional landlord function in order to be able to use its policy instruments – such as investments (in physical and knowledge infrastructure), land allocation policy and regulation (see Hollen et al., 2015, for illustrations of these policy instruments) – in such a way that it can truly contribute to the creation of strategic value of a port on different levels. For instance, as has been done by the Port of Rotterdam Authority, participating in the commercial exploitation of a new pipeline bundle in the port or participating in other ports. These are examples of new activities that are beyond the scope of the traditional landlord port authority.

The development of such activities that extend the traditional landlord model implies renewal of the port authority's business model. Levers of business model renewal that are distinguished in the literature (see, for instance, Hollen et al., 2013; Verhoeven, 2010) include the internal organisation and the management of a port authority as well as the co-creation of activities with external stakeholders, such as firms in the port. This means that, in order for port authorities to be able to enhance their contribution to the creation of strategic value for a port's country, they need to take into account a possible need for changes in these levers of business model renewal. For instance, port authorities may have to collaborate more with established firms in the port in order to create additional strategic value, which emphasises the need for more co-creation.

3.5 Conclusion

This chapter revolved around the following research question: how do ports create strategic value for their country? This question was first examined from a theoretical lens to conceptually clarify what ports' strategic value consists of. By assessing and subsequently communicating a port's strategic value in addition to its mere economic value, external stakeholders will be more willing to continue to provide resources and 'licenses to operate and to grow' to the port. Then, it was explained how strategic value for the country can be created at three levels: port-level (by improving the four determinants of a port's international competitiveness as well as the interactions between the determinants), national level (by improving the strategic connectivity of the port with other ports within its country) and international level (by improving its strategic connectivity with ports abroad). Different organisational entities, such as firms, governments and, in particular, port authorities, can contribute to the creation of strategic value on one or more of these levels. Illustrative examples have been provided from the context of the port of Rotterdam. Based on the concepts and conceptual framework provided in this chapter, readers are encouraged to critically reflect on the current strategic value of a port in their environment for the country, and on how port authorities and other entities can further increase this value.

References

Dooms, M., Van der Lugt, L., and De Langen, P. W. (2013) International strategies of port authorities: The case of the Port of Rotterdam authority. *Research in Transportation Business & Management*, 8, 148–157.

Hollen, R.M.A. (2015) *Exploratory Studies into Strategies to Enhance Innovation-driven International Competitiveness in a Port Context: Toward Ambidextrous Ports*. ERIM PhD Series Research in Management, No. 372. Retrieved from http://repub.eur.nl/pub/78881. ISBN 978-90-5892-422-3.

Hollen, R.M.A., Van den Bosch, F.A.J., and Volberda, H. W. (2013) Business model innovation of the Port of Rotterdam authority (2000–2012). In: Kuipers, B. and Zuidwijk, R. (eds.) *Smart Port Perspectives: Essays in Honour of Hans Smits*. Rotterdam, SmartPort, 29–47.

Hollen, R.M.A., Van den Bosch, F.A.J., and Volberda, H. W. (2015) Strategic levers of port authorities for industrial ecosystem development. *Maritime Economics and Logistics*, 17(1), 79–96.

Huybrechts, M., Meersman, H., Van de Voorde, E., Van Hooydonk, E., Verbeke, A., and Winkelmans, W. (eds.) (2002) *Port Competitiveness: An Economic and Legal Analysis of the Factors Determining the Competitiveness of Seaports*. Antwerp, Editions De Boeck.

Moon, H. C., Rugman, A. M., and Verbeke, A. (1995) The generalized double diamond approach to international competitiveness. *Research in Global Strategic Management*, 5, 97–114.

Porter, M. E. (1990) *The Competitive Advantage of Nations*. New York, The Free Press.

Robinson, R. (2002) Ports as elements in value-driven chain systems: The new paradigm. *Maritime Policy & Management*, 29(3), 241–255.

Schwab, K. (ed.) (2015) *The Global Competitiveness Report 2015–2016*. Geneva, World Economic Forum. ISBN-10: 92-95044-99-1.

Van den Bosch, F.A.J., and Hollen, R.M.A. (2015) *Insights From Strategic Management Research into the Port of Rotterdam for Increasing the Strategic Value of Shanghai Port for China: The Levers of Strategic Connectivity and Institutional Innovation*. Asian Economic Transformation: System Design and Strategic Adjustment, 113–126. Shanghai Forum 2014 Publication. ISBN 978-7-309-11417-1.

Van den Bosch, F.A.J., Hollen, R.M.A., Volberda, H. W., and Baaij, M.G. (2011) *The Strategic Value of the Port of Rotterdam for the International Competitiveness of the Netherlands: A First Exploration* (Accessible through ResearchGate; www.researchgate.net). Rotterdam, INSCOPE. ISBN 978-90-817220-2-5.

Van der Lugt, L. M., Dooms, M., and Parola, F. (2013) Strategy making by hybrid organizations: The case of the port authority. *Research in Transportation Business & Management*, 8, 103–113.

Verhoeven, P. (2010) A review of port authority functions: Towards a renaissance? *Maritime Policy & Management*, 37(3), 247–270.

Suggestions for further reading

Besides the papers provided as references in the text, we recommend the following texts as suggestions for further reading.

Frenken, K., Van Oort, F., and Verburg, T. (2007) Related variety, unrelated variety and regional economic growth. *Regional Studies*, 41(5), 685–697.

Haezendonck, E. (2001) *Essays on Strategy Analysis for Seaports*. Leuven, Garant.

Ng, A. (2009) *Port Competition: The Case of North Europe*. Saarbrücken, VDM Verlag.

Vonck, I., and Notteboom, T. E. (2015) *Strategic Evaluation of the Belgian Port Sector and Accompanying Services*. Research report commissioned by ING Bank.

4

PORT DEVELOPMENT COMPANY

Role and strategy

Larissa van der Lugt

4.1 Introduction

This chapter applies a strategic management perspective on the port authority and introduces a paradigm shift in thinking about the particular organisation: from port authority to port development company. Port authorities were traditionally organised in the public, governmental domain but have developed into mere market-oriented organisations; they are often still of a public or semi-public nature, but act in a highly competitive environment and more frequently change into more commercial undertakings. Port authorities or actually port development companies are the organisations that are primarily responsible for the development, management and control of the seaport. In most ports in the world, they act as landlords. This means that the port development company of a particular port is primarily responsible for the provision and management of the port area and the nautical access. The commercial or service-related activities are done by private companies located in the port.[1] But this strict landlord model is increasingly being blurred. Induced by changes in the port development companies' environment, economically and also institutionally, many of them face new challenges. As a consequence of these challenges many port development companies have had to reconsider their role and position or are in the midst of doing so. This chapter sheds a light on the transformation of port development companies – institutionally and in terms of their role and strategy.

4.2 Port authority defined

Before diving into the strategy and management issues of port authorities or port development companies (PDC, as we would like to call them from now on) first the terms port and port development company need to be defined. The strategic choices and strategic management options for PDCs are co-determined by port characteristics, such as location, infrastructure (both natural and created), functions and activities, and competitive and market environment.

For this reasoning we first define the term 'port'. The definition that is used here is derived from Notteboom and Winkelmans (2001) and is sufficiently broad to incorporate a wide range of possible strategic issues of port authorities (PAs): *'The port is a land area with maritime and hinterland access that has developed into a logistics and industrial centre, playing an important role in global industrial and logistics networks.'* The port defined as such is not an entity, but rather a collection of a diverse set of economic activities. Value is created both by providing the transport node function to port users as well as by providing land for industrial and logistics companies that are located within the port areas.

Within such a port, we define the PDC as a common institution found across the various ownership models, as a land manager or land developer with the responsibility for a safe, sustainable and competitive development of the port (De Langen and Heij, 2014). The basic functions of such a PDC include the development, management and control of the port area, including not only nautical access and basic port infrastructure but also taking into account safety and environmental issues. Operational activities in the port such as transport activities (including transshipment), logistics activities and industrial activities are in this case the domain of private companies.

PDCs generally have two 'meta' goals (Van der Lugt and De Langen, 2007):

- To facilitate a competitive, sustainable and safe economic development of the port as a whole;
- To become an efficient and effective organisation that generates income to cover costs, to make investments and, in some cases, to return to shareholders' investment.

Both goals are related, but they are also clearly different and sometimes even conflicting. On one hand, the development of a competitive and attractive port enhances the possibility of generating revenues for the PDC. On the other hand, a focus on generating revenues for the organisation itself might sometimes conflict with the broader goal of developing a competitive, but also sustainable and safe port. The tension between these goals thus serves as a challenging aspect of PDC strategy-making.

The basic business model of a PDC builds upon its two value propositions of providing a transport node function to port users that are involved in maritime transport chains and providing a land plot for logistical and industrial companies that in order to do their business are best located within a port area. The two main revenue drivers of PDCs are thus port dues and land rent.

Some other specific attributes of PDCs that are relevant for their strategy-making include:

- Ownership is in general public; it is increasingly in the form of corporatisation and only in limited cases private.
- PDCs are privately funded through direct user charges; there is a decreasing trend of the use of additional subsidies although in some countries there is additional public funding through subsidies.
- Social and economic control is partly via the market (port users, stakeholders, etc.), but PDCs are very unlikely to go bankrupt.
- There is a complex environment (global companies, civilians, political institutes), a still limited but increasing freedom to act, and also a still limited but increasing financial autonomy.
- Decision making within the organisation is often multilevel and multi-actor, which makes it complex.
- There is strong interdependency with other companies located in the port.
- PDCs are still subject to public scrutiny.

4.3 Port development company institutional reform

The institutional reform of port development companies as can be seen in practice and described in literature can be approached by two angles. In the first place, there has been a reform of the governance models of ports, which has been led by an increased involvement of private actors in ports. This has resulted in a variety of models that exist throughout the world. First there is the traditional 'service port', which is completely owned and run by the government. Next there is the tool port, whereby all assets are owned by the government, but management and operations are undertaken by private companies. In the 'landlord port', the (public) port authority (or actually port development company) is responsible for the nautical access, the port land and the basic infrastructure, while private companies invest in the assets and provide the services and operations. A 'private port' is owned and run by one or more private companies.[2] In most places in the world there has been a reform towards a landlord type of port.

Apart from the reform of the division of ownership and responsibilities in ports between governmental bodies and private companies, the port development organisations have also undergone a reform. The PDCs are often or have been in a transition process towards more autonomy and a more commercial identity. Pressure for this transition comes from port users, who are looking for a business partner; governments that reduce government spending on ports, implying that port development companies should become more financially self-sustaining; and non-governmental organisations (NGOs) that demand transparency and inclusion in decisions that affect stakeholders (Verhoeven, 2010). Port development companies are also increasingly regarded as commercial organisations in court cases, for instance in the European Union.

The reform of port authorities around the world has taken place at different paces, resulting in various outcomes. Sometimes even within the same country different institutional structures exist. In The Netherlands, the port of Rotterdam was corporatised in 2004 with the public influence through shareholding and the combination of public and commercial goals. Before 2004, it was an autonomous public entity with substantial autonomy and merely private goals. The other Dutch ports followed this reform only later, between 2011 and 2013. In other regions of Europe there is also a mixture of institutional structures in place (refer to Verhoeven (2010) for further assessment and explanations), as in the rest of the world.

Starting from a landlord port setting, the reform of PDCs, resulting in more autonomy and a more commercial identity, has led to a reconsideration of its role and position in the port. Strengthening the performance of the port, both in economic and social terms, but at the same time demonstrating a healthy financial performance (e.g., return on investment) and securing the *licence to operate* may ask for a scope of activities that goes beyond a pure landlord function.

4.4 Paradigm shift

Before going into more in detail about the strategy of the port development companies we would like to discuss a paradigm shift and explain our proposed shift to using the name port development company (PDC) instead of port authority (see also De Langen and Van der Lugt, 2017). The traditional thought on the governance of port authorities (PAs) was that PAs emerge logically out of a governmental domain and that increasingly private involvement takes place. The reasons for the traditional government-based thinking were as follows: ports are key to regional economic development; ports are, like other basic infrastructure, quasi-public goods; and ports generate negative externalities. Our paradigm shift lies in the fact that we argue that ports provide merely private goods, they act in a competitive environment and the existence of externalities is not a sufficient reason for governmental ownership and control. PAs are thus actually not authorities, but merely commercial undertakings, development companies so to say, whereby there are arguments for governmental involvement. This paradigm shift puts the PA, which we consequently will refer to as a PDC from now on, more central in the discussion on how to organise and manage ports.

Thus, PDCs are commercial undertakings, in most cases in governmental ownership, in one way or another, that are more than pure landlords and require strategies also for the own organisation. They are led by the specific characteristic of a port, a complexity of many economic activities and actors, a key issue is the scope of activities of the PDC related to the position and scope of the companies located in the port and its stakeholders around it. The next section will expand on this.

4.5 Strategic scope changes: possible directions

The strategic scope changes of PDCs can be derived from general firm expansion literature. In general firms can expand along three dimensions: the functional perspective, the geographical perspective and the organisational perspective. These dimensions also apply to the PDC.

4.5.1 Functional perspective

A functional perspective on the role and position of the PDC was introduced by Goss (1990) and enhanced by Baird (2000) and Baltazar and Brooks (2007). Baird (2000) distinguished between the landlord, the operator and the regulator function. In the landlord model, the PDC performs the landlord function, it has responsibility for a part of the regulator function and leaves the operator function to the private companies. Baltazar and Brooks (2007) specify in more detail the different functions that belong to the ports' domain and also delineated between the responsibilities of actors who are private, public or a mixture of both. In the functional perspective it is this division of the functions and responsibilities between PDC and other public organisations and private companies that is questioned. De Langen in 2004 considered an extent function of the PDC with port cluster management activities. Verhoeven (2010) extended the potential function base of the landlord PDC by adding entrepreneurial functions in addition to cluster or community manager functions. Community managers or cluster managers (also see de Langen, 2004, 2008) refer to investments in hinterland access and in facilitating activities like information technology, promotion and marketing, and training and education. Entrepreneurial PDCs undertake activities outside of their primary functions with the aim of developing their businesses.

4.5.2 Geographical perspective

A next generic development direction for firms and organisations is geographical expansion. This is also the case for PDCs. PDCs are tied to a specific port (or perhaps more in one country) and therewith to a specific geographical location. If we take a geographical perspective on the possible development direction for the strategic scope of a port development company, we are able to identify two directions. The first one is internationalisation, which is common to private/for-profit firms, and it is also a possibility for PDCs. PDCs could expand their activities to other ports in the world. In doing this they follow an internationalisation strategy in which they may export their services, for example port management, traffic management, port communication services or consultancy, or they invest with equity in overseas ports. The second direction is in the hinterland. The natural domain of port development companies is the port area. But ports are part of international transport networks, with physical elements like infrastructural connections and inland terminals that have a specific geographical location. PDCs might expand their domain within this transport network into their geographic hinterland. They may invest in multimodal connections or inland terminals.

4.5.3 Organisational perspective

In a generic approach regarding the development directions of firms, the organisational perspective should not be excluded. Firms expand not only through adding activities within their own businesses, but also through the development of their strategic networks. The scope of an organisation is a combination of its strategic core, which consists of the core assets and competencies of the firm and its strategic networks (Reve, 1990). The relevance of these strategic networks in relation to a firm's performance has received a lot of attention in strategy research, in economics, and especially in institutional economics. The economic organisation that goes beyond the boundaries of the firm has been addressed as a core issue in relation to the strategic scope question of a firm. Firms and organisations can develop strategic activities between the market and their own hierarchy to improve overall performance (Williamson, 1985). Such strategic activities include contracting and introducing incentives and alliances. Port development companies can develop their activities not only by investing in new activities, but also by actively allying with firms within their ports, in their hinterland networks and in other ports. Port development companies can also expand their strategic activities between their own organisation and other companies by developing rules, making contracts or giving incentives to the relevant stakeholders.

4.6 Strategic scope changes: further explained

Thus, along the generic ways for firm expansion, PDCs are developing core activities beyond their landlord function. These activities can also be of a different nature. What we are looking for is to give, in a normative way, theory-based arguments for these strategies. A first understanding comes by looking further at the basic business model of the PDC. Figure 4.1 gives an illustration.

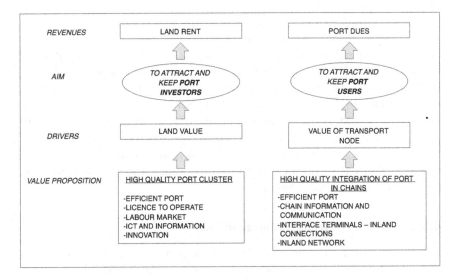

FIGURE 4.1 Business model of PDC: value proposition and revenue drivers

Figure 4.1 illustrates the key role that PDCs actually play, but also partly the logic behind PDCs developing activities beyond the landlord function. Related to its goal of increasing the land value, PDCs might develop activities that enhance the quality of the port cluster. Related to its goal of attracting more port users, PDCs might also get involved in developing activities to increase the quality of the transport chain.

Confronting the three generic dimensions for firm expansion with the actual business model of the PDC, we distinguish and underpin three new roles for PDCs beyond their landlord function:

- PDC as network/chain manager
- PDC as cluster/community manager
- PDC as (international) entrepreneur.

4.6.1 PDC as network/chain manager

Why should a port development company get involved in the hinterland? The answer is in the fact that ports are not competing as sole entities but as parts of complete transport and logistics chains (Notteboom and Rodrigue, 2005; Notteboom and Winkelmans, 2001; Notteboom, 2007; Robinson, 2002). Port users are increasingly looking at the performance of the whole chain if they select their port of use. If PDCs express their responsibility for strengthening the performance of the port, they cannot neglect the hinterland that is related to the port. A smooth working inland transport system contributes to the performance of the port and is therefore of interest to the PDC. In addition to this, for their revenue (port dues) PDCs are dependent on the performance of the port and thus also of the port's inland transport system. For both goals, at port level and at port authority level, port authorities have an interest in the ports' hinterland.

The logic behind a PDC developing activities in port-based transport chains extending to the hinterland can be found in the existence of coordination problems that result in excessive coordination costs. Actors in port-based transport chains have a strong interdependency, but each actor optimises its own operations, with the result that coordination is missing or underdeveloped. This results in chains that are not performing as efficiently and effectively as they should. PDCs cannot neglect this fact and should seriously consider effective ways to get more involved in the port-based chains. Activities they could consider are investing in inland terminals (as a landlord; the port of Barcelona is an example); investing in inland infrastructure, like railways; financing the start-up of inland transport services; and providing for information exchange platforms.

4.6.2 PDC as cluster/community manager

Resource or pooled dependency concerns situations in which 'each partner renders a discrete contribution to the whole and each is supported by the whole' (Gulati and Singh, 1998, p. 801). It exists when organisations pool their resources to achieve a shared strategic goal, the common benefits arise from combining resources into a shared pool, and each partner uses resources from the shared pool. Firms in a cluster or network show potential resource dependency in their supporting activities. By putting together resources – for example, for marketing, technology developments or human resource development – firms might find economic benefits. However, this 'collective action', even when the collective benefits of co-operation clearly exceed collective costs, in many cases does not arise spontaneously. The reasons for this are the risks of free-rider behaviour (Olson, 2002) or uneven division of benefits related to costs among the participants. Within ports these collective action problems also occur, which are related to the supporting activities of firms in the port cluster.

Thus, collective action must be stimulated or initiated in order to result in positive common benefits. PDCs are in the right institutional position to create and stimulate collective action (De Langen, 2004). PDCs can do this by creating platforms that facilitate collective action, or by joint investment in collective action activities. PDCs, if they have the appropriate institutional structure, can invest in collective goods for the port cluster and recover costs from the port users. Since different port users (shipping lines, tenants, shippers) have to pay charges for the investments of the cluster manager, such investments do not involve subsidies and do not distort the 'level playing field' for the competition between ports. Such a role as cluster manager is not required in single-user ports. Most of these ports are fully private (e.g., South Africa, Brazil, The Netherlands, United States); in ports like these, it is not effective to separate terminal ownership and port authority functions. In such ports, there is no port cluster, but one large corporate hierarchy that operates the port. Consequently, the issue of cluster management is not relevant.

Most large ports worldwide consist of a substantial and diverse number of firms and have a landlord type of port management organisation. Examples include Rotterdam, Hamburg, Vancouver, Kobe, Hong Kong, Colombo and Barcelona. For all of these ports, the role of the PDC as a cluster manager is relevant for the performance of the port.

4.6.3 PDC as (international) entrepreneur

One possible scope development direction for PDCs is to follow an internationalisation strategy that ultimately may lead to investing in overseas ports. This is referred to as equity-based strategy. Other internationalisation strategies for PDCs that do not go as far in terms of investments and risks are management or consultancy.

Motives underlying an equity-based internationalisation strategy are:

- Offering customers the possibility of entering new markets, therewith generating cargo flows;
- Setting up preferred partnerships with leading logistic and industrial players, thereby creating a competitive advantage;
- Strengthening the PDCs reputation.

Besides contributing to strengthening the port and its own organisation, PDCs can also have a more financial argument for developing international activities. PDCs have over the years invested in the development of assets and competencies that contribute to the performance of the port. PDCs try to cover these investments with the revenues from land rent and port dues in their aim to result in a financially healthy business. Actively selling their competencies to other potential markets provides PDCs with two potential benefits:

- Ex post: PDCs can generate additional revenue on the investments that they make for the development of the assets and competencies. This revenue can benefit the port.
- Ex ante: PDCs might develop a higher potential to invest in a specific facility because they can increase potential revenues by reselling the developed facility, related assets and competencies.

Another perspective is that PDCs increasingly have the need to develop capabilities and resources within their own organisation to face the requirements from the dynamic, globalising and highly competitive environment in which they operate. They can develop their existing capabilities or create new ones by actively developing activities in new places, either by themselves or by teaming up.

But what makes a port development company successful in developing an internationalisation strategy? The answer to this relates to the specific assets and competencies a PDC can offer, which are as follows (Dooms et al., 2013):

- The offer of scalable models of participation to international partners, including the appropriate investments;
- The capabilities and know-how on port and industrial area development and management, developed over a long period;
- A strong brand creating credibility for private investors;
- A global business network, spanning several industries (transport, logistics, energy, mining, petrochemical industry).

Internationalisation is an emerging strategy that only a few PDCs have pursued, such as Rotterdam and Antwerp. Researching into the intentions, the reasons why PDCs do or do not concern an internationalisation strategy provides an insight into expected future developments. One of the reasons that at the moment only a few ports follow such a strategy is due to institutional limitations in host countries and a reserved attitude of the port's managers and stakeholders. One interesting question is whether the increasing private investment in port infrastructure will negate these hurdles. Naturally, private investors want a return on their investment, which is best guaranteed by the best management of these assets, but that the manager is not necessarily the local PDC. Thus, private investors may begin offering incentives that can attract world-class port managers to the port they invest in, ultimately laying the foundation for a global port management market. From this possibility, two interesting research questions arise: Will private investment in port infrastructure increase? Will this lead to a global market for port management services?

4.7 Implementation

Finally the individual PDCs want to decide on their specific set of activities that best fit their port context and best lead to an achievement of their goals. This requires a portfolio approach. Looking at a port we can distinguish a diverse set of segments that together form the port. This can vary between container or bulk transshipment, offshore services, vessel construction or energy production. Each of the segments in the port has a certain life cycle development. The actions of the port authority depend on the phase of the product life cycle of each of the segments. For a new, upcoming segment the PDC might be involved rather as a stimulator, enabler or financer. A segment that has grown and is in a mature stage only requires good facilitation, but overall freedom to act.[3] A segment which faces a decline needs to transform. Depending on the behaviour of the firms in this latter segment, the port authority might need to discipline these companies to enhance the transformation or even might work towards phasing out the segment.

Thus, the PDC should see its port as a portfolio of economic activities or segments, each potentially in a different stage of its life cycle development and each requiring a different positioning and action of the port authority. Figure 4.2 illustrates the relationship of the PDC's acting and the potential life cycle development of actual port segments.

Figure 4.2 is an example of a possible set of port segments, with each segment its potential life cycle development phase and related acting of the PDC. But ports differ in their layouts, their contexts and thus in their functions, activities and segments.

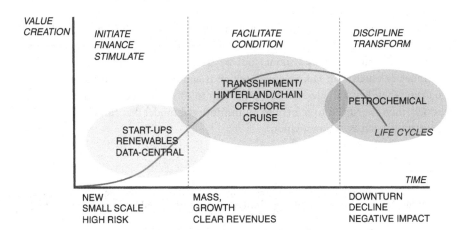

FIGURE 4.2 Relationship between the PDC's acting and the life cycle development of port segments

As the acting of the PDC is port dependent and thus port specific, PDCs will differ in their precise roles, positions and strategic activities.

4.8 Conclusion

PDCs are relevant economic actors that are important to the development and performance of seaports. However, it is no longer the norm for landlord PDCs to simply act as pure landlords and to limit their actions to the development and exploitation of the port's land and basic (nautical) infrastructure. The strategic reorientation of PDCs is induced, not only by external pressures, but also by a reform towards a more business-like organisation with greater emphasis on entrepreneurial values.

As organisations, PDCs are not similar to private companies. PDCs have a combination of private and public goals which they must continuously align. In principle, private companies face this same challenge, but on average, PDCs have a larger societal impact than private firms. Therefore, in most cases their public goals and the public interest to preserve carry more weight than those of private companies. Consequently, the two goal types, private and public, should both guide their increased value orientation.

A challenge for PDCs is keeping their strategy aligned with their direct strategic context, which is to a large extent determined by the companies in the port cluster. Due to its position, the PDC's actions in the port's value network are strongly intertwined with the actions of the port companies. Because these actors, PDC and private port companies, share a common goal, namely to bolster the port's value, they are increasingly interdependent. The interdependence between the PDC and the private port companies causes the strategy change of one actor to impact the context of the other, requiring constant adaptation on the part of both actors. A portfolio approach seems to fit here, whereby port segments are assessed upon their phase in their life cycles and upon their strategies applied. The action of the PDCs is then determined accordingly.

The new paradigm for PDCs (i.e., changing their business models from landlords to key developers and value creators in their ecosystems) requires an appropriate institutional structure. PDCs should not be seen as governments: they are more customer oriented than task oriented, which is apparent from both their strategic goals and their key activities. And if PDCs want to adopt new positions and business models, then they need to reform to an institutional model that fits a business-like orientation, such as corporatisation.

Notes

1 Over the last decades port reform programs have been implemented, based on the thought that involvement of the private sector brings more efficiency and better performance for a port.
2 Private ports exist in the UK and increasingly in Australia and New Zealand.
3 A reminder of the limited possibility that PDCs have in securing and directing cargo flows. This is the responsibility of terminal operators and maritime transport companies, that increasingly show horizontal and vertical integration, resulting in global operating

companies that make their business decisions in sometimes faraway headquarters, leaving little room for PDCs to influence this.

References

Baird, Alfred J. (2000) Port privatisation: Objectives, extent, process, and the UK experience. *Maritime Economics & Logistics*, 2(3), 177–194.

Baltazar, R., and Brooks, M. R. (2007) Port governance, devolution, and the matching framework: A configuration theory approach. In: Brooks, M. R. and Cullinane, K. (eds.) *Devolution, Port Governance and Port Performance*. London, Elsevier, 379–403.

De Langen, P. W. (2004) Analysing the performance of seaport clusters. In: Pinder, D. and Slack, B. (eds.) *Shipping and Ports in the 21st Century*. London, Routledge, 82–98.

De Langen, P. W. (2008) *Ensuring Hinterland Access: The Role of PAs*. Discussion Paper of OECD International Transport Forum, Paris, France.

De Langen, P. W., and Heij, C. (2014) Corporatisation and performance: A literature review and an analysis of the performance effects of the corporatisation of Port of Rotterdam authority. *Transport Reviews*, 34(3), 396–414.

De Langen, P. W., and Van der Lugt, L. M. (2017) Institutional reforms of port authorities in the Netherlands: The establishment of port development companies. *Research in Transportation Business & Management*, 22, 1–6.

Dooms, M., Van der Lugt, L., and De Langen, P. W. (2013) International strategies of port authorities: The case of the Port of Rotterdam authority. *Research in Transportation Business & Management*, 8, 148–157.

Goss, R.O. (1990) Economic policies and seaports: Strategies for PAs. *Maritime Policy and Management*, 17(4), 273–287.

Notteboom, T. (2007) Strategic challenges to container ports, in devolution, port governance and port performance. *Research in Transportation Economics*, 17, 29–52.

Notteboom, T., and Rodrigue, J.-P. (2005) Port regionalization: Towards a new phase in port development. *Maritime Policy and Management*, 32(3), 297–313.

Notteboom, T., and Winkelmans, W. (2001) Structural changes in logistics: How will port authorities face the challenge? *Maritime Policy and Management*, 28(1), 71–89.

Olson, M. (1971) *The Logic of Collective Action: Public Goods and the Theory of Groups*. Cambridge, MA, Harvard University Press.

Reve, T. (1990). The firm as a nexus of internal and external contracts. *The Theory of the Firm: Critical Perspectives on Business and Management*, 310–334.

Robinson, R. (2002) Ports as elements in value-driven chain systems: The new paradigm. *Maritime Policy & Management*, 29(3), 241–255.

Van der Lugt, L. M., and De Langen, P. W. (2007) *Port Authority Strategy: Beyond the Landlord, a Conceptual Approach*. Proceedings of IAME Conference, Athens, 2007.

Verhoeven, P. (2010) A review of port authority functions: Towards a renaissance? *Maritime Policy & Management*, 37(3), 247–270.

Williamson, O. E. (1985) *The Economic Institutions of Capitalism: Firms, Markets, Relational Contracting*. New York, The Free Press.

Suggestions for further reading

Besides the papers provided as references in the text, we recommend the following texts as suggestions for further reading.

Bird, J. (1963) *The Major Seaports of the United Kingdom*. London, Hutchinson.

Notteboom, T., De Langen, P., & Jacobs, W. (2013) Institutional plasticity and path dependence in seaports: Interactions between institutions, port governance reforms and port authority routines. *Journal of Transport Geography*, 27, 26–35.

Van der Lugt, L. M. (2015) *Beyond the Landlord: A Strategic Management Perspective on the Port Authority*. Doctoral thesis, Amsterdam, VU.

Verhoeven, P. (2009) European ports policy: Meeting contemporary governance challenges. *Maritime Policy and Management*, 36(1), 79–101.

Verhoeven, P., and Vanoutrive, T. (2012) A quantitative analysis of European port governance. *Maritime Economics & Logistics*, 14(2), 178–203.

5

PORT MANAGEMENT AND INSTITUTIONAL REFORM

Thirty years of theory and practice

Hercules Haralambides[1]

5.1 Global forces driving port reform

During a recent meeting of the UNCTAD *Intergovernmental Group of Experts on Ports*, port commercialisation, corporatisation and privatisation were subjects that not only evoked exceptional interest and enthusiasm, but also provoked concern among delegates.[2] Such discussions would have been unimaginable 20 years ago, when most governments were considering their port sector as one requiring massive public investment for port development; of strategic interest to the nation; or a service sector crucial to the *common interest*.

However, during the last decade, there has been a worldwide trend of institutional reform within the public sector. In some developed and developing countries this has taken the form of commercialisation or privatisation of public enterprises. Globalisation and fierce national and international competition have been the major motors of such changes.

Globalisation could be described as the increase in cross-border interdependences and integration, which has resulted from the greater mobility of factors of production and of goods and services (Campbell, 1994; Milanovic, 2012).

The need for reform in developing countries' economies is as much the result of their own precarious economic and social situation as of the fact that – without having been adequately prepared – developing countries have been exposed to the relentless forces of globalisation and intensified international competition. This exposure has been taking place simultaneously with the opening-up of their internal markets, so that they could take advantage of the recent developments in the liberalisation of international trade, particularly of the many favourable "developing country provisions" of the General Agreement on Tariffs and Trade/World Trade Organization (GATT/WTO). Most developing countries are now well aware of the tremendous potential benefits from the opening-up of their internal markets and the liberalisation of their external trade.[3] These benefits are, of course, the result of their comparative advantage, due to their still low-basis growth in industrialisation (and thus their potential of achieving significant economies of scale) and their inexpensive labour force.

It has often been argued that high port and transport costs hurt developing countries' exports, which are already little diversified and over-dependent on the very volatile international commodity prices. For that reason, developing countries have often refuted the principle of comparative advantage as one that leads to a worsening in their terms of trade, creates balance of payments bottlenecks and thus hinders their efforts to grow through diversification. Nowadays, there is another equally important factor that compounds this problem. This factor, or rather series of factors, consists of the complex developments in multimodal integrated transport, logistics networks and electronic data interchange.

Governments are increasingly realising that the poor performance of their public ports and their high costs are hampering trade and the national economy. This is especially true for most countries in the stage of economic development. The proportion of port costs in the final price of traded goods varies significantly from 0.2%, for cargo of high value per ton, to over 20% for low-value cargoes. The trade structure of most developing countries shows that their export products are mainly of the latter type (low value). Although more and more developing countries enter the world market with manufactured goods, they have in fact been providing low-end products, competing through price rather than quality. Consequently, port performance plays a bigger role for them than for developed market economies.

Differences in international transport costs (including port costs) between developed and developing countries, but also among developing countries themselves, are substantial. Developing countries pay at least 50% more for the transport of their imports than developed countries; but even within developing countries themselves, differences are pronounced: Africa pays twice as much as Asia and Latin America for its imports.

The high elasticity of the international demand for developing countries' exports and the low short-run elasticity of supply of most agricultural and mining goods they produce often leave developing countries with very slim profit margins that can be easily swallowed by increased transport and port costs. In view of the high elasticity of demand and low elasticity of supply, the *incidence* of transport costs is unfavourable to developing countries, and any transport cost increases will have to be absorbed in the FOB[4] prices exporters are enjoying.

The export of soybeans can serve as a good example. In 1991, the international FOB price for soybeans was $230 per ton. However, loading the cargo on board a ship cost $65 per ton in the port of a South American country, while it cost only $20 per ton in a North American port. Although the production cost of soybeans per ton was $165 in South America, $30 cheaper than in North America, the result was that by selling soybeans at the international market, the South American producers made no profit at all, while their North American rivals were realising a

$15 per ton profit. Poor port services were thus not only taking profits away from the national exporters, but in fact they were squeezing the country out of the world market.

The dramatic improvements in cargo handling operations that were brought about by the introduction of containerisation have enabled general cargo ships to spend hours or days now in ports rather than the weeks or months that were customary before. The reduction of port time and the corresponding increase in time at sea have eventually led to the substitution of the previous multipurpose general cargo ships with specialised high-speed container vessels of substantially (and ever-increasing) larger dimensions that can take advantage of the economies of scale afforded by the shorter port turnaround times. Port efficiency and productivity have thus been the main drivers behind the increase in containership sizes, particularly since 1988 when American President Lines introduced the post-Panamax vessels. However economically justified investments in containerisation might have been in the industrialised countries facing the north Atlantic and Pacific oceans (where the bulk of general cargo traffic is concentrated), some developing countries have reacted to the necessity of this type of investments with varying degrees of scepticism. Their legitimate worries concerned the suitability of capital-intensive techniques in countries with abundant and inexpensive labour, their lack of financial resources together with other pressing investment priorities in the country, and also the fact that the vast majority of their exports (primarily raw materials and agricultural produce) were not "containerisable."

Furthermore, the capital-intensive nature of liner shipping and the consequent "operational arrangements" within this industry, in the form of consortia and similar types of co-operation, frustrated many developing countries' plans to get actively involved in liner shipping, despite their cargo-sharing entitlements secured mainly through the provisions of the UNCTAD Code of Conduct of Liner Conferences.[5] For many developing countries, the result of this situation was that they were often seen to be played off against each other by major liner operators who had been convincingly arguing that if adequate port investments in container-handling facilities and equipment were not timely made, ports would be bypassed by major lines and thus become *backwaters*.

5.2 Public involvement in the port industry

Public involvement in ports, usually through a statutory body known as the port authority, can take various forms, ranging from the mere ownership and leasing of port land and basic infrastructure (landlord ports), to the provision of all port-related services, notably cargo-handling (service ports or comprehensive ports).

In the research I carried out for the International Labour Organisation (ILO, 1995a) on the reforms taking place in the world port industry, it came out that in most ports around the world, wet areas (63%) and quays (76%) were in public ownership without competition, while the operation of quays was fairly evenly distributed between the public and the private sectors.

5.2.1 Military protection

Many major seaports are located close to national borders and are especially vulnerable to attacks from the sea. In older times most ports were thus military protected areas. Even nowadays, a number of commercial ports around the world still dedicate a part of their infrastructure to naval bases.

5.2.2 Expropriation of sites

In many cases, ports have to extend into the water where, usually, there is no provision for the expropriation of sites. In most countries, people can acquire legal rights to territory or land and can subsequently exclude others from their use. This is never the case for water or aquatory, the more so when most countries recognise a general *right of passage*, or free navigation, to which unauthorised port structures could be considered as obstructions (Goss, 1993).

5.2.3 Economic protection

As major ports are usually *gates* to international trade, they may afford governments a convenient means to implement import restricting policies, aimed at protecting domestic markets. Import restrictions can be effected by the erection of tariff and/ or non-tariff barriers. The latter can take many forms and are usually more difficult to detect and quantify. High port tariffs, long turnaround times and inefficient ports in general could be seen as constituting effective non-tariff barriers to trade. It has sometimes been argued that import-competing domestic producers have strong vested interests in the continuing existence of inefficient ports, as this offers them effective protection. These producers could also be effective lobbyists and influential members of pressure groups that resist port reform.[6]

5.2.4 Natural monopoly

Ports are often referred to as the classic example of the so-called natural monopoly case, in which possible market failures can justify government intervention (Baumol, 1977; Shirley et al., 1991; Mosca, 2008). Under certain conditions (a given level of demand, cost structures and technological factors), a market with two or more firms can produce sub-optimal economic outcomes (for example, a certain port may be too small to have two tug operators), whereas a single firm might produce the required output more efficiently. For this reason, governments may, at times, decide to move away from a multi-firm competitive environment, towards a monopolistic, albeit regulated, situation. This can be achieved by explicit legislation, allowing only one operator; by discriminatory subsidies, finally resulting in the withdrawal of potential competitors; or, finally, by turning a blind eye to the restrictive business practices of incumbent operators.

5.2.5 Public goods

Among the many functions of public port authorities, whether (semi)autonomous or centralised, is the provision and maintenance of the port's basic infrastructure, such as breakwaters, approach channels, turning basins, rail/road connections within the port and navigational aids. Apart from the general public's interest in the safety of ports, many of the port services can clearly be considered as falling within the domain of *public goods*, in the sense that no particular user can be excluded from their use if he is not agreeable to share in the cost of their production – a situation often referred to as the *free rider problem*. Furthermore, services such as those provided by, say, breakwaters and navigational aids can be considered as *collective consumption goods* (Shoup, 1969) in which case, and up to a point, the total cost of producing them does not vary in relation to the number of their consumers. Finally, a number of port services can be considered as *non-rival in consumption* (Musgrave, 1969), given that user A's demand does not reduce (compete with) that of user B. Those port services that qualify as "public goods" ought to be provided by some public authority, although *provision* should not be confused with *production*: the latter could be entrusted either to the public or to the private sector, depending on considerations of economic efficiency. This is notably the case of the nautical-technical services (towage, pilotage, line-handling), which are increasingly brought in the ambit of the private sector, albeit under strict regulatory control.

5.2.6 Financing

The rapidly changing cargo-handling technologies; the increase in the size of modern container vessels; the limitation in the number of direct port calls, coupled with the expansion of main-line/feeder networks; and the growth of international trade have resulted in numerous port expansion/modernisation programmes, generally requiring substantial capital outlays and invariably leading to regional over-capacity. Regional port competition and the need of ports to turn ships around as quickly as possible is an additional contributing factor to regional port overcapacity (Haralambides, 2002).

These investments often exceed the financial capabilities of the public sector thus making the case for private sector involvement; a development that would make sense anyway, given that, increasingly, ports, particularly container terminals, are losing the *public good* character, becoming private goods, whose services ought in principle to be paid by their user.

However, often, in spite of its abundant financial resources, particularly those of global terminal operators (GTO),[7] the private sector may be reluctant to invest in ports, particularly when capital outlays have to be made in uncertain institutional and regulatory frameworks, or through usually frontloaded agreements (concession contracts) that cannot guarantee positive financial returns in the longer term. On the other hand, as government objectives usually extend beyond considerations of short-term financial profitability and towards the maximisation of long-term *economic welfare*, a number of port development projects that might be deemed unprofitable by the private sector can be of cardinal importance to the government. Thus, the success of any public-private partnership (PPP) is in being able to strike the right balance between the two. In other words, affording the private sector an acceptable return on investment (RoI), on an opportunity cost basis – that is, a return as good as that of the investor's second best investment – while keeping the bulk of the generated *economic rent*[8] in the hands of the public sector.

5.2.7 National/regional economic development

In addition to their main functions as interface, storage and distribution points, efficient ports also function as growth poles, attracting new activities and stimulating trade (Rimmer, 1984; Haralambides, 2012). In this way, and apart from their obvious direct contribution to GDP growth and regional development, the indirect contribution of ports to the economy is also substantial, given their importance for the competitiveness of the country's export industries. State intervention is thus often justified on the grounds of these "not solely commercial" objectives of ports. The costs of such *macroeconomic* objectives of the State, however, often manifested in departures from economic efficiency in a strict sense, ought to be borne by the State itself, and it would be unrealistic to expect private investors (only interested in RoI) to share in them. Again, if private investment is deemed desirable, concession contracts should be drafted in such a way so as to strike a balance between public *welfare* objectives and private *profitability* ones.

However, as I argue later, the influence of this *statist* approach to port management is fading out today in most parts of the world. Its problem is in *limiting* port management *autonomy* at a time when ports are expected to become new-business developers and compete for traffic internationally. This new role of ports, and related governance models, is finding strong support at least in Europe (ESPO, 2011), vis-à-vis earlier (public) port governance models that tended to limit the functions of a port authority to those of the harbour master's office.[9]

The realisation of indirect *macroeconomic* objectives may indeed generate numerous benefits for the region or the country by and large, but these do not necessarily produce visible financial rewards for the ports concerned. Thus, the efficiency and productivity of the latter might, at first sight, be considered as disappointing and inferior to that of comparable privately owned enterprises with clear-cut financial objectives (e.g., UK ports), or compared to ports which have been *centrally prioritised* on the basis of macroeconomic objectives. Moreover, such a "central" prioritisation is not void of strong resistance by local communities, whenever the fortunes of one (prioritised) port are at the cost of another. Few would argue against the benefits from looking at ports as a "port system"; these have been described earlier, in the case of Japan. But the objectives of such a *regional* approach to port development are to enhance regional cohesion, thus leading to a more balanced development across the region. Thus, together with prioritisation, a system ought to be found, and agreed upon, to distribute the fruits of this *planning* policy, as equitably as possible, also among the "unlucky" ports, their cities and citizens. Otherwise, local political opposition could be so strong as to frustrate any *central* attempt to "allocate roles."[10]

Finally, a *statist* approach to port management reduces port authorities to mere administrators, and if this is the objective of the State, then it should be spelled out clearly in the relevant port laws, rather than expecting ports to function under commercial management principles, often with financial targets imposed on them. There is no worse thing than *responsibility without authority*, and governments, as well as public organisations, fortunately start to realise that *you can't have your cake and eat it too*.

5.3   Government retrenchment and major issues of concern in the port industry

It is sometimes argued that policies of public sector retrenchment, together with the encouragement of more private sector initiatives, are rooted in ideological origins. However, regardless of how true this opinion may have been in the past, current economic and political developments worldwide can no longer support its validity. Instead, the reasons for explaining the widespread popularity of the various divestiture programmes are to be found, among others, in the increasing economic interdependency among nations and the trend towards the globalisation of all forms of economic activity.

Regardless of ideological postures and doctrines, an increasing number of governments (and ordinary citizens) realise that it is no longer possible to isolate their economies, or insulate them from external economic influences and shocks. Even if this was still possible, such a policy's effectiveness towards increasing growth and industrialisation would be more than doubtful, at least today.

In many countries, governments have become painfully aware of the inadequacy of their state-owned enterprises (SOE) in an environment of increasing international interdependence and global competition. Market-oriented policies are thus becoming more and more popular in order to reap the benefits of higher efficiency and productivity, and to reduce the financial and administrative burden SOEs often impose on their owner, the State.

High levels of staffing,[11] together with the absence of risk in economic activity; the lack of accountability for economic performance (staff assessments); the impersonality of operational structures; and a missing sense of belonging and achievement can very effectively remove employees' natural drive for initiative, innovation and higher efficiency, consequently resulting in very low (and sometimes even negative) labour productivity. The low productivity of the public sector is one of the major driving forces behind the various divestiture programmes throughout the world.

However, in the case of ports in particular, it would be fundamentally wrong to believe that these are the only factors accounting for the low labour productivity of the public sector. Comparisons between different countries or between different sectors of the same economy should, therefore, be contemplated with extreme care.

Labour productivity ought not to be measured only as "output per man/hour" or "tonnes handled per gang-shift," as is sometimes the practice in many ports, but as *output per man/hour produced with a certain stock of fixed capital of a given technology and operational characteristics*. Thus, differences in labour productivity between the private and the public sector could be explained equally well by the fact that the level of fixed capital investment in the latter sector is frequently inadequate or obsolete, due to scarcity of financial resources, budgetary constraints and the economic priorities of the government.

Notwithstanding this, employment in most state-owned ports, and to that effect in the wider public sector by and large, is usually characterised by high levels of staffing.[12] Many times, this is not only the result of the government's employment creation policy – particularly in developing countries with rapidly growing populations and an anaemic private sector – but also of the fact that, through its permanency of employment, fringe benefits and stability of income, employment in the public sector is often an arduously sought-after objective, many times pursued through systems of "political clientelism."

However, large-scale employment in the public sector creates inelastic government expenditures, it increases the public sector's borrowing requirements (PSBR) and it may lead to inflation and high interest rates. In their turn, the latter can hinder the private (domestic and foreign) sector's propensity to invest and subsequently result in less output, employment and growth. Additionally, inelastic government expenditures can reduce the effectiveness of fiscal policy as a tool of economic stabilisation. The latter (at least nowadays) is almost invariably a precondition for the successful implementation of structural adjustment programmes and often the reason for the divestiture plans of the government.

5.4 Management issues

The capital intensity of liner shipping and the need for maximum capacity utilisation and fast turnaround times, in order to achieve adequate rates of return on investment, have increased pressures on ports for further improvements in labour productivity and operational efficiency. In its efforts to adjust to the new demand requirements, the port industry has also become a capital-intensive one, requiring massive investments in port infrastructure and sophisticated cargo-handling equipment. In this way, containerisation and the cargo-handling techniques it induced have had an equally profound impact on port employment. As with all other capital-intensive innovations, containerisation substituted capital for labour and thus resulted in substantial redundancies, accompanied however by remarkable increases in labour productivity.

However, port performance and labour productivity measures obtained from various ports around the world still demonstrate substantial differences from one port to another, even within the same region. Evidence from the US West Coast (US West Coast Longshoremen's Union) shows that port labour productivity increased by more than ten times. In 1960, 29 million work hours resulted in the handling of

29 million cargo tons while, in 1990, the same amount of working hours produced 175 million cargo tons.[13]

High costs, poor services and low efficiency and productivity appear, however, to be only the symptoms of the problem. An UNCTAD survey carried out in four African countries (Ivory Coast, Ethiopia, Kenya and Senegal) showed that port problems were not of a technical nature and that investment in modern port facilities had been universally good; apart from minor omissions, there were no cases of serious infrastructure defects.

It was thus evidenced that although many ports are in possession of the right infrastructure and necessary equipment, what they lack is effective management, or modern management know-how. In many instances, basic management principles such as those of clear description of objectives and area of authority and responsibility; accountability and control; adequate rules and regulations; good statistical and information systems; financial accounting and cost control; quality control; human resource development; and so forth appear to be amiss.[14]

Yet, the management ability of port managers, including those in developing countries, should not be underestimated. A cursory look into the management techniques of most ports today will immediately show that the aforementioned managerial skills are rather well-known to most port managers and many of them have already been in place. Modern port management knowledge has in fact been well spread in many developing countries through various training activities during the last decades, and it is not uncommon today to find many port managers in developing countries that have been trained abroad in modern management techniques. In many ports, the problem seems not to be the lack of modern management techniques, but rather the lack of their effective implementation. Managerial measures do not thus touch the root of the problem which, in most cases, seems to be institutional.

More often than not, the interface between the government and the port is too heavy. As a result of unnecessary bureaucracy and state intervention, ports have many times been prevented from carrying out their management streamlining efforts and reacting to the needs of the market. Furthermore, the lack of competition often results in a negative service attitude within the port. Because of the "soft budget constraint" and the frequent low-interest government loans or subsidies, the "opportunity cost of capital" is a principle virtually unknown to many port managers. This may explain why *cost control* is often a low-ranked priority in many public ports. Besides, port tariffs are often state-controlled and do not correspond to market prices, something that affects adversely the management's motivation to seek cost reductions.

Thus, investments are not always made in time and when they do they are not market-oriented or cost-effective. Decision makers may be more responsible for political or administrative priorities rather than commercial ones. The difficulties connected with the quality of port decision-making are often due to the excessive distance between the place where the problem arises and the place where the solution is worked out. Centralised public port administrators rarely make decisions without consultation at a ministerial level, and they often have a very relaxed attitude regarding commercial matters.[15] In the UNCTAD study mentioned earlier, it was shown that good intentions to improve port performance had, in most cases, run into problems of implementation or were over-laden with subsequent controls[16] combined with a distinctive unwillingness of the middle ranks of central government to delegate authority.

5.5 Labour issues

In the earlier days (up to the beginning of the 1960s) *general cargo*, carried by liner shipping, was transported, in various forms of packaging (pallets, boxes, barrels, crates, slings), by relatively small vessels, known as general cargo ships. These were twin-deckers and multi-deckers, that is, ships with holds (cargo compartments) in a shelf-like arrangement where goods were stowed in small pre-packaged consignments (parcels) according to destination. This was a very labour-intensive process and, often, ships were known to spend most of their productive time in port, waiting to load or discharge. And although seafaring was great fun in those days, the same cannot be said for *casual* port work which was rather ill-considered and looked down by society.[17] Also, congestion was a chronic problem in most ports, raising the cost of transport and hindering the development of trade. Equally importantly, such delays in ports made ship arrivals (and consequently port work) erratic and unpredictable,[18] obliging manufacturers, wholesalers and retailers to keep large stocks. As a consequence, warehousing and carrying costs were adding up to the cost of transport, making final goods more expensive and, again, hindering the development of international trade.

Containerisation and the introduction of the new cargo-handling techniques in ports have changed all this. Around the world, the port industry has invested a lot in order to cope with the new technological requirements. Modern container terminals – and corresponding cargo-handling equipment – have been built and new, more efficient, organisational forms (including privatisation) have been adopted in an effort to speed up port operations. Operational practices have been streamlined; the element of uncertainty in cargo flows largely removed; forward planning has been facilitated; port labour regularised; and customs procedures simplified. These developments took place under the firm understanding of governments and local authorities that ports, now, constitute the most important link (node) in the overall door-to-door supply chain and thus inefficiencies (bottlenecks) in the port sector can easily whittle away at all benefits derived from economies of scale in liner shipping and in global supply chain management.

As said, these developments, notably the predictability of ship arrivals and of port work – and consequently of port labour requirements – have provided an important stimulus for the registration of port workers.

Moreover, the capital intensity of the new technologies, together with carriers' ability to now plan a reliable shipping network, have resulted in a need for more intensive port capacity utilisation. This has been achieved mainly through the extension of working hours, which in many ports was done through the introduction of shifts. As the new technology also required a skilled workforce, the need for the regularisation of employment relations was apparent, as there was no way casual labour could provide the adequate, responsible and skilled manpower necessary to move cargo safely and efficiently through a modern port using advanced equipment (Couper, 1986: 53). Regularisation of employment, finally, provided casual workers with some form of guaranteed employment or income, and it was thus strongly supported by trade unions, who often made it an explicit objective in their negotiations concerning the social impacts of the introduction of the new cargo-handling methods.

The necessary adjustment of manning levels, however, has often been prevented or delayed due to pressure from the affected labour, often represented by powerful trade unions. On the one hand, port workers had an interest in the introduction of modern cargo-handling techniques, as this reduced their hard physical work and afforded them regular employment. Besides, unions realised that the introduction of the new techniques was necessary to secure the competitive position of the port, which directly affected their long-term employment prospects. On the other hand, however, workers feared that the new technologies would lead to a considerable reduction in employment (in which they were right), and this has brought many of them to resist technological change. Already in 1969, there were refusals to operate new types of equipment, and shift systems and gang sizes were not reduced in line with the changed needs (Couper, 1986: 2).

An additional reason for the resistance of port labour to change relates to the "through-transport" concept and the door-to-door possibilities that containerisation now affords. In other words, a considerable part of what was previously considered as "port work" today shifts to areas outside the port domain. From the point of view of port labour, this development further exacerbated workers' misgivings with the new technology: The well-proclaimed advantages of containerisation were not localised but dispersed throughout the regional/national economy and, thus, not immediately visible or directly beneficial to the workers who had contributed to their accomplishment. In the end, as port management needed the co-operation of workers in order to implement new technology successfully, certain promises regarding job security and financial compensation were made to unions, applicable to the fortunate workers who would remain employed after the necessary redundancies.

In many countries, all work falling under a certain definition of "dock work" and taking place within a certain statutorily defined "port area" is restricted to registered workers who sometimes have the sole legal right to carry it out, often organised in *labour pools*, even when they do not have the necessary skills. This situation often leads to *ghosting*, where non-registered dock workers, some employed directly by port operators, carry out whatever work is necessary, while registered dock workers are paid in effect to watch the non-registered employees with the necessary know-how actually carrying out the work.[19] This relatively protected position of registered port workers can be seen as one of the reasons why they often enjoy higher wages than those paid for comparable jobs elsewhere in the economy. Some observers argue that this privileged position has finally resulted in a negative attitude of the public and other unions towards port workers.

Sometimes, pressure to maintain old-fashioned manning levels comes also indirectly from the government, which is reluctant to face the financial and political consequences of labour force reductions that can lead to substantial compensation payments to those leaving the industry, or even disruptions to foreign trade. Furthermore, and contrary to most developed countries where one of the prime objectives of management is to improve port efficiency, many developing countries see port "efficiency" as a matter of only secondary importance; in the absence of social safety nets, keeping people "working" is considered to be at least of equal importance.

This often leads to an additional labour problem facing many ports, which is the age structure of the workforce. The continuous surplus in the number of registered dock workers and the "job for life" basis on which they are in practice employed can discourage employers from recruiting new, younger staff. That was the case in the UK, where the average age of registered dock workers increased from 44.2 years in 1980 to 47.1 years in 1988.[20] Regularisation of port employment has also created large numbers of different job categories. Often, strict demarcation lines between different jobs and different activities exist, a fact that severely limits, and in many cases totally prohibits, the transferability of workers from one activity (job category) to another. These labour rigidities often lead to large gang sizes, excessive overmanning, little labour mobility and high port user costs. In many ports around the world, the inflexible and monopolistic supply of port labour has effectively discouraged intended private sector activities around the port and has, thus, deprived the latter from one of its main functions, that of being a "growth pole" for the region and the country.

5.6 Measures (and degrees) of port reform

Considerable confusion and uncertainty surround the term *port reform* and its various manifestations, which may range from a simple reorganisation of the internal management procedures of a port authority, to the outright sale of port land, that is, *privatisation*. For the purposes of this chapter, port reform should be taken to mean *a process of change and transformation, through the introduction of private sector characteristics in public port administration, aimed to improve port efficiency and performance*.

Given the undoubtedly complex economic and legal nature of the issue, this confusion is in most cases unintentional. Indeed, differences between the concepts of, say, commercialisation and corporatisation, or between a *lease licence* and a *concession contract*, are often not easily discernible, even among experts. But the complexity of the issue of port reform has often been used strategically in order to resist change, notably by declared statists or public sector employees. The most common method of doing this is through the use of the word *privatisation*, a word that admittedly carries with it a lot of negative connotations, as synonymous to reform.[21] In what follows, the various types of port reform are discussed in order of increasing need for change and private sector involvement, compared to the traditional situation of a publicly owned and operated port which serves as a starting point.

5.6.1   Improving port administration

The improvement of port management and administration within the current organisational structure and without changes in law or national policy can be seen as a first option of port reform. As can be seen from Table 5.1, the need for such improvements is widely felt in most ports. Surprisingly, however, carried away by the well-publicised merits of more radical port reforms, ports and governments often tend to neglect the sometimes substantial benefits that can be reaped by improving the port's organisational structure, information systems, managerial techniques, financial management, setting clear objectives, training, empowerment of staff, teamwork, and the development of a corporate culture – attributes that should be considered as prerequisites to more radical reforms anyway. Above all, management restructuring requires strong and competent leadership which – more often than not, unfortunately – cannot be found among the retired civil servants, or navy admirals, who customarily frequent the chair of a port authority's president in many countries.

5.6.2   Liberalisation (deregulation)

Under liberalisation, the private sector is allowed to provide port services, sometimes in competition with the public sector. Liberalisation entails the removal of statutory restrictions limiting entrance of the private sector to the port services market, and of discriminatory rules discouraging competition. Eventually, these restrictions are replaced by regulations that encourage or even require competition. For some countries, the advantage of liberalisation is that the introduction of some form of

TABLE 5.1 Types of port reform programmes

	Global (%)	Europe (%)	South and Central America (%)	South East Asia (%)
Improving port administration	43	37	25	85
Liberalisation/deregulation	28	11	37	40
Commercialisation	45	37	37	25
Corporatisation	17	21	37	30
Privatisation	16	11	37	45
Other	19	21	–	30

Source: ILO (1995b).
Note: Percentages may not total 100% due to multiple answers.

competition in port services leads to efficiency improvements, while the overall regulatory control over the (strategically important) port remains completely in the hands of the port authority or the relevant government department.

Obviously, decentralisation is a *sine qua non* for greater port management freedom and autonomy. However, decentralisation alone cannot solve the problem of lacking management incentives or competencies; having the power does not necessarily mean using it as well, and in many cases, doing nothing is considered much safer than doing something (Gordon, 2016, p. 45).[22] For instance, reformers in China, and in many other countries as well, were caught in the decentralisation/recentralisation cycle: once decentralised, power and authority were quickly abused, disorder occurred, control was called for, power and authority were taken back by the *centre* and the situation was back to its original state. Then another cycle started with re-decentralisation and the old scenario repeated itself. This is a quite common situation in many developing countries, where not only the necessary legislation for decentralisation is inadequate, but the mechanisms and institutions (banking, judicial, business ethics, market culture, etc.) of a market economy are not in place. Old control has been given up, while new (regulatory) control has not been created.

A possible disadvantage of deregulation is the potential danger of *cream skimming*. The private sector will only be interested to provide *those* port services that are potentially profitable, that is, container terminal operations. In a statutory monopoly port, the sometimes unprofitable port services – albeit required, such as security – can be cross-subsidised by the profitable ones (cargo handling). However, as a result of liberalisation, the public sector may be losing revenues from profitable port activities, having at the same time few possibilities for cross-subsidisation. This issue should be seriously considered when leasing out port facilities to private operators: if the port authority is to continue providing commercially unprofitable services, and in the absence of central/regional government support, lease payments of private operators should be determined at a level that would allow the efficient provision of the various port services entrusted to the port authority.[23] Such an arrangement is also in the interest of the private operators, given that their efficiency improvements in cargo handling can be easily nullified by inefficient dredging, mooring, pilotage, towage, engineering, security, fire protection and similar operations entrusted to the port authority.

Furthermore, ports in many countries have been run for a long time as administrative entities with both infrastructure and superstructure belonging to, and often operated by, the port authority. In such cases, deregulation does not automatically bring in new competition because the latter is restricted not only by regulations, or market size, but also by a lack of competitors, due to financial incapability or lack of management know-how. It may well be then that after deregulation measures have been put in place, and efforts made to restore competition, no new entrant is found to complement and compete with the old monopoly and force it to change.[24] It could thus be easily realised that the old organisation is too strong to be changed by (unassisted) market forces alone, and some more proactive reforms may be required together with deregulation.

5.6.3 Commercialisation

Commercialisation implies the introduction of a commercial, business-like environment, in which the port management is accountable for its decisions and performance. In the earlier stages, ports still retain their status as quasi-government departments. In the commercialisation stage, the status of a *state-owned enterprise* is justified, as the previous "government department" now changes into a public company.

The main objective of commercialisation is to increase management autonomy and accountability (World Bank, 1994: 9). If port managers in public port organisations are not held responsible for port performance, they will not always take all the necessary steps for securing cost reductions or improvements in productivity. Furthermore, as the management of commercialised ports is still public, it often hesitates to consider, in time, possible reductions in employment. Port labour contracts are usually not governed by regular labour law, but they have a quasi-civil service status. Solutions to these situations could be found in an increased accountability for port managers and workers, or in the contracting out of certain port functions to the private sector. Several approaches are in use to achieve this: performance agreements, management contracts, service contract/contracting out, lease and concessions.

5.6.4 Corporatisation

Corporatisation requires the transformation of public sector organisations (SOEs) into publicly listed private companies, the shares of which are held in majority by the public sector (central and/or municipal government). Although enterprises in the commercialisation stage are introducing more private sector characteristics in their operations, they still lack the legal corporate independence often needed to ensure efficient operations. Corporatisation affords the enterprise a status of independence and subjects it to the same legal requirements with those of a private firm. A whole new company is thus established, enjoying administrative and financial flexibility and autonomy, enabling it to close agreements, and make decisions on pricing, investments, and human resources, without continuous reference to the government. All land and moveable and fixed assets are transferred to the new company as paid up capital.

A significant advantage of corporatisation is to be found in its commercial accounting procedures, which make financial cost control more transparent, thus facilitating the identification of sources of inefficiency. As the government does not exercise direct control over port management, corporatisation is in general a more attractive alternative to foreign investors than the other stages of port reform discussed earlier.

5.6.5 Privatisation

Privatisation is the most radical and possibly most complex exercise in port structural adjustment programmes. It could be defined as the conditional, and often transitory, transfer of port ownership from the public – to the private sector. However, although this definition serves a methodological purpose, *pure* privatisation such as this is rarely found in practice. In many cases, the increasing private sector participation in the management, operations and development of ports (described above as commercialisation/corporatisation) would also be often defined as "privatisation." Privatisation can take various forms:

- *Public offering*: In those cases where the shares of the port company are quoted, even partly, on the stock exchange and can be freely traded, the government may decide on a public offering. It may also decide to retain a major part of the stock (corporatisation) in order to exercise control over future port activities.
- *Management/employee buy-out*: In this situation, the government decides to divest its shares to the employees, so that the latter assume ownership of the port. A buy-out would be more appropriate whenever the employees are highly motivated and keen on buying the company. Demand prospects have to be stable and the size of the port should be rather limited.
- *Private placement*: Through a process of competitive tendering, various potential private investors can submit a quotation. By negotiation, the government can then decide which offer is the most attractive. It is possible that offers are made by a consortium of companies, banks or even a group of employees.
- *BOO/BOT*: In this case, a private company Builds, Owns and Operates an asset for a certain period. Under a BOT arrangement, at the end of the period the asset is Transferred back to the government. If privatisation takes place in this way, the private sector is given an exclusive concession to operate an infrastructural project, such as a bridge or a port, and it assumes the risk of completing it. BOO/BOT is a form of non-debt financing of public sector activities, in which the private sector finances the construction and the costs are recovered through user fees. Depending on the project, incentives may include guaranteed purchase of output, tariff support in the early years, concessionary rates of income tax, free repatriation of dividends and capital, and exemption from customs duties, turnover tax and excise duties.
- *Sale of assets*: This alternative can be considered when private investors are not interested in acquiring the whole of the company, or when better results can be expected through a partial rather than an outright sale.
- *Joint venture*: A joint venture represents an enterprise in which two or more private companies, or an SOE and private investor(s), jointly own the equity of the port company.

Most countries actually experiencing port privatisation have adopted public-private joint venture options. Port joint ventures are often attractive to both government and the private sector. The former can thus reduce administrative and financial burdens, improve efficiency and promote competition. The private sector views this arrangement favourably whenever the magnitude of the required investments, and associated commercial risks, are beyond its capabilities, or when complete owner-ship of assets and operational control are not allowed.

5.7 Interim port reform authority

The structural adjustment of ports is a complex process, with many interests involved and a significant impact on port management workers and employees. The fact that many ports are natural monopolies makes such adjustment even more complex. The existence of an *interim authority*, which controls and directs the structural adjustment process, can facilitate the smooth and effective implementation of this process. Several recent port privatisation efforts have made use of such an arrangement. The "Steering Committee" in Thailand and the "Waterfront Industry Reform Authority" (WIRA) in Australia are two examples.

The interim port authority ought to include representatives of the relevant government departments, often supported by a team of experts. The latter is usually multidisciplinary in nature and it includes representatives of the private sector. An interim authority has several tasks, the most important of which is the selection of an appropriate strategy for privatisation. Thus, the evaluation of the suitability and/or desirability of the different privatisation alternatives would be one of this authority's main challenges. Another important task is related to the establishment and control of a *tendering procedure*. The interim authority can further assist with the negotiation process and the evaluation of the various offers.

5.8 Concluding considerations

It has already been noted that, nowadays, the increased internationalisation of all forms of economic activity, mass media, foreign experts and modern telecommunications intrigue developing countries to attempt comparisons with other nations, western ones included, many of which at a completely different stage of economic and social development, with institutional frameworks that were set up years ago. If superficially attempted, such comparisons can be extremely dangerous and misleading, particularly when successful economic reforms in other countries are taken *prima facie,* without a thorough understanding of all their implications, and without adequate comprehension of the simple fact that, if proper *institutions* are not in place, the future of privatisation, and to effect the country's economic development by and large, cannot be taken for granted.

To give a simple example, the listing of a privatised port's shares in the country's stock exchange would be next to pointless, if the latter is not functioning properly, the volume of transactions and its liquidity are low, the dissemination of market information inadequate and if capital markets, in general, are inefficient. In situations such as these, the real value of the port will be far from being reflected in the nominal value of its shares and, thus, domestic and foreign investors' interest could not be expected to be significant.

To enhance the possibilities of survival in a competitive environment, the government can improve the institutional environment of the port, thereby enhancing its ability to respond adequately and promptly to the changing market conditions. Several well-documented divestiture experiences show that certain prerequisites regarding the port's economic environment have to be met, if the full benefits from divestiture are to be realised. A hospitable and efficient business environment has, thus, to exist, distortions that hinder domestic and foreign competition eliminated, and an efficient capital market with considerable absorptive capacity developed.

In addition, the retrenchment of the economic role of the state and the encouragement of greater private sector participation should constitute a careful long-term social cost-benefit analysis, undertaken by the government. The results of this analysis should form the government's basis for designing and implementing programmes of economic reform. Its strategy, once decided, should be firm, with clear and transparent objectives, and it should be widely explained through a process of extensive consultation, particularly with those parties that are adversely affected by the proposed reforms. The importance of consultation in structural adjustment could not be over-emphasised, not only in securing labour's co-operation, but also in convincing the latter that the attempted reforms aim at enhancing the country's general economic welfare, which should be every government's utmost objective. This strong message has to be successfully and timely conveyed to trade unions and employees.

Notes

1 A much earlier version of this chapter appeared as H. E. Haralambides, S. Ma and A. W. Veenstra (1997) World-Wide Experiences of Port Reform. In: H. Meersman and E. v.d. Voorde (eds.) *Transforming the Port and Transportation Business*. Acco, Leuven (Belgium). The original publication was based considerably on work the authors had undertaken for the United Nations (Professor Haralambides for ILO and Professor Ma for UNCTAD). All views and opinions expressed there (and consequently here) are of the author(s) only and in no way commit the United Nations or any of its agencies. The current version is a thoroughly revised and updated chapter, including my more recent experiences as president of the Italian port of Brindisi. Readers are strongly advised not to skip the frequent notes which I am sure, for most, will be most elucidating! Many of the current updates are included in these notes but, for me, the most gratifying thing is that almost everything I was predicting 20 years ago has come true in 2013.

2 I often use the three terms indiscriminately, under the overall characterisation 'port reform'. Here the term is meant as the gradual retrenchment of the public sector from economic activity and the introduction of commercial principles in port management.

3 And, 20 years later (2012), they have been proven right. On the contrary, the Western world has been losing out in this zero-sum game, as it has proven to be after all. Our initial enthusiasm with globalisation and trade liberalisation was based on a false premise: that is, that our saturated economies and increasing returns to scale industries could only survive if and only if we could expand the international market for our exports. Unfortunately, this didn't happen. Instead of producing 'here' and exporting 'there', foreign direct investment (FDI) started to flow 'there', producing 'there' and, often, re-importing back to Europe. Profits of European multinational companies have not been repatriated in a way that would allow us to sustain our welfare systems and way of life, developed over decades with the taxes of our fathers and forefathers. These systems are now being unravelled in

the pursuit of the holy grail of cost competitiveness, and as a result of a 'cheap consumerism' that is signing the death certificate of, at least, Europe. The bad news is that these trends are no longer reversible.

4 FOB (Free On Board) is a term in international commercial law that specifies at what point the seller transfers ownership of his/her goods to the buyer.

5 The United Nations Conference of Plenipotentiaries on a Code of Conduct for Liner Conferences, better known as the UNCTAD Code, came into force in October 1983. The Code, a result of political pressure on the side of developing countries, was nothing more than a protectionist cargo-sharing arrangement, on the basis of the infamous 40-40-20 formula, whereby trade was reserved for the ships of the two trading nations, leaving only 20% open to the ships of third countries. The United States never ratified the Convention, whereas the European Union did so reluctantly and conditionally on the basis of the Brussels Package. In short, the latter ensured that the restrictions of the Code did not apply among EU member states as well as between them and other OECD countries.

6 During my brief spell as president of the Italian port of Brindisi, I had the opportunity to witness firsthand another type of resistance to reform. In the supervisory bodies of port-stakeholders, known as port committees (Comitato Portuale), sat such persons as representatives of shipping agents, shipowners, port operators and others – that is, 'clients' of the port who often had not only a conflict of interest between their supervisory role and that of a port client, but also every interest in maintaining the *status quo*, which was affording them effective protection from outside competition. As the *red line* between "supervision" and "management" is not always as clear-cut in Italy (and of course in other countries of the European south) as it is in northern Europe, this situation often frustrated my decisions, bringing them to a grinding halt, to the detriment of the general interests of the city and of the wider port community these stakeholders were supposed to safeguard. Two cases are typical and I would be amiss not to mention them here. When I brought Grimaldi Lines to Brindisi (by far Europe's biggest short-sea-shipping and multimodal operator), certain local agents of competitor lines mounted a war against the newcomer, lobbying travelers not to use Grimaldi ships because he would eventually monopolise the traffic and this would result in higher ticket prices in the future! When I decided to develop the cruise business in the port, tendering a beautiful berth overlooking our magnificent medieval castle (Castello Alfonsino), a local cruise agent (and member of the port committee) went live on television, expressing his concerns about the health of our American and Canadian visitors, who would disembark in the vicinity of a coal berth!

7 Such as Hutchinson Port Holdings (HPH); PSA; DP World; APM Terminals; ICTSI.

8 It should not be forgotten that, in spite of intra-port competition, a port will always have a "captive audience," and thus a significant market power.

9 I hope the exaggeration of the witticism would help in driving the point home.

10 My "Brindisi experiences" could help, here too, in order to drive this point home. It seems that due to some "unwritten law," going back for decades, Brindisi – maybe one of the most backward cities in Europe – has been condemned to the handling of coal, also with cargo handling and inland transportation techniques which have a lot to be desired, compared to modern cargo handling, storage, and coal distribution practices. The environmental impacts from the (improper) handling of coal are only too well-known to be repeated here. As a result, the citizenry is constantly expressing strong concerns, often quite vociferously. At the same time, the more "sexy" types of port traffic, such as cruise and ferry, have moved almost exclusively to its new competitor, the port of Bari, located just 100 kilometers to the north, and the seat of the regional government. Thus, for decades, the port of Brindisi has been in a state of heart-breaking decay. As soon as I moved in, I made it clear to all that Brindisi was *open to business*. The first big success came with the arrival of Grimaldi Lines. Soon after this, I was "summoned" to the regional headquarters in Bari for a "meeting." The punch line of that meeting was *to tell me* that I should be "careful" with my decisions and with talking to shipowners, for Bari was living from its port and they were not as lucky as Brindisi to have so much coal!

11 If one is unfortunate enough to have to found his way through the corridors of a government ministry, somewhere in Spain, Portugal, Italy or Greece, he couldn't fail to notice filing cabinets placed in corridors, often against fire-fighting regulations, so that more space is created inside the offices to accommodate an increasing number of seemingly working people.

12 I witnessed with my own eyes, at a major port of a certain Asian country, 10 dockers washing a container, all holding on to the same water hose! Let it be noted too that they were casual workers, while the registered ones were watching them from the bridge, blissfully smoking their cigarettes.

13 Comparisons of labour productivity should however be attempted with utmost caution. Labour productivity is a function of the existing capital equipment and if the latter is inferior or obsolete, as it often is in developing countries, labour can produce *only that much* with it.

14 As soon as I moved in at Brindisi, the standard question each and every staff member of the port authority asked me was *could you please tell me what is my job?* People were being reallocated from one department to the other without notice or motivation, at the spur of a moment; the port's statistical information system was run, quite independently and without any control, by an external private company; quality control was an unknown concept; and a human resources officer didn't exist. Immediately, I started a process of internal reorganisation, including a system of staff assessments. Interestingly, it was the very same person who asked me to define his job that fought the most against staff assessments!

15 All of my predecessors at Brindisi were quite versed in finding their way to the right ministerial office, or pick up the phone and call the Minister, but they wouldn't speak a word of English. A *marketing and communications* department did not exist at the port authority at the time I stepped in.

16 The construction of the new passenger terminal of Brindisi was delayed for months, because the director of the regional branch of the Ministry of Cultural Heritage thought that the logo design on the roof of the terminal was a bit too high and this would spoil the historical landscape view of Brindisi! The local landscape architects objected too, albeit for a different reason. Forty years ago, in the area of the intended terminal, nowadays right in the middle of the port, there used to be a beach and, thus, they thought we should keep it this way notwithstanding the fact that, today, no one would even dream of swimming there even if one could!

17 Given the unpredictability of port work, port management could not possibly employ permanent staff, having them idle and waiting for a ship to arrive. Labour was thus *casual*, that is, employed for as long, as much, and whenever required. Recruitment was very different too. Each morning, a number of labourers would show up to a foreman waiting at the gate and he, on the basis of certain "criteria" that had more to do with *natural selection* than anything else, would thumb-in the youngest and the strongest. The rest would return to their "locales" to indulge in whatever it was they were indulging in.

18 Delays in one port propagate onto others with a cascading effect.

19 The white paper "Employment in the Ports: The Dock Labour Scheme" gives a clear example of this.

20 As a measure of comparison, before the Australian Waterfront Reforms came into force, the average age of the workforce was over 50 years.

21 I vividly remember Greece's socialist ex-prime minister, George Papandreou, at that time leader of the main opposition party PASOK, in the middle of the tear gases of the riot police, proclaiming that the government, by offering a concession of a terminal of the port of Piraeus to the Chinese company COSCO, was selling off the crown police, proclaiming that the government, by offering a concession of a terminal of the port of Piraeus to the Chinese company COSCO, was selling off the "crown jewels"!

22 What William Baumol called the "security of the management team" has a big role to play here: focusing on performance and the need for change may require some rather

hard-nosed decisions, for example, forced redundancies that could threaten the "survival" of the management team. The latter would thus opt for passivity rather than action. As R. A. Gordon (2016, p.45) has so succinctly put it, "the management of corporations do not receive the fruits which may result from taking successful action, while their position in the organisation may be jeopardised in the event of a failure."

23 At Brindisi, the port dues we were charging to Ro-Ro ferry operators were roughly €1 per passenger, while *security* only was costing us €3 per passenger. What was even more interesting was that some ferry operators and their local agents would refuse to pay even this €1, claiming that the port is a public facility, thus to be used by anyone without an obligation to pay! This was indeed a "world first," lasting for years, with debts to the port authority accumulating year after year. Finally, even the thought of increasing port dues, in order to cover costs, would be immediately met by the threat of ferry operators to leave the port for the neighboring port of Bari, which would be more than happy to welcome them.

24 This is particularly the case where tenders are only advertised locally and/or in the national language which, naturally, cannot entice foreign competition.

References

Baumol, W. J. (1977) On the proper cost tests for natural monopoly in a multiproduct industry. *American Economic Review*, 67, 809–822.

Campbell, D. (1994) Foreign investment, labour immobility and the quality of employment. *International Labour Review*, 133(2), 185–188.

Couper, A. D. (1986) *New Cargo-Handling Techniques: Implications for Port Employment and Skills*. Geneva, International Labour Office.

ESPO (2011) *European Port Governance: Report of an Inquiry into the Current Governance of European Seaports*. European Sea Ports Organisation (ESPO).

European Commission (2013). *Proposal for a Regulation of the European Parliament and of the Council Establishing a Framework on Market Access to Port Services and Financial Transparency of Ports*. European Commission.

Gordon, R. J. (2016). *The Rise and Fall of American Growth: The US Standard of Living since the Civil War*. Princeton, NJ, Princeton University Press.

Goss, R. O. (1993) *Port Privatisation: The Public Interest*. Proceedings of the 9th Port Logistics Conference, Alexandria, Egypt.

Haralambides, H. E. (2002) Competition, excess capacity and the pricing of port infrastructure. *International Journal of Maritime Economics*, 4, 323–347.

Haralambides, H. E. (2012) *Ports: Engines for Growth and Employment*. European Ports Policy Review: Unlocking the Growth Potential. European Commission, Brussels, 25–26 September 2012.

ILO (1995a) *Social and Labour Problems Caused by Structural Adjustment in the Port Industry of Selected Countries in the Asia-Pacific Region*. Geneva, International Labour Organisation.

ILO (1995b) *Social and Labour Effects of Structural Adjustment Programmes in the World Port Industry*. Geneva, International Labour Organisation.

Milanovic, B. (2012) *Global Income Inequality by the Numbers: In History and Now. An Overview*. World Bank Policy Research, Working Paper 6259, Washington, World Bank. Retrieved from http://elibrary.worldbank.org/doi/pdf/10.1596/1813-9450-6259.

Mosca, M. (2008) On the origins of the concept of natural monopoly. *The European Journal of the History of Economic Thought*, 45(2), 317–353.

Musgrave, R. (1969) Provision of social goods. In: Margolis, J. and Guitton, H. (eds.) *Public Economics*. New York, St. Martin's Press.

Rimmer, P. (1984) Japanese seaports: Economic development and state intervention. In: Hoyle, B. and Hilling, D. (eds.) *Seaport Systems and Spatial Change.* Chichester, John Wiley & Sons Ltd., 99–134.

Shoup, C. S. (1969) *Public Finance.* Chicago, Aldine.

World Bank (1994) *World Development Report 1994: Infrastructure for Development.* New York, Oxford University Press.

Suggestions for further reading

Besides the papers provided as references in the text, we recommend the following texts as suggestions for further reading.

Acciaro, M. (2013) A critical review of port pricing literature: What role for academic research? *The Asian Journal of Shipping and Logistics,* 29(2), 207–228.

Adam, C., Cavendish, W., and Mistry, P. S. (1992) *Adjusting Privatization: Case Studies From Developing Countries.* London, James Curry Ltd.

Baird, A. (1994) *Port Privatisation in the UK, A Flawed Process.* Proceedings of the World Port Privatisation Conference 1994, London, World Port Privatisation Conference.

Barros, C. P., Haralambides, H. E., Hussain, M., and Peypoch, N. (2011) Seaport efficiency and productivity growth In: Cullinane, K. (ed.) *International Handbook of Maritime Economics.* Cheltenham, Edward Elgar, 363–382.

Beplat, K. (1989) *Mega Trends in Containerization: Directions and Projections.* Hamburg, Reinecke & Ass.

Bureau of Transport and Communications Economics (1988) *Economic Significance of the Waterfront.* Information Paper 29, Canberra, Australian Government Publishing Service.

Dicken, P. (1992) *Global Shift: The Internationalization of Economic Activity.* London, Paul Chapman Publishing Ltd., Chapter 2, 4, 6, 13.

Haralambides, H. E., and Acciaro, M. (2015) The new European port policy proposals: Too much ado about nothing? *Maritime Economics and Logistics,* 17(2), 127–141.

Haralambides, H. E., and Gujar, G. (2011) The Indian dry ports sector: Pricing policies and opportunities for public-private partnerships. *Research in Transportation Economics,* 33, 51–58.

Haralambides, H. E., Verbeke, A., Musso, E., and Benacchio, M. (2005) Port financing and pricing in the EU: Theory, politics and reality. In: Lee, T. W. and Cullinane, K.P.B. (eds.) *World Shipping and Port Development.* Basingstoke, Palgrave-Macmillan, 199–216.

Harding, A. S. (1990) *Restrictive Labour Practices in Seaports.* Washington, The World Bank.

Kikeri, S., Nellis, J., and Shirley, M. (1992) *Privatisation: The Lessons of Experience.* Washington, World Bank.

Kotwal, M. (1992) *Indian Country Report Presented at the Asia/Pacific Rim Dockers' Seminar.* International Transport Workers Federation, Yokohama, 5–7 October 1992.

Notteboom, T., Verhoeven, P., and Fontanet, M. (2012) Current practices in European ports on the awarding of seaport terminals to private operators: Towards an industry good practice guide. *Maritime Policy & Management,* 39(1), 107–123.

OECD–DAC (1993) *Development Co-operation: Aid in Transition.* Report by A. R. Love, Development Assistance Committee (DAC), 39–42.

Oxford Economics (2011) *The New Digital Economy: How it Will Transform Business.* A Research Paper Produced in Collaboration with AT&T, Cisco, Citi, PwC & SAP.

Port Development International (1988) *Thailand, the Privatisation Process.* November 1988, 16–21.

Port Development International (1993) *Privates on Parade*. December/January 1993, 33–53.

Ports and Harbours Bureau of Japan, Ministry of Transport (1993) *Ports and Harbours in Japan 1993*. Japan, Ministry of Transport.

Rosenstein-Rodan, P. N. (1943) Problems of industrialisation of Eastern and South-Eastern Europe. *The Economic Journal*, 53(210/211), 202–211.

Shirley, M., and Nellis, J. (1991) Public *Enterprise Reform: The Lessons of Experience*. Washington, World Bank.

Singh, A. (1994) Global economic changes, skills and international competitiveness. *International Labour Review*, 133(2), 167–183.

Slack, B., and Starr, J. (eds.) (1994) Containerization and the load centre concept. *Maritime Policy and Management*, 21(3), 185.

Thomas, B. (1994) The privatization of United Kingdom seaports. *Maritime Policy Management*, 21(3), 181–195.

World Bank (2001). World Bank Port Reform Toolkit. www.worldbank.org/transport/ports/toolkit.

6

THE PORT-CITY INTERFACE

Olaf Merk

6.1 Introduction

Ports are often located in cities, close to consumers and producers. This proximity of port and city can bring conflicts, but can also provide unique development opportunities. This chapter provides an overview of this balancing act and applications of effective management of the interface of port and city. How to manage port-city conflicts, how to arrange for "peaceful co-existence" between ports and cities, and how to create synergies between ports and cities, by coupling their respective strengths? Such are the main questions covered in this chapter.

The chapter will start by identifying the four main challenges and opportunities of the port-city interface (Section 6.2), related to land use, environmental impacts, traffic and the economy. The subsequent sections will develop these further and provide in each field policies that have been applied and proven to be successful. Section 6.3 will focus on land use in port-cities and will cover port relocations and redevelopment of port areas. The urban areas most closely located to port areas suffer in many cases from environmental impacts from ports; Section 6.4 will treat main instruments to mitigate these impacts, in particular air pollution and noise. Ports generate hinterland traffic that intermingles with urban traffic; Section 6.5 focuses on the policy instruments that can be applied to avoid port-related traffic exacerbating urban congestion. Ports not only present challenges to their cities, but also economic opportunities; Section 6.6 assesses how the port can be mobilised as a driver of an urban economy. The chapter will end with a conclusion and expected future developments.[1]

6.2 The interface of port and city: challenges and opportunities

Ports are often at the origin of cities. Many cities started as trading posts, allowing small towns to become cities, and fuelling urban development, thanks to the

TABLE 6.1 Main non-urban ports

Type of port	Reason non-urban location	Examples
Some bulk ports	Closeness to natural resources	Port Hedland, Corpus Christi, Novorossiysk
Some transshipment ports	Closeness to shipping routes	Salalah, Freeport, Gioia Tauro, Algeciras, Marsaxlokk
Non-urban gateway ports	Decongestion of urban ports	Felixstowe, Laem Chabang, Lianyungang

Source: Adaptation of Hall and Jacobs (2012).

prosperity associated with trade. Ports are often still closely connected with their cities, in particular in emerging economies. A striking example in recent history is the case of Shenzhen (China), a small fishing village that became one of the world's largest metropolises and ports in a few decades, thanks to export-driven growth made possible by a free-trade zone and extensive port development.

Many of the largest cities in the world have the largest ports in the world. This is particularly the case in Asia, with Shanghai and Osaka-Kobe as examples of very large metropolitan areas and very large ports. Other Asian metropolises with very large ports include Guangzhou, Shenzhen, Tianjin and Hong Kong. The link between metropolitan size and port size can also be seen in North America (New York and Los Angeles), and to a lesser extent in Europe, where London and Barcelona are examples of large cities with large ports.

There are large ports that are not located in cities, but there are usually very specific reasons for this: because they are close to natural resources or global shipping routes, or because of a deliberate decision to decongest urban ports (Table 6.1).

Different types of port-cities can be identified, which are dependent on port size and city size, ranging from coastal port towns to world port-cities (Figure 6.1). *World port-cities* are large cities with large ports; examples include New York, Hong Kong, Tokyo and Singapore. In a *port metropolis*, the urban function is large, whereas the port function is smaller but nevertheless considerable, as in Cape Town and Buenos Aires. When the port function is even smaller in a large metropolis, it can be considered a *coastal metropolis* (Stockholm, Baltimore and Tunis). However, the opposite phenomenon also exists; in these cases, the size of the port is relatively larger than the city. These could be called *major port-cities*, such as Rotterdam, Le Havre and Genoa, and *major port towns*, such as Freeport, Gioia Tauro and Laem Chabang. These port-cities could be considered to have different characteristics, challenges and development perspectives.

Despite the differences among port-cities, they also have common challenges. In this chapter we discuss four types of such generic challenges of port-cities, related to land use, environment, traffic and the economy (Table 6.2). The sections below discuss these challenges, as well as the various policies that can be applied to resolve these challenges.

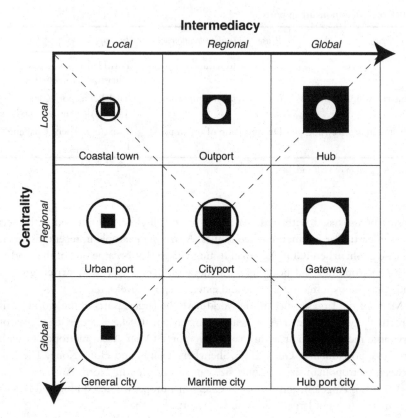

Intermediacy

FIGURE 6.1   Typology of port-cities

Source: Ducruet and Lee (2006).

Note: The circle represents the city; the square represents the port. The larger the circle, the larger the urban population. The larger the square, the larger the port volume.

TABLE 6.2 Four port-city challenges and the policy strategies to solve these

Type	Challenge	Policies
Land use	Resolve conflicting views on using land	Port relocation Redevelopment of old port land
Environment	Mitigate environmental impacts of ports on cities	Information, regulation, incentives, clean technologies
Traffic	Reduce impact of port traffic on urban congestion	Port gate policies, dedicated freight routes, dry ports
Economy	Create local benefits of ports for cities	Industrial development, maritime clusters, waterfronts

Source: Adaptation of OECD (2014).

6.3 Balancing land use of port and city

6.3.1 Port-city land use challenges

Ports are traditionally located in the centre of cities, close to consumers and economic activity. This made sense in times when both cities and ports were relatively small and could fairly easily exist together, but this is no longer the case. The co-existence of city and port has become more complicated, which is rooted in different interests that express themselves in different perspectives on land use.

Modern ports are space intensive. They require land to transfer goods from sea to land and the other way around. This requires space for cargo handling, storage, and related activities carried out in port areas. These can be logistics activities, but also activities that transform the goods. Since the Industrial Revolution, many ports have developed into industrial estates, a practice which has survived until now (see Chapter 18). As a result, many of the largest ports of the world now occupy very large areas. For example, the land surface of the Belgian port of Antwerp is equivalent to a third of the surface of the city of Antwerp.

Maritime transport over the last decades has been characterised by increasing ship size; for example, the average capacity of a container ship doubled between 2005 and 2015 (ITF/OECD, 2015). This increased ship size means that ports become more land intensive; their quays need to be longer (because ships have become longer) and their yards need to be bigger because big ships bring more cargo at once into a port.

Ports need more space, but space is something that is scarce in cities. Many cities are expanding and have a shortage of land for housing and other functions. This is especially the case in developing countries with rapid urbanisation (attracting people from rural areas) and in successful global cities (attracting people from other countries). Both instances intensify the need for urban land to accommodate population growth. The increasing scarcity of land translates itself into higher land prices, which means that port functions in these cities will be difficult to sustain, considering the opportunity costs of port land.

So, there is an inherent conflict for land in port-cities. This conflict can be more or less explicit and has generally resulted in two different developments: relocation of the port, or part of the port, away from city centres; and the redevelopment of former port lands.

6.3.2 Port location and relocation

Ports have evolved spatially over time, generally away from city cores and towards the sea. This is the case of many estuary ports, such as Rotterdam, which originated upstream but which expanded downstream. This is a long-term development that started with industrialisation, requiring larger ports and larger urban populations. This spatial evolution of ports has been amplified by containerisation and the related increase in economies of scale in shipping.

Gradual port relocation can take place if there is vacant land available for port development adjacent to the current port. This is not the case for all ports. Various ports are located in bays and are completely surrounded by the city. In that case, the only place where new adjacent port development is able to take place is in the sea, via offshore terminals developed via landfill. An example of such a port that expanded via an offshore terminal is Genova.

A more *radical* form of port relocation consists of creating whole new ports at considerable distance from the city centre. In this way, the port could be located where maritime access (depth) is favourable and where large land surfaces are vacant and cheap. In many cases, this development liberates valuable port land for urban waterfront development. Table 6.3 provides an overview of urban ports that have to deal with such new ports, in addition to the existing city-port. Two options could be distinguished:

1   New ports that form part of (and are created by) the old urban port (e.g., Busan, Shanghai and Marseille);
2   New ports that are created independent of and in competition with the old urban port (Rio de Janeiro–Sepetiba, Kolkata–Haldia, St. Petersburg–Ust-Luga).

In practice, many of these new ports manage to capture a lot of the port traffic from the urban port, which could subsequently become more focused on port functions that are more compatible with urban activities, such as passenger terminals, marinas, fishing ports and various non-port functions.

The process of a new port capturing the cargo of the old port could be planned and anticipated as such (e.g., in the case of Tangiers-Med), but this is not always the case. In some cases, there is a transition period in which the two ports operate at the same time, which could give rise to problems. For example, there is considerable

TABLE 6.3 Urban ports and new ports

Old city port	New port site
Busan	Busan New Port
Shanghai	Yangshan
Rio de Janeiro	Sepetiba
Marseille	Fos
Kolkata	Haldia
Bremen	Bremerhaven
St. Petersburg	Ust-Luga
Tangiers	Tangiers-Med

Source: Adaptation of OECD (2014).

truck traffic between the old and new port in Busan (contributing to urban congestion), because the new port has important transshipment functions, whereas some of
the feedering activity still takes place at the old port. Many ports consist of multiple
port sites, one site close to the city centre and one located at some distance from the
urban core (the out-port). A classic example of this is Marseille with its non-urban
port site at Fos, which has been well documented in a seminal article by Hoyle
(1989). Multi-site ports have to face the fundamental challenge to find a balanced
division of functions between the different port sites. They often face recurring
discussions on the future of the urban port, in particular on the question if all or
most of the port activity should be moved to the out-port.

A radical relocation is relatively unproblematic if the old port is no longer very
active or if the terminal equipment has already been written off. Relocation of ports
can be facilitated by land swaps, with ports giving up some of their land in urban
cores in exchange for new land for port development. This can become a source of
conflict if urban and port interests are not aligned, and if the port fears that firms
located in the current port might move to other sites than the new port sites.

6.3.3 Waterfront regeneration

The emergence of container shipping accelerated the abandonment of old port
areas, mainly due to the fact that their piers had too little space to deal with containers. As a result, port functions, especially deep-sea shipping, started moving out of
the historic port areas. Ports were faced with the challenge – and opportunity – that
surrounds the redevelopment of abandoned port land areas in the heart of the urban
core (Hoyle, 1989). The many waterfront developments of recent decades have
used different ways to reinvent the old port areas, including commercialising the
proximity to water (with marinas, fisheries and aquariums), using the port function
for tourism (cruise ship passenger terminals), promoting a port's maritime heritage
(with the preservation of historic buildings), and organising major events to attract
people and tourism.

Successful waterfront projects, such as those indicated in Table 6.4, have managed to combine diverse functions that make the waterfront area economically
vibrant, such as port functions, recreational and cultural activities and food-related
businesses such as food markets or restaurants. Port Vell in Barcelona, which attracts
more than 16 million visitors per year, is an exemplary case, where the old port area
has been transformed into a successful waterfront area with an interesting mix of
functions.

Achieving the right mix of functions can be a challenging task, due to the
difficulties in financing a project. The land use of waterfront projects typically
includes residential, commercial, tourism and recreational functions; yet the cities
or redevelopment agencies are often obliged to include residential developments
because low-density land uses – such as park or recreation-based anchorage – do
not generate the revenue to cover the cost of buildings or preparation of the sites.
Finding the right mix is closely related to how the project is going to be financed

TABLE 6.4 Famous waterfront development projects

Waterfront	City	Country
Port Vell	Barcelona	Spain
HafenCity	Hamburg	Germany
Abandoibarra	Bilbao	Spain
Puerto Madera	Buenos Aires	Argentina
Porto Antico	Genova	Italy
Canary Wharf	London	United Kingdom
Victoria and Albert Waterfront	Cape Town	South Africa

Source: Adaptation of OECD (2014).

and the financial capacity of the public sector involved, which must ensure that the waterfront serves local economic as well as social interests. It is crucial to balance the functions that help finance the project (e.g., residences) and those that do not, yet are nevertheless essential to develop a vibrant waterfront (e.g., leisure or recreational sites).

6.4 Mitigating environmental impacts of ports

6.4.1 Environmental challenges of port-cities

The environmental impacts of ports can be related to the *shipping* activity in a port, the activity on the *port* land itself and the environmental impacts of *hinterland transport* to and from ports (OECD, 2011). The main impact falls on air emissions, water quality, soil, waste, biodiversity and noise. These environmental impacts can have severe consequences for the health of the population of the port-city, especially for the poorer parts of port-cities.

Shipping *emissions* can present a large share of the total emissions in the port-city, for example, up to half of the SO_2 emissions in Hong Kong and Los Angeles/Long Beach. The main air emissions impact from ports to their cities come from SO_x, NO_x and particulate matter (PM) emissions. *Soil pollution* from the maritime transportation sector is mainly linked to the terrestrial activities in port areas. Port activities produce waste, especially from oil terminals, fuel deposits and dry-dock operations, which produce oily and toxic sludge. Ports' impact on *biodiversity* is due mainly to air emissions, waste and ballast water (the transfer of ballast water could lead to the introduction of non-indigenous marine species). *Dredging* may have an impact on the ecosystem, but in most dredging projects the impact is temporary and often limited through environmental monitoring and compensatory measures. *Noise* impact from ports can derive from ships, cranes, trucks, trains and industrial activity, which can affect a significant number of urban residents. The next sections will concentrate on the impacts that are most relevant to most urban citizens: air emissions and noise.

6.4.2 Improving air quality in port-cities

Air quality in port-cities can be improved by four different types of measures: information, regulation, incentives and clean technology. These measures have been defined on different intervention levels. Information on air emissions in many cases is dealt with at the local level (port or city level). Air emissions from maritime transport are regulated at the global level, but there can be additional regulations at the national, regional and even port levels. Incentives and clean technology has in practice mostly been implemented at the local (port) and regional and national levels. Mitigating negative port impacts requires the interplay of different levels of intervention, ranging from the local on up. Given the nature of the shipping industry, some environmental impacts of shipping are best tackled at the global level. Self-regulation of ports can work, but in most cases, external pressure is needed.

Information is an important policy measure to mitigate air emissions, for example in the form of emissions inventories. These inventories seek to identify emissions levels that occur within a given area, according to their source. Sources can be mobile (i.e., ships and vessels entering and leaving, cranes, trains) or stationary (i.e., energy production facilities). Various ports have established such emission inventories; some ports have integrated these into larger sustainability reports that provide measures of a diverse range of environmental impacts and mitigation initiatives. Such efforts at quantification are essential, as they provide a baseline against which subsequent progress and performance can be measured. Methodologies employed and main indicators covered vary from port authority to port authority.

The main *regulations* on air pollution are expressed in Annex VI of the Marpol Convention of the International Maritime Organisation (IMO). This Annex establishes a maximum limit on the sulphur that is allowed in ship fuels, and defines limits on the NO_x emissions from ships. Part of this regulation consists of emission control areas (Box 6.1), in which stricter rules on air emissions from ships are applied. Both

---

BOX 6.1 WHAT ARE EMISSION CONTROL AREAS?

Emission control areas (ECAs) are sea areas in which stricter controls are established to minimise airborne emissions from ships. There are currently four ECAs: the Baltic Sea, the North Sea, the North American ECA (covering most of the US and Canadian coasts) and the US Caribbean ECA (including Puerto Rico and the US Virgin Islands). Since 1 January 2015 the maximum limit of sulphur content in ship fuels within ECAs is 0.1%; outside ECAs this is currently 3.5%.

TABLE 6.5 Port incentive schemes to mitigate emission impacts close to cities

Focus	Incentive	Examples of ports
Vessel speed	Lower port dues for vessels with speeds below 12 knots	Long Beach
Clean fuel	Lower port dues for ships using cleaner fuel	Ports in Sweden
Alternative fuel	Subsidies for vessels using LNG	Rotterdam, Gothenburg
Clean ships	Lower port dues for cleaner ships	Most large ports in OECD countries
Clean trucks	Grants for replacement of old port trucks	Los Angeles, Long Beach, Vancouver
Hinterland mode	20% discount on port dues for containers by rail	Ports in Spain

Source: Adaptation of OECD (2014).

the global emissions cap and the maximum limits in the ECAs have become more restrictive over time and will continue to become more restrictive in the future.

In addition, many ports have introduced incentives to stimulate reductions in emissions from shipping, in particular shipping taking place in the vicinity of the port area. These incentives often consist of reductions of port dues to stimulate the behaviour that is desired from shipping companies or other transport actors, and in some cases of subsidies. They have a variety of angles to achieve the intended reduction of emissions, via lower vessel speeds, cleaner fuel, alternative fuel, cleaner trucks and more hinterland traffic by trains (Table 6.5).

Ports can also reduce emissions by stimulating *clean technologies* and equipment. Among the many experiences worldwide, here are a few examples:

- The port of Houston has implemented several energy-efficiency measures, including improved lighting control systems and the installation of new window systems to reduce dependence on artificial lighting.
- In the ports of the State of Virginia, suppliers were instructed to provide only the cargo-handling equipment with the lowest emissions engines on the market.
- The port of Busan (Korea) has switched from fuel-driven rubber-tired gantry (RTG) cranes to electricity-driven RTG cranes.

In addition, various ports have implemented shore power facilities in their ports, in order to reduce air emissions from ships (Box 6.2).

BOX 6.2 WHAT IS SHORE POWER?

Shore power facilities enable ships to use power generated by the local grid when docked in a port, instead of keeping their auxiliary engines running. This reduces emissions, as the energy source of the grid is in most cases cleaner than ship fuel. Alternative names for shore power are "cold ironing," shore connection, on-shore power supply, and alternative maritime power supply. Especially in Europe and North America, an increasing number of ports provide shore power to different types of ships. One of the first ports to provide shore power facilities was the port of Gothenburg (Sweden). Shore power not only requires an on-shore power connection, but also ships that are able to connect to this power source. For this reason, shore power is most feasible for point-to-point connections, such as ferries. Increasingly though, issues of compatibility are being resolved, and other ship types are connected to shore power in ports. Shore power connections are mandatory in the State of California, and will be mandatory in EU countries from 2025.

6.4.3 Mitigating noise impacts

Noise impacts from ports are subject to international regulation, including an IMO resolution on noise limits from ships and the Environmental Noise Directive of the European Commission, which requires noise mapping for industrial port areas near urban areas. As a result, various ports have invested in noise measurement. In addition, three types of measures can be identified to reduce noise impacts of ports: technical possibilities for source mitigation, port design and barriers, and adaptations in residential areas (Table 6.6).

TABLE 6.6 Main instruments to mitigate noise impacts from ports

Focus	Possible measures
Source of noise	Exhaust silencers on ships, shore power, electrification of port equipment
Transmission	Port layout, port planning schedules, sound barriers, differentiated port dues
Reception	Adaptations in residential areas: insulation, building codes, noise measurement

Source: Adaptation of OECD (2014).

6.5   Avoiding port-related traffic congestion

6.5.1   Urban congestion challenges of port-cities

The presence of a port can lead to urban congestion, caused by the hinterland traffic to and from the port area. A large share of freight transport between a port and its hinterland is by truck, which adds to road traffic volumes, and often to congestion costs in metropolitan areas that are struggling with congestion. The issue is particularly pronounced in developing countries and emerging port-cities, with rapid urbanisation and ports often located in city centres.

Urban congestion due to port-related traffic originates at the port-land interface: the port gates and the arteries from the port to main hinterland destinations. High truck volumes also contribute disproportionately to traffic accidents and ensuing delays (Giuliano and O'Brien, 2008). Congestion on urban road networks not only reduces urban quality of life, but also port competitiveness.

6.5.2   Port gate strategies

One of the main port-related traffic mitigation measures relates to reduction of idle trucks at port gates. This presents highly relevant challenges in many port-cities, leading to urban congestion and environmental impacts. Two main policy instruments in this respect are terminal appointment systems and extending gate opening hours. The following section assesses these instruments as applied in ports and port-cities.

The goal of *truck appointment systems* is to reduce road congestion at port terminals by giving a preferential treatment to trucks that choose to schedule an appointment. The idea is that an appointment system allows terminals to spread truck movements more equally over the day. Terminal gate appointments are usually voluntary, but have in a few cases also been imposed on terminals by law, for example, in California. The results of terminal gate appointment systems have been mixed, but have proved positive in the ports of New Orleans and Georgia (US), improving traffic flows and reducing truck turnaround times in ports.

Extended gate hours attempt to redistribute the arrival times of trucks to port terminals throughout the day. The idea is that offering incentives to use off-peak hours will reduce congestion at port terminals, as well as nearby roadways. The most well-known example of extended gate hours is the PierPASS programme implemented in the ports of Los Angeles and Long Beach, which have managed to significantly reduce the share of peak hour traffic to the port, and with that the congestion in the urban area.

6.5.3   Dedicated freight routes in cities

Some cities have dedicated freight lanes, which facilitates fast and uninterrupted freight transport, allowing for limited intermingling of freight with urban passenger transportation. One of the most well-known examples of such an urban freight corridor is the Alameda Corridor, a railway connection of 32 km connecting the ports of Los Angeles and Long Beach to the transcontinental railways in the US; the corridor is partly underground. Similar dedicated freight corridors exist for trucks. Since 2009, the port of Valparaiso (Chile) is connected to a distribution centre via a dedicated tunnel for trucks, de-congesting major city artery roads. Dedicated infrastructure for trucks will not lead to a modal shift, but might still be effective in reducing port-related congestion in port-cities.

6.5.4 *Extended gates and dry ports*

Regional approaches towards freight transport can help to create enough critical mass for non-truck transportation, in the form of dry ports and extended gates. Trucks generally have a competitive advantage for shorter distance transport; only as distances are longer does freight transport by train generally become a competitive transport mode. Large economies of scale can be reaped, but a certain logistical organisation is required for this in the form of distribution centres in which large amounts of containers and cargo can be grouped before being dispatched to individual destinations. Such a system of selective dry ports or distribution centres has made it possible for relatively small container ports such as Gothenburg (Sweden) to achieve high railway shares in total hinterland traffic. A related approach is that of extended gates, which basically has relocated part of the port closer to the hinterland by displacing cargo handling, customs and other procedures towards an inland port, allowing for a de-congestion of the port. Such a concept is well-developed in the port of Antwerp (Belgium), which has a large set of partnerships creating a network of inland extended gates. Ports have generally become more aware of the need to be better linked to hinterlands, with various ports taking stakes in inland terminals and distribution centres, creating dry ports, merging with inland ports and facilitating part of the hinterland transportation. More information on dry port development is provided in Chapter 15 of this book.

6.6 Mobilising the port as driver of an urban economy

6.6.1 *Industrial development in ports*

In many port-cities, industrial development and port development have traditionally gone hand in hand. These first forms of port-city industrialisation were more or less spontaneous, but since the late 1950s, a wave of planned industrialisation related to ports has taken place. These policies were in most cases driven by national states supporting "champions" as a means of developing economically disadvantaged areas. The fundamental reasons for their development lie within the sphere of maritime transport, namely the development of very large bulk carriers, which have dramatically reduced the costs of long-distance ocean transport (Vigarié, 1981). This heavy industrial development in coastal areas, frequently referred to as Maritime Industrial

TABLE 6.7 Examples of ports with large-scale industrial development

Ports	Industrial development
Rotterdam	Petro-chemical
Antwerp	Chemical
Amsterdam/IJmuiden	Iron and steel
Marseille-Fos	Petro-chemical
Dunkirk	Coal, energy
Taranto	Steel

Development Areas (MIDAs), was land intensive, with requirements for sites of at least 2,000 hectares. Major MIDA projects took place in Europe (see examples in Table 6.7), the United States and Japan.

Originally concentrated in heavy industry, policies gradually shifted to lighter industrial activities, after the economic crisis of the mid-1970s. New oil refining capacity and production of primary chemicals and steel in developing countries meant a rationalisation of the industries that underpinned MIDA development, with a refocusing of port development projects. At the same time, increased population pressure in port-cities such as Rotterdam, Hamburg and Yokohama led to pressure to limit pollution and diversify economic activity. Larger areas in ports became devoted to warehousing, commercial activities and light industries.

Port-industrial planning projects like these have had mixed success rates. In many cases, they have led to rapid increases of population, employment and economic growth. They have in some cases increased the industrial potential of nations and facilitated the restructuration of postwar economies. At the same time, there have been many partial failures as a result of over-ambitious projects or a lack of continuity in planning. In southern Italy, no effective MIDAs were developed apart from Taranto (Vigarié, 1981).

One of the main challenges related to port-industrial development is the creation of linkages with the local economy. This often proves challenging, because most of the industries that have invested in MIDAs are multinational companies whose development strategies are often not aligned with those of regions and cities. This lack of economic linkages within the region may enforce vulnerability of regions related to one-sided economic development and path dependency. Such one-sided development can increase a port's economic vulnerability, cutting off other possibilities for development. The economic vulnerability of industrial development in ports is underlined by the current global industrial restructuring. With the prospect of industrial rationalisation looming, many ports and port-cities are assessing new industrial opportunities that could build on existing assets and infrastructure, including industrial ecology and renewable energy in ports. Chapter 16 presents a detailed coverage of industrial activities in seaports.

6.6.2 *Maritime clusters*

Successful maritime clusters enhance the port's contribution to its surrounding city and region. For this reason, the formation of maritime clusters has been seized upon as a policy objective in many parts of the world, and governments now have at their disposal a diverse range of instruments that may help embryonic maritime clusters to emerge and consolidate, and enhance mature clusters. The success of a given instrument for encouraging maritime clusters is context dependent; policy cannot create clusters out of nothing. In most instances, clusters emerge through path-dependent and market-induced processes, meaning that not all maritime clusters can be encouraged in the same manner. The port cluster can thus be composed of very different subsectoral components. Table 6.8 summarises some of the more famous examples of maritime clusters around the world with the main subsector in which these clusters excel.

Governments are increasingly choosing to support and stimulate cluster growth. Policies should be tailored to suit the needs of the cluster's specific comparative advantages and needs. Broadly defined, they can be grouped into four different types: *developmental support* instruments that support the emergence and maturation of embryonic clusters through the formulation of broad development strategies and the provision of basic facilitating infrastructure; *fiscal and financial incentive* instruments that seek to spur or renew growth in existing clusters, by providing fiscal relief or financial transfers to strategic aspects of the cluster; *co-ordination and information-sharing* instruments that aim to improve cluster governance and overcome collective action problems; and *human capital matching* instruments that seek to better

TABLE 6.8 World leading maritime clusters and their main subsectors

Maritime cluster	*Main subsectors*
London	Financial services, marine insurance, ship registry, ship owners, ship classification, ship agency and forwarding, legal services, shipbuilding and repair, research, maritime regulators
Singapore	Financial services, marine insurance, ship registry, ship owners, ship agency and forwarding, legal services, shipbuilding and repair, research
Hong Kong	Financial services, ship registry, ship owners, legal services, shipbuilding and repair, research
Hamburg	Financial services, ship registry, shipbuilding and repair, research
Piraeus	Ship owners, ship registry, ship brokers, financial and legal services, research
Oslo	Ship classification, ship owners, ship registry, financial service, marine insurance
Shanghai	Financial services, ship agency and forwarding, marine personnel, research

Source: Adaptation of OECD (2014).

embed the cluster locally, by improving the match between the local labour pool and the cluster's human capital requirements.

6.7 Conclusion

The interaction between ports and their cities is subject to policy dilemmas. Port authorities and city governments do not necessarily have the same interests, goals and perception of challenges and policies needed. The challenge for port-cities is to find synergies between the two perspectives, for example, by introducing smart, selective goals for port growth, attracting high value added port employment, using the port as a site for green businesses and developing mixed urban waterfronts with room for port functions.

Concrete impacts and implications differ depending on local circumstances, on the character of the port-city interface and the functional composition of the port and its city. Large-scale industrial development on or close to port sites requires a huge amount of bulk goods, generally associated with fairly limited job intensity, a variety of environmental impacts and strong local economic linkages. Container traffic has similar low job intensity, fewer local economic linkages and environmental impacts related to shipping and hinterland traffic, but overall less polluting impacts, because the connected economic activity is less industrial. Maritime business services generally generate high value added and limited environmental impacts, but are connected to large ports or large metropolitan areas. Cruise shipping is less space intensive than most other port functions, but the economic value it generates is fairly limited unless it is linked to a port-related waterfront. However, it can have relatively severe environmental impacts (emissions, noise) especially if terminals are close to city centres, which is frequently the case.

The mix of port-city policies should be coherent: policy instruments should neither overlap nor work at cross-purposes. Many of the policy choices made depend on the local situation, but the most convincing examples of policy performance involve a coherent package of inter-related instruments. Alignment between local and national policies is particularly important in this regard. Much depends on the situation in a specific port-city: some ports are owned and controlled by their cities, whereas others are owned by a national government, and yet other ports are completely privatised. These ownership patterns evidently change the dynamics between the city and its port. Whatever these institutional differences, port-cities are generally faced with a need for policy alignment on at least two levels: between the port administration and the city administration, and between the city and higher levels of government (central and regional/state).

6.8 Expected future developments

Urbanisation is expected to dramatically increase over the next decades. This urbanisation will take place in emerging and developing countries, where most cities are port-cities. At the same time, many of these emerging port-cities will become more engaged in global trade, maritime transport and port development. This seems to suggest that the port-city interface and its main challenges will only become more relevant in the years to come. The challenge will be to reflect on how policy experiences of highly developed port-cities could be adapted to be of use to the emerging port-cities.

The shipping industry will probably continue to be dominated by a search for economies of scale and cost reductions. This will affect port-cities in many ways. The desire of ports to accommodate the new mega-ships will intensify the spatial relocation process of ports and various negative impacts. Local economic development related to ports might face increasing pressure due to continuing reduction of labour intensity of cargo handling functions, localisation of logistics functions outside ports and global restructuring of the industrial sectors that have so far been located close to ports, such as the petro-chemical sector. New economic opportunities for port-cities could be related to the global energy transition, utilising the locational advantage that ports could have with respect to industrial ecology and bio-based economic functions.

Note

1 This chapter draws extensively from the OECD publication "The Competitiveness of Global Port-Cities" (OECD, 2014).

References

Ducruet, C., and Lee, S. W. (2006) Frontline soldiers of globalisation: Port-city evolution and regional competition. *Geojournal*, 67(2), 107–122.

Giuliano, G., and O'Brien, T. (2008) Extended gate operations at the ports of Los Angeles and Long Beach: A preliminary assessment. *Maritime Policy and Management*, 35(2), 215–235.

Hall, P.V., and Jacobs, W. (2012) Why are maritime ports (still) urban, and why should policy-makers care? *Maritime Policy and Management*, 39(2), 189–206.

Hoyle, B. (1989) The port-city interface: Trends, problems and examples. *Geoforum*, 20(4), 429–435.

ITF/OECD (2015) *The Impact of Mega-ships, International Transport Forum*. Paris, OECD.

OECD (2011) *Environmental Impacts of International Shipping: The Role of Ports*. Paris, OECD Publishing.

OECD (2014) *The Competitiveness of Global Port-cities: Synthesis Report*. Paris, OECD.

Vigarié, A. (1981) Maritime industrial development areas: Structural evolution and implications for regional development. In: Hoyle, B. and Pinder, D. (eds.) *Cityport Industrialization and Regional Development; Spatial Analysis and Planning Strategies*. Oxford, Pergamon Press.

Suggestions for further reading

Besides the papers provided as references in the text, we recommend the following texts as suggestions for further reading.

Hein, C. (Ed.). (2011) *Port Cities: Dynamic Landscapes and Global Networks.* London, Routledge Publishing.

Hesse, M. (2008) *The City as a Terminal: The Urban Context of Logistics and Freight Transport.* Aldershot, Ashgate Publishing.

Vroomans, J., Kuipers, B., and Geerlings, H. (2016) *Understanding Governance Perspectives and Handling Dynamics: The Role of Cultural Dynamics in the Port-City Relationships.* Paper ID41, Proceedings IAME Conference Hamburg.

Wang, J., Olivier, D., Notteboom, T., and Slack, B. (Eds.). (2007) *Ports, Cities and Global Supply Chains.* Aldershot, Ashgate Publishing.

7

PORT PERFORMANCE

Shmuel Yahalom and Changqian Guan

7.1 Introduction

A port is a complex site along a coast where ships dock to transfer cargo or passengers to or from land. It is a hub for multimodal transportation network activities, including cargo/passenger terminal facilities for freight movement and other economic activities. A port is a gateway of international and regional trade. Port performance indicators (PPIs)[1] are important statistical tools to measure port performance.

PPIs are used in three focus areas: economics and finance, operations, and development needs. The measurement instruments are operational efficiency, cost efficiency, and economic effectiveness to serve various objectives, such as internal management control, communication to stakeholders, marketing tools to attract business, and a basis for comparison between ports.

PPIs could be designed for each facility and activity that takes place in the port[2] and used for analysis. Even though there is not a formal measure used to determine the overall performance of a port or a terminal, an important measure that is widely used for this purpose is throughput volume. Globally, ports are ranked by the cargo volume that they handle.

Because most of the activities in a port are related to cargo handling, the majority of PPIs are focused on operational efficiency of the cargo terminal operations and target major function areas – that is, vessel movement, loading and discharging, cargo storage, and receipt/delivery of cargo. More than 90% of general cargo is containerised. Container terminal operations are the most complicated and physically challenging, involving globally hundreds and thousands of container movements daily. The PPIs discussed in this chapter will follow the container movement through the terminal.

This chapter provides a general description of major functional areas of container terminal operations and it discusses what activities are measured, how these measurements are developed, and where they can be used. Additional explanations have also been provided to demonstrate some aspects of the complexity of container terminal operations. The chapter addresses PPIs with respect to social-economic and environmental issues and the role of government.

7.2   Container terminal logistics

A port is a key node in the global supply chain network and container terminals, providing the critical interface between water transport and surface transport.

The port is the starting and end point of surface transport for import and export goods respectively. It functions as a cargo transfer facility. Therefore, port performance data collection starts with information about the port and terminal facilities and the data generated from freight handling for analysis.

The port also functions as a storage/warehouse where containers are picked up and delivered. Import and export containers are moved in opposite directions. The terminal's yard has two exits and two entrances (pier and gate); operation and performance always addresses these unique features. Figure 7.1 shows a generic configuration of a container terminal. There are three major operations taking place: vessel loading/discharging, container yard, and gate operations. If the container terminal has an on-dock intermodal yard, then intermodal operation is included.

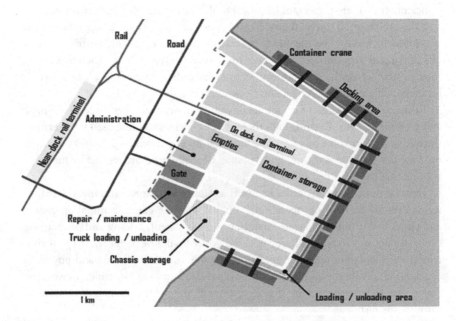

FIGURE 7.1   Generic configuration of a container terminal

Source: Jean-Paul Rodrigue, Hofstra University, https://people.hofstra.edu/geotrans/eng/ch4en/.

To facilitate import and export container movements, there are different types of container handling equipment and systems, depending on operations needs and various trade-offs regarding capability, cost, and efficiency. For vessel loading and discharging operations, there are three types of cranes:

- Ship-to-shore gantry crane (most common)
- Conventional shore crane (quay crane)
- Mobile crane.

For horizontal transfer between the container yard and berth, the following equipment is used:

- Terminal truck/chassis
- Straddle carrier
- Shuttle carrier
- Automated guided vehicle.

For container yard operations, the following equipment is used:

- Top loader or reach stacker
- Straddle carrier (SC)
- Rubber tire gantry (RTG)
- Rail-mounted gantry (RMG)
- Automated stacking crane (ASC).

The selection of a container yard handling system (main type of equipment) depends on a set of criteria:

- Equipment cost
- Manpower requirement
- Size of the container yard
- Equipment capability (stacking density, lifting height, speed, etc.).

On the water side, the pier is designed and built to accommodate vessels that call at the port. The pier is equipped with ship-to-shore (STS) gantry cranes for loading and discharging. The latest generation of STS cranes is capable of handling ships with 23 containers across and eight to nine containers high on deck (maximum 65 tons). The number of STS gantry cranes used on a ship should be maximised in order to minimise port time – as it is said in the industry, "Ships make money at sea, not sitting in port."

Once a container is discharged from the ship, it is placed in a container stack in the yard. See the example of the port of Yang Shan, Shanghai. A typical stack of loaded containers could be four or five high and three to ten across depending on the availability of land and equipment used. Empty containers can be stacked up eight to ten high. As technology changes, so does the configuration of holding containers.

There are three principles of container terminal operations: planning, controlling, and keeping track of every container movement. Advance planning is necessary to allocate yard space, equipment, and manpower to meet operational requirements. Every container movement should be controlled to ensure that no unauthorised movements occur and to avoid confusion and waste of resources. Because there are thousands of containers going in and out of the terminal gate, loading and discharging from and on to the vessel, and going through the yard on a daily basis, proper record keeping of container movement is of utmost importance for data collection and management purposes. All the information on container movements in a container terminal is kept on a real-time basis, providing management with the most up-to-date information. The management of container terminal operations is done with the help of a terminal operating system (TOS), an essential IT software. For example, a spreader is equipped with a container number reader that transmits the container number to the operation centre upon container pickup off the vessel or from the yard. Similarly, once a container is placed in the yard, its location in the grid is submitted to the operations centre for record keeping. The opposite process takes place for container pickup in the yard.

Due to its vast physical expansiveness and operational complexity, data on ports can be divided into the following categories:

- *Physical characteristics*: A port has distinct physical characteristics: draft, pier length, yard size, yard design, number of gate lanes, types and number of equipment, number of reefer slots, space for empties, and so forth. These characteristics are *fixed* in the short term.
- *Operational characteristics*: Operational characteristics are *administrative*, such as hours of operation, labour start time, and number of people that make up a work gang, and so forth; and are determined by operation need and by contract negotiations, which include labour cost, equipment cost, and other fees.
- *Operations information*: Operations data is collected and transmitted to the TOS. A large amount of operations information is associated with *time*. The data includes the time it takes to move a container from the vessel to its parking spot in the yard, the time it takes to hoist a container on and off a vessel, the time it takes to find a container in the stack and load it on a truck, truck turnaround time, and so forth.
- *Financial information*: The management office generates financial information about revenues and costs (revenues from fees, demerge fees, charges, etc.; costs from labour, operations, etc.).

These key internal data sources and other external sources are the foundation for determining port performance and the development of PPIs. Time and cost are the most important components in port performance assessment.

7.3 Economics

International trade is a leading sector in the global economy; countries specialise and export. A country such as China, for example, designed its economic growth and development for more than a decade on growing its exports to the rest of the world. In this effort, 90% of international trade is carried out by vessels. Therefore, the port is a key transportation link along the global supply chain.

A port is the gateway of trade for a country. All coastal countries have ports. Landlocked countries seek access to ports and therefore, frequently develop relationships with coastal countries to use their ports for international trade. Transport by vessel is economical because it offers the lowest cost option due to vessel size. The cost of moving containers by sea in real and nominal terms has declined over time. As global trade grows, the demand for vessel transport increases, vessel size increases, and the demand for port services increases.

Ports play an important role along the supply chain. As service providers for ship owners/operators, ports try to provide fast turnaround service at a low cost. The port's competitive environment keeps in check a vessel's discharge and load operations cost and the time it takes. The competition between ports is frequently also about port productivity, which is instrumental in minimising vessel port time. In this respect, we note that a vessel owner/operator is mobile and may choose any port for its operations.

The direct and indirect economic effects of a port on the local, regional, and national economy are significant.

7.3.1 Economic development

Port planning and development are forward-looking, taking into consideration industrial and technological changes. A port, like any other business, seeks to increase market share and profitability. The plan of a port's economic development and growth (public or private) is to demonstrate its comparative advantage via specific features and attributes in order to attract more business from the shipping industry and the shippers. Thus, a port highlights its location, productivity, technology, services, and value-added services on the premises and nearby. The private ports are adamant in getting new business. A number of PPIs are instrumental in this effort.

7.3.2 Employment

A port is a labour-intensive work environment. The port-related employment is divided between direct, indirect, and induced. Direct employment includes stevedores. The stevedores are usually organised labour, which is defined as people who are employed in loading and unloading ships. In some ports stevedores also move containers in the port. Unorganised labour are all the others.

Stevedores work in gangs of 18 to 24 individuals per gantry crane (gang). One large vessel could employ more than ten gangs simultaneously for a duration of

40 to 80 hours per gang. The number of hours depends on the number of containers handled on the ship.

The container terminal also employs individuals responsible for handling containers in the yard and for loading them onto some type of surface transportation mode. Another group of employees carries out the traditional jobs of accounting, management, IT and terminal operations jobs (monitoring, tracking, checking, etc.). The other transportation mode operators (truck, rail, barge, etc.) work directly with the port management to complete the supply chain services. Different types of direct jobs include tug support services, pilots, chandelling, insurance, finance, government, regulators, security, fuelling, and more.

Indirect and induced port jobs

Port-related employment includes jobs of various kinds that are normally available in any industry. However, these jobs are port specific, including insurance, legal, finance, consulting, management, repair, computer services, and others.

For example, in 2014 the number of individuals directly employed in deep-water seaport employment in the US was almost 542,000 (Table 7.1). The commercial port activity employed an estimated 23.1 million individuals. This group generated an estimated $4.6 trillion.

7.3.3   Competition

Ports compete with each other for business. The competition is for more business from shipping lines and also for more business from customers who prefer Port

TABLE 7.1 US deep-water seaport employment 2014

Jobs	Total
Direct[1]	541,946
Induced[2]	822,884
Indirect[3]	372,017
Subtotal	1,736,847
Importers/exporters (Direct/induced/indirect)	21,380,000
Total jobs	23,116,847
Total economic value (in millions)	$4,557,104

[1] Direct jobs are directly related to the business operation.
[2] Induced jobs are jobs created due to business expenditure generated by the direct and indirect jobs.
[3] Indirect jobs are created due to purchases by a business.
Source: AAPA, American Association of Port Authorities, www.aapa-ports.org/Industry/content.cfm?ItemNumber=1032.

A to Port B. The competition is by highlighting the quality of services the port offers using PPIs. The quality of service is usually associated with the efficient use of time. A port emphasises turnaround time, the time from unloading to delivery, waiting time outside the terminal, operating hours, productivity, dwell-time, and others.

7.3.4 Technology

Port performance is closely associated with the technology used in tracking the loading and unloading of containers and the technology used to handle the containers in the yard and gate. The adoption of new technology is designed to reduce operating time, reduce the number of employees, and better utilise port space. A demonstration of the impact of technology is through performance indicators.

7.3.4.1 Operating cycle time

Technology is designed to reduce operating time. For example, a spreader can pick up two or four containers in one lift. Other technology is designed to optimise yard operations after each movement of a container, thereby minimising container search time and pickup time for the next container. This technology, for example, is based on programs using sophisticated algorithms. In advanced ports yard operation is handled from a climate-controlled back office, where an operator uses cameras and a joystick to move containers back and forth from the pier to the yard or from the yard to loading a truck or a rail car. PPIs are instrumental in this effort. More advanced technology is expected with time.

7.3.4.2 Employees

Technology sometimes substitutes for employees. Information technology is extensively used for record keeping, transactions, communications, and the development of PPIs. In the past there were clerks whose main job was to keep records manually and communicate with port customers; today those functions are performed by technology, thereby reducing the number of employees, reducing errors, and operating around the clock. For example, a typical terminal has one clerk in the gate complex responsible for a few entry booths. The communications with the truck driver for pickup or delivery documents is by camera, submitted by suction tubes and discussed via speaker with the clerk. The container's exterior image is also inspected by cameras. The exterior image is transmitted for review of damages before clearing it to enter or to leave. However, in low labour cost regions one could expect a larger number of employees. The degree of substitution between labour and technology depends on the trade-off costs between the two. The trend is expected to continue towards more technology and less labour.

7.3.4.3 Space utilisation

Ports that are fixed in location in the short term are usually also fixed in ground space (acres). Some ports can expand the ground space in the long term. The growth in global trade implies that many ports require more space. Since ground space is fixed, space can be obtained only by increasing density by increasing the height of container stacking. A comparison of a contemporary container terminal with a terminal of 20 or more years ago highlights the technological differences. *The increase of space utilisation is by using new technologies.* PPIs demonstrate the utilisation of space.

The new technologies of different designs have a common denominator of higher and wider container stacks. The yard itself went through a transformation to multiple rows of lined-up containers. In a modern port there are no more trucks next to container rows picking up or dropping off containers. Trucks wait in a designated space for pickup or delivery. Furthermore, truckers are assigned pickup and drop-off times as well. Improvements in space utilisation and pickup by appointment are expected to continue. The container stacking limits are due to weight limits. PPI tools are used to track this performance.

Indirectly, space utilisation is also achieved by the number of days a container can stay in the container terminal for free. A common period in the US is four days. New technologies provide for better communications, and thereby the potential reduction of the number of free days, that is, increased space utilisation and throughput. A PPI for container dwell-time would determine the container's duration of stay.

7.3.5 Performance

Terminal performance assessment must address the three key physical subsystems that make up a terminal: berth, yard, and gate. These facilities, starting with the pier, are the site for vessel discharging and loading of containers. The yard is a holding facility (or a warehouse) of containers. The gate is exactly what its name means – a cross point where containers either leave or enter the terminal. All the three facilities are involved with container movement. The movements are recorded manually or electronically and therefore provide data for measuring and/or determining terminal performance. The information available in each of the three areas could be used for the development of indicators and operation analysis. The analysis is performed in each predetermined category and cross categories.

Container terminal performance (Figure 7.2) is based on statistical indicators. *Statistical indicators are quantitative measures that state the behaviour of a variable.* Container terminal performance determination starts with obtaining data. Data collection in the container terminal includes internal and external data. The internal data of a facility can be divided into two types, the facility itself and the activities that take place in the facility. External data could be associated with the terminal location. For example, external data, in addition to data from the vessel files, could be the size of the population or business community in a given radius of the port

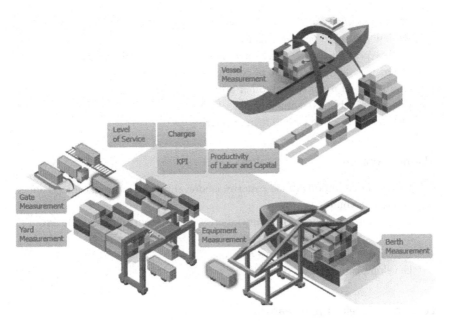

FIGURE 7.2   Container terminal operations

Source: Container and Port Analytics, www.Prognoz.com.

that potentially uses the port, the number of warehouses in the port and the sur-rounding area, access roads, and so forth.

Facility data is fixed in the short term. Port activity data is imbedded in the ongoing terminal activities. Internal data is generated and collected continually by the terminal. Depending on data collection design, the data could potentially include every move that takes place in the facility.

Once the data is available, it should be categorised for analysis. The categories of data collected are facilities, operations, cost, revenue, productivity, and finance. Obviously, the more data, the more analysis.

Container terminal performance data in the form of PPIs is a fundamen-tal method used to determine the state of the terminal with respect to its past and present. PPIs are also used to compare other terminals for planning further developments and investments. A sample list of reasons for using PPIs, divided into categories, is as follows:

General reasons

- Identify trends
- Identify demand for port services
- Provide insight to port management, including feedback
- Develop management tools to monitor performance
- Implement new technology and determine their impact

- Identify bottlenecks
- Review and improve performance
- Review port design (piers, terminal, yards, and value-added services)
- Determine the need/update of computerised systems
- Compare performance with other ports
- Prepare for labour negotiations
- Identify status of repair and maintenance.

Planning reasons

- Changing conditions (trade patterns, labour, ship size, routing, cargo changes, costs, accessibility)
- Scarcity of management personnel (performance, reporting standards, training)
- Scarcity of capital resources
- Determine needs and improve control
- Develop planning tools (simulation).

Economic development reasons

- Determine port development needs and goal setting, including deviation from goals
- Determine and target port investments
- Obtain information for negotiations of contracts
- Determine port fees, including congestion fees, tariffs, and others
- Indicate economics of scale/diseconomies of scale
- Identify competitive forces.

Investment reasons

- Identify public/private investments (strategic investors, landlord requirements)
- Improve port performance (quantitative tools, data mining).

As indicated, container terminal performance has many aspects, measures, and uses. Generally, every *activity* should be quantified with respect to time, revenue, and growth. The results should be used for cross comparison with other terminals, locally, nationally, and internationally and identify best practices. Similarly, all the data of the terminal facilities should be quantified and compared for the same.

The number of *potential indicators* that could be generated from available data in a container terminal is very large. However, the key performance measures used are as follows.

7.3.5.1 Productivity

This is a measure of output per unit of time. In general, this measure of output is with respect to *mobile* assets such as gantry crane and *fixed* assets (facilities) such as terminal, pier, yard, and gate. The productivity measure could also cross between the two in order to obtain an aggregate indicator such as throughput. Throughput by itself is not a PI. Common productivity measures that are used include:

- Output per gantry crane

  The output per crane can be measured in terms of gross output or net output. Gross output includes moving hatches, moving the gantry crane, and so forth. The output can be further divided, taking into consideration weather conditions, seasons, holiday cycles, and so forth. Thus, the traditional measure of output per gantry crane includes the following sample:
  - The number of container *moves* per crane per hour. For example, there are reports of 40 to 80 moves/hour.
  - The number of containers *lifts* per crane per hour.[3] There are reports of 25 to 40 lifts/hour.
  - The number of *double moves* per crane per hour.
  - Total number of hours a crane operates per vessel.
  - The number of cranes assigned to a vessel.
  - Other measures could be added to address other issues; for example, hoist time, time to move hatches, time to move crane, and so forth. Additional measures could be designed to address specific issues of interest of the container terminal management.
- Other indicators or variations of them that are used for productivity analysis include:
  - Output per gate: the number of inbound or outbound container trucks per time period (hour, day, week, month, year, etc.).
  - Turnaround time: the amount of time (in hours) a truck takes from check-in at the gate to check-out at the gate.
  - Throughput: the number of containers a terminal could process, usually in a year. Other time periods could be used.
  - Queuing outside the gate: the amount of time a container trucks waits before it is processed at the gate (measured in hours).
  - Gate hours: the time a gate complex is open for business. For example, 6 AM to 6 PM.
  - Real-time operation: a container terminal operation time that excludes the time of breaks for lunch, change of shifts, bad weather, equipment down time, and so forth. Productivity could be measured against real-time operations instead of operation time in general.
  - Dwell-time: The average time an import or export container stays in the terminal before being picked up (measured in hours or days).

Productivity measures could be many. The preceding list includes traditional measures/indicators. There is no limit to the number of indicators that could be developed in order to address a specific issue of interest to terminal management.

Other measures that impact performance directly or indirectly include the following.

7.3.5.2   Port location and geographic considerations

Port structure and space layout impact performance. This is usually associated with the port location and geographic considerations. A port with insufficient space and/or poor layout will not be attractive to stakeholders and/or customers. The equipment will not operate efficiently and it might prohibit automation, which in turn will reduce productivity and performance. For example, correlating port structure and automation to performance could provide insights into this relationship.

7.3.5.3   Bottlenecks

Bottlenecks could be internal or external to the terminal. In either case there is an impact on performance. For example, an external bottleneck due to permanent infrastructure limitations impacts traffic flow from and to the terminal, which impacts delivery and pickup schedules and trucker's waiting time and location. Internal bottlenecks impact operation timing and performance. The larger the number of external trucks in the container terminal, the larger is the probability of breakdown, bottlenecks, or congestion. An indicator of equipment breakdown or truck breakdown could be correlated with bottlenecks and delays.

7.3.5.4   Support services

A port with insufficient port support services of piloting, fuelling, repair, and so forth, is not an attractive facility. In case of a need of assistance that cannot materialise, operations will slow down or stop. A port needs to determine the key support services and assure their existence. An indication of the time it takes waiting for a support service could be a PI of disturbance and a catalyst for change.

7.3.5.5   Dwell time and demurrage

Demurrage fees are imposed on delays in picking up containers. Allowing a container to stay in the container terminal beyond the allotted time disturbs terminal operations and space utilisation. The terminal administration is responsible to move the containers on and off in order to keep operations moving. An indicator of the average container stays in the terminal together with demurrage collections due to delay in pickup could be an indicator of disturbance to terminal operations.

7.3.5.6   Gate complex

A smart gate is a gate with technical equipment for video checking with the driver, transmission of documents, and container inspection, replacing an individual sitting at a booth. This new gate allows fast transactions at a lower cost which increases output and terminal performance. A comparison of various PI activities at the gate can be compared in order to determine optimum methodology.

7.3.5.7   Pier planning

Export containers are prepared on the pier in advance of the vessel arrival. Moving containers around the terminal in advance of a vessel's arrival in order to accommodate a stowing plan is common practice in container terminals in order to save time. There is a need to measure the integrity of the system and identify the frequency of delays. The reasons for disruption could be many, including the stowing plan, on board a vessel or on the pier. This type of measure will identify the time lost and the cause. In addition, there is a need for pier planning due to changes in ship size, technology, and shipping line services.

7.3.5.8   Yard planning

Yard planning is a complex process. The objective is to minimise the number of times containers are moved. Because the yard activities simultaneously include loading and unloading one or more vessels and loading and unloading trucks and rail cars, keeping track of these activities and analysing them for performance are critical in order to improve performance.

The design of the yard operation is in blocks to reduce interference between various types of traffic flow. The blocks could be defined in each terminal differently, but all container terminals should have blocks for standard containers, empty containers, and refrigerated containers. Developing a host of PPIs designed to address different segments of the operation and blocks will provide important assistance in tracking and/or improving performance. Other issues for consideration include safety and security and their indicators.

7.3.5.9   Pickup by appointment

A new trend is developing of pickup by appointment. A trucker is given a window of time to come and pick up a container. Complying with the allotted time improves terminal performance and space utilisation. Missing the window does the opposite. Once implemented or simulating the operation could identify the container terminal performance differential with the alternative system. The pickup by appointment indicator would identify its success and/or deficiencies, which could lead to proposing remedies.

7.3.5.10   Terminal operation synchronisation

A container terminal is a complex work environment. All the subsystems must work well together. An aggregate operations measure of throughput, for example, could provide an indication of overall terminal performance together with aggregate earnings and costs. However, a delay in one part or area spills over to another. Disaggregate operation indicators, including spill overs which impose delays from one region to another, are important to have. Good management is aware of the complexities and designs the terminal operations accordingly while reviewing the operations for improvements. Frequently terminal operations software is designed to optimise the comprehensive system's performance. The comprehensive system might be oversimplified and does not include some important processes and variables. An aggregate and disaggregate review of the system's operation is important.

7.3.6 Indicators[4]

PPIs are statistical instruments developed to determine performance, usually in order to improve management and control. The use of indicators could be very wide. However, the fact that massive amounts of data are collected and available via records of activities does not imply that it is used.

Internally the indicators are analysed in order to improve performance in terms of time and money. The internal indicators developed by ports and terminal are frequently *not* shared with the rest of the world. They are usually for internal use only, but it could change.

Externally the number of indicators shared with the public, regulators, or investors are small, including throughput, turnover, and profits. They are usually used in order to attract more business and/or investments.

The data collection methods and definitions of indicators by port or terminal are not *standardised*. There is no uniformity in definitions. Therefore, the use of indicators, especially for comparison, should be used with caution.

The data could be used with general economic and/or industry indicators in order to forecast change and provide trends and comparisons. A large variety of indicators could be developed from the port data, operations and others. The indicators should be targeted in order to answer specific questions. The indicators could be *leading, lagging*, or *coincidental* indicators.[5] A terminal with an extensive database could develop numerous indicators for analysis addressing all aspects of management interest (Table 7.2). An efficient use of data incudes a monthly review of predetermined indicators.

TABLE 7.2 Introduction of PPIs in the port of Rotterdam

Year/Period	Indicators
Beginning twentieth century	Number of ships
	Throughput volume
1990s	Port-related employment
	Value added
	Port value added as % of regional GDP
2002	Development in turnover
	Profitability of firms in port
2003	Investment level of private firm in port area
2004	Establishment of (new) companies in port area

Source: Port of Rotterdam (2006), Rabobank (2003), ECORYS-NEL (2003), Rebel Group/Buck (2005). From: Peter de Langen et al. (2007) "New Indicators to Measure Port Performance," *Journal of Maritime Research*, Vol. IV, No. 1.

Port indicators are categorised as:

* General
* Operations
* Finance and/or economic.

7.3.6.1 General indicators

General indicators are categorised as indicators that provide general information about the port. The sample can be expanded to include other indicators.

* Ship ownership and flag of registration
* Duration of contract with shipping line (time in years)
* Characteristics of ships calling the port: type, size, length, beam, height, gross ton, etc.
* Number of ships calling the port or terminal (number/terminal or port)
* Ship arrival time/departure time (time/ship)
* Turnaround time of ship in port (hours/ship)
* Time ship started/completed operations (time/ship)
* Time berthing (time/ship)
* Gang hours worked (time/ship)
* Total of hours ship was worked on and idle (net hours, gross hours)
* Containers discharged/loaded (number/ship)
* Tons discharged/loaded per ship (ton/ship)
* Tonnage handled (tons)
* Value-added indicators.

Some indicators fit into more than one category.

7.3.6.2 Operations indicators

The potential number of terminal operation indicators are many. There is no limit of what could be measured as an indicator. As indicated before, the indicators are designed to collect different information to determine port performance.

The operations indicators should be blocked by ship size and type in order to generate consistency in analysis and comparisons. A sample of indicators includes:

* Quantity of cargo worked per ship hour in port or berth (ton/hour, containers/hour).
* Number of gangs employed per ship per shift. This is equal to total gross gang time, divided by total time that berthing ships were actually worked (time/berth).
* Average tonnage per ship: total tonnage worked for all ships, divided by the number of ships (ton/ship).
* Number of gangs per ship (gangs/ship).

- Tons/containers per gang hour (tons/gang-hour or containers/gang-hour).
- Berth utilisation.

As stated before, some indicators could be in more than one category.

7.3.6.3 Analysis indicators

The analysis uses the indicator data obtained to determine performance. The potential analysis of scenarios is endless. However, a sample of common issues analysed includes:

- Waiting time: total time between arrivals and berthing for all berthing ships, divided by the number of berthing ships (hours/ship).
- Arrival rate (from arrival time): number of ships arrived per month/number of days per month (ships/day).
- Service time (from service time): total time between berthing and departure for all ships, divided by the number of ships (hours/ship).
- Turnaround time: total time between arrival and departure for all ships, divided by the number of ships.
- Fraction of time berthing ships worked: total time that berthing ships were actually worked, for all ships, divided by the total time between berthing and departure of all ships.
- Tons per ship hour in port: tonnage worked, divided by total time between arrival and departure.
- Tons per gang hour: total tonnage worked, divided by total gross gang time.
- Fraction of time gangs idle: total idle gang time, divided by total gross gang time.

7.3.6.4 Financial/economic indicators

All activities in the container terminal have monetary aspects. The financial and economic indicators are designed to provide information about revenues produced by each service and the costs of each service. All cost aspects are differentiated between fixed and variable costs.

There are times that it is hard to distinguish between categories. These cost and revenue indicators are listed in various manners (including duplicate). In general, financial indicators include:

- Total costs (labour, facilities, utility, offices, etc.)
- Labour cost (supervision, management, review, etc.)
- Direct labour cost
- Ship revenues
- Capital equipment cost
- Return on capital cost

- Fixed cost and variable cost
- Intermodal facilities cost/revenues (rail, track on dock rail, etc.)
- Berth revenues
- Cargo revenue related to the cargo handling operation (ship to storage, ship to ship, etc.)
- Landlord port contracts.

Examples of specific financial indicators include:

- Berth occupancy revenue per ton of cargo: total berth occupancy revenue produced, divided by tonnage worked ($/ton).
- Berth revenue per ship ($/ship).
- Container slot occupancy revenue ($/slot).
- Cargo handling revenue per ton of cargo: total revenue produced from transferring cargo to or from ships, from or to storage areas, divided by tonnage worked ($/ton).
- Labour expenditure per ton of cargo: total direct labour expenditure for transfer of cargo to or from ships, from or to storage areas, divided by tonnage worked ($/ton).
- Capital equipment expenditure per ton of cargo: total amortisation and interest allocated to and maintenance and operating costs incurred for the berth group, excluding the costs of transit sheds and warehousing, divided by tonnage worked ($/ton).
- Total contributions: the berth occupancy and cargo handling revenues minus labour and capital equipment expenditure.
- Revenues per ton of cargo: total contribution divided by tonnage worked ($/ton).
- Return on investment: 100 times the gross revenues (the sum of working expenses, staff and miscellaneous expenses and capital charges) divided by the sum of working capital and present value of assets.

As indicated before, additional measures are used in the industry. The measures are developed to address a specific issue or need. There is no limit to developing measures.

7.4 Socio-economic issues

The proximity and the interaction of the port with a port-city and surroundings has its positive and negative aspects. Close proximity provides opportunities of employment and its multiplier effect. A key port could generate these benefits for a long period of time. For example, in an extreme case, the community growth is linked to the container terminal, the key industry in the area, and would become dependent on it, such as the port of Kaohsiung in Taiwan. Being dependent on the port, in the long term, could become a liability if things change due to the development of a competing port. For example, the port of Shanghai has an impact on the port of Kaohsiung. Other key gateway ports with an overwhelming impact on a region are the port of Singapore and the port of Hong Kong.

During the period of container terminal growth, the community grows and develops in every possible aspect: infrastructure, housing, schools, services, support services, repair, small business, entertainment, shopping, and so forth. However, should there be a turnaround towards a decline, the same community will exhibit a decline in all the same sectors and additional socio-economic problems associated with social programs: family pressure, divorces, psychological issues, and so forth. Thus, a community should try to diversify its industrial base; it should preferably not be dependent on only one sector.

Port performance should be credited and take into consideration social issues during the good times and the opposite at the bad times. Port management should engage and/or partner with community leaders and address issues of growth and community investments and raise concerns of liability if the wave turns. *Transparency* is the instrument of cooperation between the two.

7.5 Environmental issues

A port generates a large amount of traffic from both the water side and land side. This traffic, which usually uses fossil fuel, causes air and water pollution, noise and congestion. These *externalities* are perceived as negative to the community and positive to industries that measure the externalities' impact to find solution to the problems.

Communities address the negative externalities by regulations. Some regulations are local, others are regional and even national. The degree of imposing regulations depends on location and the severity of the negative impact of the externality. For example, less developed countries or regions are willing to tolerate more environmental damage than others.

The community and the terminal management should cooperate and collaborate in identifying environmental issues and work together to mitigate their impact. Environmental PPIs could be used to determine the status of the environmental changes and impact. Environmental indicators could include air, water, and noise. Indirect data of the impact on traffic congestion should also be collected.

Collaboration mitigates the potential of one side impacting on the other, which could lead to imposing a one-sided regulation.

7.6 Government issues

The government is engaged with the port on different fronts. One of the fronts is its role as *landlord*. Another front is its role as a *regulator* and *inspector*. In some places there could be a conflict of interest between the two responsibilities. Some of the activities require space, take time, and require resources. As a part of the government's role, it generates data associated with its activities such as accidents, the number of inspections performed by customs and security, the time that the inspections took and their associated delays, and others. If additional regulations are added, data to determine their impact on time and cost should be collected and analysed.

Government is engaged in various forms of PPIs. Government resources fund some of the indicators. The resource allocation for indicators could include dredging, inspections, and others. This data can be analysed against other performance measures, to determine its impact and lead to policy changes.

The government role is to oversee the port activities in order to assure that the port does not impose negative externalities on society. In case the port does not self-control its negative impact, government agencies could impose regulations. The regulations are designed to look for the best interest of the public using best practice standards. However, before regulations are imposed, data must be collected and analysed. The data collected is used as an indicator for analysis.

7.7 Conclusions

Port performance indicators are instrumental in determining the health of a terminal/port and the standing with comparing ports. Indicators are instruments used to identify the need for new technology, investments, regulations, terminal growth, change of leadership, and many others. The data for making these types of decisions is continually collected and available from the terminal operations activities. Some of the data is used; more can be used.

Every terminal can develop new indicators for internal or external use. Both the data generated and the indicators are not standardised; every terminal can collect and report data in its own way. Therefore, comparing terminals or ports is a challenge. Port data collection and reporting should be standardised. Port data should be more transparent in order to increase the awareness of port operations and efficiency, and in order to compare port performance. Data sharing between ports and stakeholders is important in order to increase cooperation, develop best practice standards, reduce negative externalities, and assist in developing more scientific tools for port performance analysis.

Notes

1 Note that the term "port" performance is frequently used even though the performance/indicator is for a "terminal."
2 There is a clear distinction between a port and a terminal even though sometimes the two are confused. For example, the port of New York/New Jersey consists of five terminals where some are identified as a port, i.e., port of Newark, Port Elizabeth, Global Container Terminal in Bayonne, Global Container Terminal in Staten Island, and port of Red Hook.
3 Moves and lifts are not the same. Lifts refers to the number of times a gantry crane generates a lift, which could be of one or more containers. For example, one lift could include two containers, that is, two moves.
4 Some of the indicators in this section are from UNCTAD (1976), "Port performance indicators," United Nations Conference on Trade and Development, New York, May.
5 Leading indicators reflect future changes in the trend due to construction, planning, purchasing, and so forth. Lagging indicators trail behind the activities reflecting commitment

already underway. Coincidental indicators move with the trend reflecting concurrent activities.

References

AAPA, American Association of Port Authorities. Retrieved from www.aapa-ports.org/Industry/content.cfm?ItemNumber=1032

De Langen, P., Nijdam, M., and Van der Horst, M. (2007) New indicators to measure port performance. *Journal of Maritime Research*, IV(1), 23–36.

UNCTAD (1976) *Port Performance Indicators*. United Nations Conference on Trade and Development, New York, May.

Suggestions for further reading

Besides the papers provided as references in the text, we recommend the following texts as suggestions for further reading.

Bichou, K., and Gray, R. (2004) A logistics and supply chain management approach to port performance measurement. *Maritime Policy & Management: The Flagship Journal of International Shipping and Port Research*, 31(1), 47–67.

Chow, G., Heaver, T. D., and Henriksson, L. E. (1994) Logistics performance: Definition and measurement. *International Journal of Physical Distribution & Logistics Management*, 24(1), 17–28.

De Castilho, B., and Daganzo, C. F. (1993). *Handling Strategies for Import Containers at Marine Terminals*. University of California Transportation Center.

Geerlings, H. and van Duin, R. (2011). A new method for assessing CO_2-emissions from container terminals: a promising approach applied in Rotterdam. *Journal of Cleaner Production*, 19(6–7), 657–666.

Kim, K. H., and Kim, H. B. (2002). The optimal sizing of the storage space and handling facilities for import containers. *Transportation Research Part B: Methodological*, 36(9), 821–835.

Marlow, P. B., and Paixão, A. C. (2003) Measuring lean ports performance. *International Journal of Transport Management*, 1(4), 189–202.

Talley, W. K. (1994) Performance indicators and port performance evaluation. *The Logistics and Transportation Review*, 30(4), 339–352.

Talley, W. K. (2006) Chapter 22 port performance: An economics perspective. *Research in Transportation Economics*, 17, 499–516.

Tongzon, J. L. (1995) Determinants of port performance and efficiency. *Transportation Research Part A: Policy and Practice*, 29(3), 246–252.

8

ACCESSIBILITY OF PORTS AND NETWORKS

Francesco Corman and Rudy R. Negenborn

8.1 Introduction

As ports are increasingly confronted with congestion (on the road as well as on rail and on the water), accessibility has become a key port performance indicator. Overall accessibility is related to the ease of realising a certain transport activity. Policies aimed at guaranteeing accessibility of ports are increasingly focused on management issues (so-called orgware and software), instead of on the building of new infrastructure (hardware).

For instance, many initiatives are being undertaken to increase the modal shift from trucks towards short sea, inland shipping, or railways. Those can be supported by building new infrastructure (such as a new railway line), improving the management of the infrastructure (for example, by providing updated information over the time windows for seamless traveling along an inland seaway), or dealing only with organisational aspects, such as changing laws or subsidies so that vehicles with larger capacity will be preferred to vehicles with less capacity, such as trucks.

The aim of this chapter is to provide an accessible insight in different perspectives on accessibility and ways in which accessibility can be improved. Hereby, perspectives and measures are motivated with examples from the ports in the Hamburg–Le Havre range.

This chapter is organised as follows. Section 8.2 introduces the issue of port accessibility. How can port accessibility be defined and what are typical indicators used for quantifying degrees of accessibility? Several indicators and directions are reviewed, which have been proposed to somehow quantify this concept. Section 8.3 subsequently emphasises the importance of addressing this issue. The different directions for assessing and improving accessibility are overall linked to a generic network representation. Actual values and examples of indicators for several port areas where accessibility is an issue are discussed. Section 8.4 provides an overview of directions for addressing the accessibility issue. Section 8.5 provides concluding remarks and challenges for the future.

8.2 Accessibility of port areas: definitions

Accessibility is a word that has many meanings and is used in a range of domains. Its origin comes from the Latin verb *accedere*, that is composed of the words *ad* – "towards" – and *cedere* – "to move."

There are many ways in which accessibility can be defined. From the eyes of an urban planner, accessibility may refer to the relation between changes in a transport system (network, production and attraction of trips) and journey length. Differently, accessibility might refer also to the difficulty of reaching a disaster area in case of relief. From a theoretical systems and control perspective, accessibility is associated to being able to reach a certain set of dynamically varying states. Several authors provide an actual definition of accessibility. For example, Hansen (1959) defines accessibility as the potential of opportunities for interacting, while Burns (1979) focuses on the individual, and defines accessibility as the freedom of individuals to decide whether or not to participate in different activities. The definition for accessibility is also related to the perspective of the actor; people with particular mobility needs have a different concept of accessibility than people without. Similarly, accessibility of a particular place can be completely different than when considering different transport modes. Accessibility is a rather vague notion – a common term that everyone uses – until faced with the problem of defining and measuring it (Gould, 1969).

In the transportation science community, accessibility is generally related to the ease with which places or activities may be reached from a given location using a particular transport system (Morris et al., 1978). This accessibility definition can be translated immediately to port areas. Nevertheless, growing competition in the international markets has shifted the focus of accessibility from activities strictly related to ports to activities related to whole transport chains. The concept of accessibility is in this sense related to the effort to reach a destination or the ability to reach multiple destinations across a complete, possibly complex, transport chain. As a result, ports are striving to enhance the quality of the hinterland transport services (Notteboom and Winkelmans, 2004). The quality of port hinterland access is the result of the interrelation of the behaviour of many actors including terminal operators, freight forwarders, container operators, and the port authority (Van der Horst and De Langen, 2008).

A structured approach to evaluate accessibility measures can be found in Geurs and Van Wee (2004). Among those, the most common ways to quantify accessibility are based on simulated or observed *infrastructure*-based performance or service level. Among the ways to quantitatively measure accessibility, the main indicators are related to a measure of spatial separation. This can be measured by one or more attributes of the links between areas that separate the origins from the destination (one of which will be the port area). These can be distance (crow-fly or network distance); travel time; travel costs; travel time reliability and variance; time for entering or exiting a particular system (access and egress, transshipment, capacity of terminal); time related to perform some needed activities within a non-transport system (such as clearance of formalities, waiting for an available slot on a transport vehicle or on a transport network); in the opportunity or convenience to couple certain activities together; to any other attribute that acts as a deterrent or constraint to access. All those characteristics can be quantified to a certain extent, based on some infrastructural measurements.

Geurs and Van Wee (2004) remark that infrastructure-based measures do not include a land-use component; that is, they are not sensitive to changes in the spatial distribution of activities if service levels (e.g., travel speed, times or costs) remain constant. Such a spatial distribution of opportunities offered is instead assessed by means of *land-use* components which, in a port context, refer to the size and overlap of different hinterlands for different modes and goods.

Geurs and Van Wee (2004) define also different directions of accessibility. Accessibility for port areas mostly develops along a *transportation* component (i.e., related to the disutility for someone to cover the distance between a supply point and a demand point) and a temporal component (related to availability in time when to perform some activities). A more exhaustive list of characteristics that should be considered when assessing accessibility refers to:[1]

Economic actors, types of goods, and companies involved: Those are associated to the location of the origins (and destinations) of the freight flows, and to the activities to be performed within the port area. For instance, availability of refinery or processing facilities will increase the accessibility of a port area with regard to the particular goods that are processed.

Spatial detail: Accessibility may concern transport that takes place at a continental, transnational, interregional, regional, municipality or neighbourhood scale. For many ports, this relates directly to the size of the hinterland and the zone of attraction of freight flows from the deep sea. Every spatial scale might require a different spatial resolution with regard to area size, the network considered, and the time scale analysed.

Travel modes: Accessibility might be measured for only one transport mode or for a multimodal or inter-modal transport system. Moreover, in a (multimodal or inter-modal) network, the route choice is also relevant; the most convenient route can be chosen, according to a variety of objectives. Those objectives typically relate to minimum cost, minimum time, shortest distance, route passing via a particular set of locations, or avoiding a particular location or bottleneck (e.g., for hazardous materials). Route choice can also involve different modes, and further considering the possibility or need for interchange, quantity of interchanges, minimum time for each such interchange, availability of guaranteed connections, or the necessity to include or avoid a particular interchange.

Time scale: Especially in the context of time-responsive logistic chains, accessibility also needs to be considered with regard to a target arrival or departure time or both, arrival or departure during a specified period, depart after, and arrive before. This directly yields the need to consider travel time reliability. Moreover, the variation of accessibility would probably show time dynamics that have seasonal, weekly and daily patterns, in the forms of peaks.

Information, offline and online: The availability of information about the network characteristics might result in different transport activities and trips planned and executed. This offline information can induce a different degree of accessibility than what is actually available on the network itself. More-over, the availability of information during the journey itself, with real-time updates and a certain reliability of information, might result in different accessibilities in practice.

8.3 The importance of accessibility

Accessibility is crucial for a port. Hinterland port access is the key to its success. In this section the importance of the accessibility issue is motivated and illustrated. Actual values of accessibility indicators for several port areas where accessibility is an issue are presented. As such, this section quantitatively illustrates the importance of addressing accessibility issues.

Figure 8.1 illustrates the size of the hinterlands of the ports in the Hamburg–Le Havre range. The figure shows, per port, the portion of the weekly shuttle services to the other major ports. The figure moreover shows the relative size of the ports in terms of total number of services. The large ranges of the Rotterdam and Hamburg hinterlands are clearly observed. It is also noteworthy that the hinterlands of these ports overlap for a significant part.

Table 8.1 shows the advantages and disadvantages of different types of hinterland transport, based on modes (Van der Horst and De Langen, 2008). As is clear from this table, although each mode has its particular advantages, there are also numerous disadvantages, each relating to accessibility (congestion, unreliability in times, not being able to physically access a particular location).

Figure 8.2 shows the trends in total transported goods (billion tons per km). The growth in total volume is quite apparent, and that is especially reflected in a direct increase of the road transport. This, combined with limited infrastructure investments, results in major congestion, undermining the accessibility of the port areas for people and goods. To this end, various initiatives are targeting increasing the share of transport over water and rail. Furthermore, also those modes, without additional measures, and looking at the growth trend, might result in accessibility problems lead to a more intense road, water, and rail use, leading to larger accessibility problems.

The growth trends of Figure 8.2 changed with the financial crisis. This crisis caused a significant reduction in the volumes of cargo transported worldwide. Although this may seem to be a solution for accessibility problems, in fact by now (2018), the decline in volume of cargo to be transported seems to have stalled and for the major ports in the Hamburg–Le Havre range an incline in transport throughput can be observed, as illustrated in Figure 8.3.

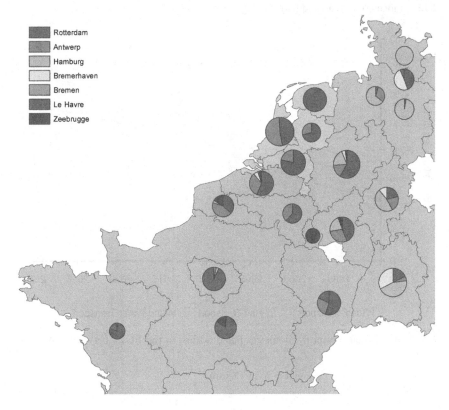

FIGURE 8.1 Hinterlands of the ports in the Hamburg–Le Havre range

Source: Based on *Kennisinstituut voor Mobiliteitsbeleid* (2014).

TABLE 8.1 Advantages and disadvantages of the different hinterland transport modes

Transport mode	Advantages	Disadvantages
Road	Extremely wide infrastructure	Congestion on roads
	Reliable in time	Most expensive form of hinterland transport
	No extra transshipments needed	
	Fast	
	Can always reach end of transport chain	
Railway	Wide infrastructure	Cannot always reach end of transport chain
	Competitive for road transportation over longer distances	Reliability in time
		Not the fastest transport mode
Waterway	Relatively cheap	Slow
	Infrastructure in large part of Europe	Unreliable in time
		Cannot always reach end of transport chain

Source: Based on Van der Horst and De Langen (2008).

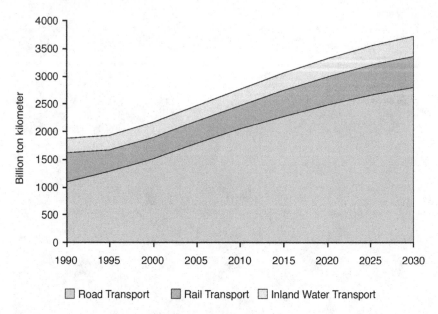

FIGURE 8.2 Total transport kilometres per modality in the EU27 between 1990 and 2030 (expected)

Source: Based on European Environment Agency (2016).

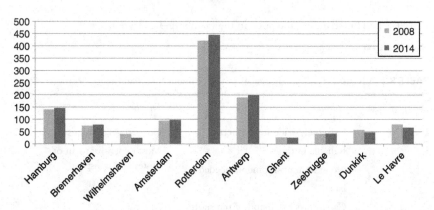

FIGURE 8.3 Total throughput, million tons of the ports in the Hamburg–Le Havre range

Source: Based on data from the Port of Rotterdam Authority.

8.3.1 Road

A study by *Kennisinstituut voor Mobiliteitsbeleid* (2014) analyzed how different measures contributed to reduction of travel time on main roads in The Netherlands. As is illustrated in Figure 8.4, between 2008 and 2013 measures that successfully contributed to a reduction of travel time involved stimulation of teleworking (working

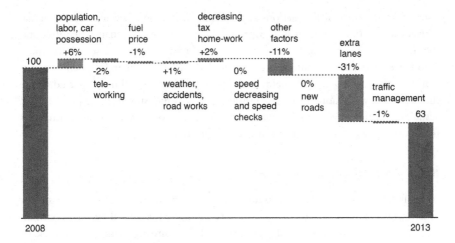

FIGURE 8.4 Travel time losses on the main roads in The Netherlands

Source: Based on *Kennisinstituut voor Mobiliteitsbeleid* (2014).

TABLE 8.2 Online arrival and departure times of trains at RSC Rotterdam

Date	Planned		Actual		Track	Train	Units	Operator
	Arrival	Departure	Arrival	Departure				
22/11	06:15	12:23	07:28	–	15	Ludwigshafen	59	Rail Chem
23/11	03:30	09:45	04:27	06:36	16	Novara 2	25	DB Schenker
23/11	02:15	17:45	00:33	07:43	17	Novara 1	34	DB Schenker
23/11	07:45	09:40	08:35	0907	1	Niederglatt CT	13	DB Schenker

Source: As provided by the website of RSC Rotterdam on 23 November 2011 at 9:22 AM.

at a distance, rather than in a fixed office), an increase in the fuel price, construction of new lanes and improved traffic management. Factors decreasing accessibility consisted in particular of a growth in car possession, road works/accidents and weather events.

8.3.2 Rail

In the current situation of the RSC (Rail Service Centre) terminal, arrival and departure times of trains are published online. When comparing scheduled and real arrival and departure times, large differences are evident. Also, the duration of the visits at the terminal show large differences between scheduled and reality. Table 8.2 shows arrival and departure times for both scheduled and real times.

The figure shows typical situations at the RSC: for example, a train with an actual arrival time of 00:33, which is 1 hour and 47 minutes earlier than scheduled, and the departure of the same train is more than 10 hours before the scheduled time. Another example is the train "Ludwigshafen" for 22 November. The screenshot is from 23 November and still, the train to Ludwigshafen has still not departed, while the departure was planned for 12:23 on the 22nd. When studying the planned and actual departure times in more detail, it can in fact be observed that such situations frequently appear, which cause clear accessibility issues.

8.3.3 Water

Accessibility to main water/sea routes might be regarded in two different aspects. One is towards the deep sea, and accessibility might be quantified by the maximum draft available, to the extent of the area reachable directly without any lock or obstacle, and moreover with the sailing time between the deep sea and the berth. From the inland perspective, the travel time along the waterways determines accessibility, as well as the presence of locks, bridges or other obstacles resulting in travel time increases.

Table 8.3 shows the expected waiting times for vessels at the locks between Rotterdam and Antwerp in 2020. Table 8.4 shows the average and maximum handling time (i.e., time between entering and leaving the locks). These tables result from experiments carried out within the project "LIVRA – Logistical Chain

TABLE 8.3 Expected waiting times for vessels at the locks in 2020

Lock complex	Average waiting time [minutes]
Volkerak	121–200
Krammer	91–120
Hansweert	31–60
Kreekrak	61–90

Source: Derived from data in Logica (2009).

TABLE 8.4 Mean service time at the locks

Lock complex	Average service time	Maximum service time
Volkerak	36 minutes	52 minutes
Krammer	43 minutes	59 minutes
Hansweert	30 minutes	48 minutes
Kreekrak	27 minutes	39 minutes

Source: Derived from data in Logica (2009).

Information Waterways Rotterdam-Antwerp" (Logica, 2009). Clearly, it is desirable that the non-effective waiting time before handling can start is kept to a minimum. In The Netherlands, Dutch authorities for waterway transport have set a norm for waiting times at locks, at 30 minutes. The norm must increase the reliability for inland waterway transport. Since the expected waiting times on the inland waterway connection between Rotterdam and Antwerp does not fulfil the norm of 30 minutes, measures need to be taken to find ways to comply with the norm.

8.4   Approaches for improved port accessibility

Issues in accessibility are observed at the physical, real-world infrastructural level, that is, those places where the actual transport takes place. The movements of the goods to be transported are the result of a number of interacting elements. Figure 8.5 illustrates three levels at which measures for improving accessibility can be categorised.

- Hardware: Closest to the actual accessibility problem are the hardware measures. Building or expanding new infrastructure to increase the throughput of the network can directly improve accessibility.
- Software: One level higher are measures categorised as software. Optimising the flow to avoid bottlenecks and maximising the utilisation of the network are typical examples. Such measures assume that a particular infrastructure and hardware solution is already in place.
- Orgware: At one level higher, organisations determine when which particular goods need to be transported from some place to another. Adjusting the flow before it enters the network can hereby be used to reduce accessibility issues at the lowest level.

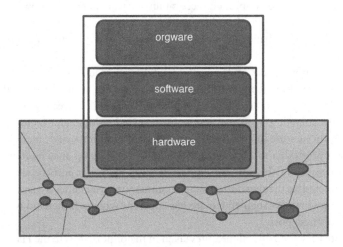

FIGURE 8.5   Relation between hardware, software and orgware as directions for improvement of accessibility

TABLE 8.5 Types of measures

	Hardware	*Software*	*Orgware*
Road	Travel time reduction	Travel time reduction	Less vehicles in the area
		Travel time variance reduction	Peak shaving
		Peak shaving	
Rail	Travel time reduction	Travel time reduction	
		More capacity along the lines	
Sea	Travel time reduction	More capacity at ports	Consolidation of small flows
Intermodal general		Transshipment time reduction	Peak shaving
			Ease modal shift
		Travel time reduction	Express documentation

By building or expanding new infrastructure it is possible to easily increase accessibility, although this comes at a high price. Better use of the existing infrastructure is usually a cheaper option. By guiding the flow, the network can be better utilised and bottlenecks can be avoided. Optimising the flow has its limitations, however, if for instance we look at the temporal component with the rush hour peak. During rush hour the total capacity of the network is too low to handle the flow; even when guided, traffic jams can occur. Actively adjusting the demand for the network is an option that cannot be forgotten. Providing a financial incentive to avoid the rush hours is an example of improving accessibility via orgware. The flow is adjusted before it enters the system. In other words, the hardware and software handle the supply, while orgware handles the demand. These types of measures are summarised in Table 8.5.

It is worth noting that improving accessibility of road links for ports competing for the same hinterland is not necessarily increasing profitability of the port. Research has shown that increasing corridor capacity will increase a port's output against the competing ports and results in increased volume and increased profit. Differently, an increase in inland road capacity only may or may not increase a port's output and profit due to extra effects. While the local delays may be reduced, and future expansions of the port are supported, extra passenger and commuter traffic will be attracted by the new infrastructure. Similar uncertain effects may result from road pricing schemes. Such effects are less apparent when a hinterland is primarily served by a single port.

Next are presented several measures taken in major port areas in the Hamburg–Le Havre range over the different accessibility improvement measure levels, organised based on the mode or modes involved in the measure (road, rail, water, intermodal).

8.4.1 Road accessibility

8.4.2.1 Hardware

The A15 is the main highway for accessing the port of Rotterdam by road, and suffers from major accessibility issues due to congestion. Figure 8.6 illustrates that during peak hours, up to on average 40% of the traffic consists of trucks.

The A15 has a number of infrastructure bottlenecks; moreover, incidents can result in severe congestion, and are quite frequent (about one every three days). To improve accessibility, a major reconstruction and expansion of the A15 is being carried out. This includes new bridges and extra lanes. It is noted that during these construction works, the accessibility issue is in fact more problematic. More at a planning stage, new infrastructure expansions include a new link between the Rotterdam area and the north of the country, the Blankenburg tunnel.

From application over the total Dutch network, the impact of *extra lanes* can be approximated by a reduction of travel time losses (travel time) of 30% (*Kennisinstituut voor Mobiliteitsbeleid*, 2014). Moreover, extra lanes result in a reduction of extreme travel time losses (reliability) by 38% (*Kennisinstituut voor Mobiliteitsbeleid*, 2014).

Similar types of hardware accessibility measures have been made in Hamburg, with infrastructural expansion of the A26, which goes along the river Elbe and connects the port of Hamburg with the German hinterland. This is not only to improve accessibility of the port, but also to reduce travel time for goods and local inhabitants; moreover, road safety in the area is expected to increase.

In addition, the road infrastructure of the Le Havre port is responsible for the largest share (85%) of accessibility to the port area. The Antwerp port area has benefitted from the construction of a series of links around the city centre, through a series of tunnels in the port area, which also increases road accessibility of the port. The accessibility improvements that are expected for this infrastructure (Liefkenshoek tunnel and Oosterweel connection) should be reached when a large highway

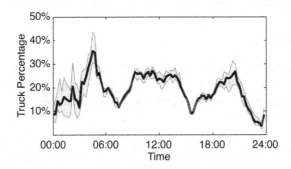

FIGURE 8.6 Average truck percentage on the A15 leading to port of Rotterdam, median and quartiles

Source: Based on Schreiter et al. (2013).

connection would continue this corridor towards The Netherlands. A third link connecting the two banks of the river Schelde is also planned.

8.4.2.2 Software

Autonomous vehicles are currently investigated by a few research groups worldwide. TNO is considering driverless trucks to improve the performance of road traffic. The concept is to have a truck follow a truck which is controlled by a conventional driver. The main argument of self-driven or intelligent cars is safety, as a faster reaction time can be reached by a computer; moreover, in-car systems can help reducing the consequence for instance by pre-crash systems which automatically brake. Moreover, various degrees of increased automation in cars can achieve coordinated control to result in intelligent operations and reduce congestion (for example, adjusting the cruise control of the vehicle). Main obstacles for the rollout of such technologies are regulations and assignment of responsibilities. The Dutch Ministry of Infrastructure and Environment pledged in the summer 2014 a removal of existing roadblocks that stand in the way of autonomous vehicles.

BOS HbR is a decision support system for *dynamic traffic management* that has been tested along the A15, with the aim of controlling the traffic flow and reducing congestion. A basic dynamic traffic management approach already implemented in other roads in The Netherlands has shown a reduced unreliability (extreme travel time losses) by 10% (Schreiter et al., 2013). A dynamic traffic management system, mainly via dynamic traffic signs, is also planned to be implemented in the Antwerp area.

At a more network-wide level, Negenborn et al. (2010) presents a number of approaches using an intelligent infrastructures perspective, in which control technology is used to improve performance, including accessibility, of road traffic networks. The idea common to the approaches presented is to use models of the traffic systems to be controlled, and use these models in a predictive setting to determine what actions lead to the best performance. Approaches are presented based on model predictive control, taking uncertainty into account in various degrees.

8.4.2.3 Orgware

Most of the accessibility problems relate to the presence of sharp peaks in traffic. *Spitsmijding* (literally "rush hour avoidance," is being developed by Verkeersonderneming, an agency with private and public stakeholders, targeting mobility at system level in the greater Rotterdam area) aims at attracting vehicles out of the congested peak hours. A first pilot to determine the economic value of peaks on passengers' traffic, has led to an application also for freight transport in the Rotterdam port area. Practically speaking, Spitsmijden is a monetary compensation for every truck that is repeatedly taken out of the peak hours on the main highway leading to the port of Rotterdam, the A15. The idea is to shave the peak and distribute traffic flows evenly across time slots.

This can be combined with limited infrastructure investment to offer parking places for trucks so that they can avoid peak hours. Such *truck parking* offers always accessible friendly places to rest for free, in a secure area, and with the possibility to compensate for variations in travel time, for a more effective and predictable transport. The initiative is gaining momentum among the truck drivers over the whole Hamburg–Le Havre range.

A more radical peak shaving initiative involves structurally moving part of the freight flows transported by truck to night times. The immediate gains are a much shorter travel time, which can reduce travel by up to 40%. The coordination of this effort is undertaken by the Verkeersonderneming and TLN.

8.4.2   Rail accessibility

8.4.2.1   Hardware

The *Betuweroute* has been a major infrastructural investment in the years 1995–2005, providing a direct, dedicated connection for freight trains from the port of Rotterdam to the German hinterland via Duisburg. Currently, about 79% of the total traffic between the port of Rotterdam and the German border is using the Betuweroute, as reported by Keyrail, although the quality of the link further into Germany is still to be improved. Completion of a currently missing dedicated link around the border with Germany is expected to allow more freight trains over the link, as currently the link used has mixed traffic operation. Completion of this link is furthermore expected to improve passenger and freight transport performance and accessibility of the country from Germany. The design of the Betuweroute project aimed for a daily traffic of 150 trains, and to be able to deal with double-stacked container trains.

A similar project concerns a direct dedicated connection by rail between Rotterdam and Antwerp, along the Hamburg–Le Havre range. Moreover, a project is aiming at improving the hinterland accessibility of Antwerp towards the Ruhr area in Germany, which is seen as a limiting factor for the development of the port of Antwerp. To this end, old railway lines that were discontinued in the past have been renewed and brought up to the current standards. A few interconnections between this line and the main railway network and The Netherlands have been planned.

8.4.2.2   Software

Container Logistics Maasvlakte is the name of a series of efforts to develop a central coordination centre for bundling containers over the Maasvlakte area, and plan and manage all rail traffic in the port of Rotterdam area. The main focus is to strengthen the sea-rail link.

A discussion on how railway dynamic traffic management can be used to resolve accessibility bottlenecks in complex and densely used networks (independent of whether this entails cargo or passenger transport) is presented in Negenborn et al.

(2010). A thorough assessment is provided of a full implementation of dynamic railway traffic management. In particular, the evaluation of the so-called ROMA system is carried out using two dispatching areas in The Netherlands where accessibility is a major issue. This system for improving accessibility is currently being prepared for possible further employment in the UK and Denmark.

A major factor hindering a good accessibility in the *far* hinterland is the current setup of multiple countries to cross with multiple regulations and technological standards, especially for *safety and signalling systems* (controlling movements of the trains) and *electric power* (for the electric locomotives). As for the former, the European train traffic control system standard ERTMS is the direction to solve the problem and allows for a seamless operation without multiple expensive multi-system locomotives. All new EU-funded projects are required to comply with this, including for instance the Betuweroute. A goal of the EU would be to cover the full TEN-T corridors, such as the Rotterdam-Genua Corridor, with standardised signalling and a safety system and a standard power system (25kV).

8.4.2.3   Orgware

The *rail incubator* aims at developing further the rail shuttle connections servicing Rotterdam. Currently, about 250 weekly shuttle services are operated. Efforts are being done to identify growing regions (such as southern Germany and eastern Europe) which have a limited accessibility to the port of Rotterdam, and co-funding the setup of rail shuttles and joint marketing, over a time horizon of two years. A similar initiative has started in Antwerp, which is called Antwerp Intermodal Solutions. A key idea is the strong collaboration with the hinterland hubs, which would allow for consolidation and grouping of freight flows over rail.

8.4.3   *Water accessibility*

8.4.3.1   Hardware

Locks allow the port of Antwerp to provide a constant level of operations on areas protected from tides by some of the largest locks worldwide. Construction and improvements of the sluices are required, which have to handle all incoming and outgoing seagoing vessels; ship sizes and characteristics need to be considered for future developments.

Floating cranes are movable equipment positioned on the seaside of the ship. These provide extra capacity to load/unload a ship, and moreover allow for a direct transshipment between seagoing vessel and inland traffic such as barges. Despite the technical interest of this project, the current surplus in capacity on terminals and international shipping means that this has not yet applied.

Other necessary works on water accessibility regard the maintenance of the *draft* on the seagoing links, by continuous dredging of the relevant main canals of access. This relates to access to Antwerp port area by the deepest container ships, or the expected deepening and widening of the Elbe channels to improve the accessibility of the port of Hamburg.

8.4.3.2 Software

NextLogic is an information exchange platform being developed mainly by the port of Rotterdam. It tries to tackle the problem of inefficient vessel calls and long waiting times. The system will act as an information exchange platform, with up-to-date and real-time information of containers, locations and destinations. Moreover, such information can be used to efficiently plan rotations and the processing of barges at terminals and depots, according to different structures that can match the different requirements and guarantee a high level of short-sea and oceangoing accessibility. Among the concepts considered are hubs consolidating streams of containers; hop barges which shuttle across multiple inland terminals; exchange of containers at the sea-land border at different terminals, and relying on an internal transport system to reach the destination terminal; and reservicing of containers on different barges due to real-time availability. A similar central point of contact is also considered for the port of Antwerp, which should also consider a barge traffic system. The main goal of such a system is to be able to request realistic time slots, and plan the port's operations based on forecast and a proactive schedule of loading and unloading operations.

The overall goal is an increased accessibility at the water level, by an increase in utilisation of quays and cranes, and increased reliability of inland container shipping.

A different aspect relates to the optimisation and harmonisation within locks, and barge speeds. This is particularly interesting in the Hamburg–Le Havre range, as the barge connection between the port of Rotterdam and the port of Antwerp includes a number of locks (see Tables 8.3 and 8.4). Locks are necessary to maintain that the water remains at a certain safe level. Unfortunately, they can also cause congestions as the lock capacity can be lower than the amount of vessels that want to use it. This results in extra travel time (up to three hours extra), and unavoidably affects the reliability of the waterway transport mode. By centrally monitoring and coordinating lock operations and controlling the speed of incoming vessels, which are tracked via AIS data, the expected waiting times can be reduced greatly and the opening times of the locks matched with the passage of the ships. Examples of such approaches are found in Ocampo-Martinez and Negenborn (2015).

Motorway of the seas is the codename for the improvement of short sea connections. Some key factors for the further modal shift to sea over longer distance are identified in efficient, regular and reliable connections and rapid administrative procedures with a high level of service.

8.4.3.3 Hardware + orgware

A variety of approaches combining hardware and orgware are dedicated to improving water accessibility, by connecting water transport with a multimodal transport network. Depending on their distance from the main port, one can identify three different categories.

A *barge service centre* is a neutral barge terminal directly at the main port, in this case Maasvlakte 2, for handling of inland containers, and delivery to multiple terminals via an internal transport system. It is expected that the terminal will be dedicated to barges, and the availability of the internal transport system will be very high, thus providing effective improvement in the accessibility of the port area by inland shipping. In fact, there will be a reduction in small calls, decreased number of hops between terminals, a backup site in case of calamities and a possibility to handle empty containers. Due to the current overcapacity at the Maasvlakte 2, the future of the BSC is not clear, it is expected to break even and feasibility might be reached around the year 2035.

A container *transferium* is a local terminal in proximity to the port. The concept is to haul containers with barges from and to the major terminals in the port. The barges ferry frequently between the transferium and the terminals. The objective is to attract modal share from the road. Currently a container transferium is in operation in Alblasserdam with a planned capacity of 200,000 TEU/year.

Similar concepts regard a *barge hub* terminal, which can be defined as a terminal more upstream from the port which bundles and sorts containers and sends them further, in a consolidated manner, by barge towards the port area. This should have the same effect as a barge service centre, that is, reducing the number of barge calls in the congested port area.

From the hub (be it a service centre, a transferium or a barge hub), the *routing on the inland links* is also relevant. This shipment to the destination terminal might take the form of a hub-and-spoke network, a roundtrip network or a combination of the two, for optimal frequency for each terminal being visited, and efficient consolidation of freight flows. To this end, *inland container terminals* can be used, possibly in combination with extended gates (discussed in Section 8.4.4.3).

Parallel efforts to reach a consolidation of small inland waterways flows and resulting in a modal shift from trucks to inland shipping in the port area have been considered in Antwerp via the concept of premium barge services. This results in shuttle barges connecting terminals and asks for coordination between multiple players.

While those efforts regard mostly the inland accessibility, a similar idea towards improving deep-sea accessibility is the one of *offshore hubs*. Those are purely transshipment terminals, meant to reduce the time spent by the oceangoing ships in the port area. Instead, freight flows might be transshipped directly to a short sea link. A proposal has been made to develop such an offshore hub in the North Sea, in the Orkney Islands and connected to the main ports in the Hamburg–Le Havre range by the motorways of the sea.

8.4.3.4 Orgware

The main design of infrastructure for sea accessibility is done by the terminal operators, which consists of mostly overseas traffic. But this has an impact on all of the other vessels which are calling, for example, feeder vessels or barges, as these have a longer handling time per amount of freight. Without any involvement of policy measures, this might result in extra handling charges which are asked of the short sea operators and inland shipping operators, if they do not load and unload a minimum number of containers per call.

8.4.4 Intermodal

8.4.4.1 Hardware + software

Dedicated intermodal information systems might be coupled with facilities such as a *barge service centre* and a *rail service centre*, which aim at bundling the demand over the available barge and rail links. With these systems, the final goal from the port of Rotterdam perspective is to reach the target goals of only 35% of freight flow going by truck.

8.4.4.2 Software

InlandLinks is a project aimed at stimulating the development of Rotterdam as a hub in a large intermodal network. This results in an online information system which helps companies by providing a comprehensive showcase of the last most accurate information about speed, sustainability, costs of transport offers, locations and details of the terminals, including available services and facilities.

Integration of expedited customs procedure, a single point of contact, and an information system for the overall port activities (*community system*) is being undertaken in the Le Havre RoRo Terminal to increase the amount of freight flows, and the accessibility towards the (mostly) French and European hinterland.

Similar efforts have been made in Antwerp by providing a *route planner* for intermodal connections, consisting of hundreds of terminals in 15 European countries.

8.4.4.3 Orgware

The concept of *synchromodality* also aims at improving the total travel time and costs for a logistic link (see Chapter 15 for a detailed analysis of synchromodality). In particular this concept involves the contractual separation between the product a logistic service provider offers – that is, the delivery of products at specified costs, quality and sustainability – from the means this product is actually fulfilled, that might for instance involve different modes, time of travel, organisation and so forth. Thus the extent by which synchromodality fosters amodal booking of transport services encourages the modal shift in general. More in detail, modes such as barge and rail could be better exploited as a reaction to updated traffic information, and the availability of assets or infrastructure.

To reach a proper coordination of different transport logistic providers, their orchestration is important. Among those efforts, *Truckload Match* is a pool of multiple transport companies that cooperate to perform logistic services all over Europe. The key idea is to share information and transport capacity, in order to provide a more time-responsive service, and avoid empty rides for improved economic performance. This is done by matching the rides of the pool of logistic providers. A side effect is also a reduced amount of vehicles in the port area.

A number of hinterland terminals are also currently working based on *extended gates*. These are inland routes under the control of the sea terminal. The latter have Authorised Economic Operators status and thus the containers do not need to be cleared by customs until the terminal. In the most common setup, the container transport is arranged by the customer from/to the extended gate at the inland terminal and the terminal operator handles the transport between the sea terminal and the inland terminal. This results in fewer trucks picking up individual containers at the port. The future sees that more terminals in the hinterland and possibly abroad might acquire the status of extended gate, which further simplifies crossing the border.

8.5   Conclusions and future research

Accessibility of ports and networks is a major issue for maintaining an attractive and competitive position in international supply chains. This chapter has reviewed the issue of accessibility by defining what it is, determining which performance indicators can be used to highlight accessibility issues, and surveying a number of approaches that have been considered to improve the accessibility. As such, this chapter provides an introduction to novices in this field, and a starting point for further research.

Accessibility itself is a vague concept that can be quantified by a set of measurements based on infrastructure characteristics. To directly improve those characteristics, improvement in the infrastructure that is available looks at the most direct way to go from A to B. On the other hand, the way the infrastructure is managed is also crucial. Intelligent control approaches can help keep optimal performance of the infrastructure, despite external influences, demand peaks, incidents and increased demand. Furthermore, the amount of roles of many actors involved in the overall port hinterland access can be translated to an important role of policies and organisational aspects in improving accessibility. To improve the accessibility of a port area at a system level, a comprehensive approach needs to be adopted, where those three main directions are coordinated and balanced. Most approaches for improving accessibility need in fact the interplay of multiple factors, to result in an intelligent infrastructure and a smart, dynamic organisational setup.

Acknowledgements

This research is supported by the STW Maritime Project "ShipDrive: A Novel Methodology for Integrated Modeling, Control, and Optimisation of Hybrid Ship Systems" (project 13276) of the Dutch Technology Foundation STW.

Note

1 As was investigated in the COST Action TU1002, Accessibility instruments for planning practice in Europe.

References

Burns, L. (1979) *Transportation, Temporal, and Spatial Components of Accessibility*. Lexington, MA, Lexington Books.

European Environment Agency (2016) *Modal Split of Freight Transport in EU 27, 1990–2030*. Retrieved from www.eea.europa.eu/data-and-maps/figures/modal-split-of-freight-transport-in-eu-27-1990-2030, accessed 2016.

Geurs, K. T., and Van Wee, B. (2004) Accessibility evaluation of land-use and transport strategies: Review and research directions. *Journal of Transport Geography*, 12, 127–140.

Gould, P. R. (1969) *Spatial Diffusion*. Washington, Association of American Geographers.

Hansen, W. G. (1959) How accessibility shapes land use. *Journal of the American Institute of Planners*, 25(2), 73–76.

Kennisinstituut voor Mobiliteitsbeleid (2014) *Mobiliteitsbeeld 2014*. Ministerie van Infrastructuur en Milieu.

Logica. (2009) *Functionele specificatie, applicatie voor LIVRA praktijkproeven Voorspelling en Samenhang*. Technical Report, The Netherlands, December 18, 2009.

Morris, J. M., Dumble, P., and Wigan, M. R. (1978) Accessibility indicators for transport planning. *Transportation Research Part A: General*, 13, 91–109.

Negenborn, R. R., Lukszo, Z., and Hellendoorn, H. (eds.) (2010) *Intelligent Infrastructures*. Dordrecht, The Netherlands, Springer.

Notteboom, T., and Winkelmans, W. (2004) *Factual Report on the European Port Sector: FR-WP1: Overall Market Dynamics and Their Influence on the Port Sector*. Brussels, European Sea Ports Organisation (ESPO).

Ocampo-Martinez, C., and Negenborn, R. R. (eds.) (2015) *Transport of Water Versus Transport Over Water*. New York, Springer.

Schreiter, T., Van Lint, J.W.C., and Hoogendoorn, S.P. (2013) *Vehicle-Class Specific Control of Freeway Traffic*. Proceedings of the Transportation Research Board 92nd Annual Meeting, 13-0585, 2013.

Van der Horst, M. R., and De Langen, P. W. (2008) Coordination in hinterland transport chains: A major challenge for the seaport community. *Maritime Economics and Logistics*, 10, 108–129.

Suggestions for further reading

Besides the papers provided as references in the text, we recommend the following texts as suggestions for further reading.

Paardenkooper-Suli, K. M. (2014) *The Port of Rotterdam and the Maritime Container: The Rise and Fall of Rotterdam's Hinterland (1966–2010)*. Erasmus University Rotterdam.

Van den Berg, R. (2015) *Strategies and New Business Models in Intermodal Hinterland Transport*. Eindhoven, Eindhoven University of Technology.

9
PORT HINTERLAND RELATIONS
Lessons to be learned from a cost-benefit analysis of a large investment project

Christa Sys and Thierry Vanelslander

9.1 Introduction

The issue of port competitiveness has been an important topic over recent years and, with the effects that ports are suffering from the recent crisis, it is only gaining importance. Understanding the determinants of port competitiveness is important for port authorities so as to determine infrastructure and investment needs.

The port competition literature reviewed in, among others, Aronietis et al. (2011) reveals a considerable range of factors that have an influence on the decision of port choice. Shipping companies comment that a decision to call at a port cannot be made without available cargo from/to that port, which is closely linked to ports' geographical location and the area that can be served through it. Hinterland transport infrastructure plays a key role in this. Meersman et al. (2010) too state that: "large seaports essentially require three elements: maritime access, goods-handling capacity, and distributive capacity, including adequate connections with the hinterland." Each port authority considers investment issues concerning smooth access to the hinterland.

Public authorities in particular invest in port infrastructure and hinterland connections, but the question arises, who benefits from the return on investment (Meersman et al., 2014): the port region or the hinterland? This chapter intends establishing a Social Benefit Analysis Framework (hereinafter abbreviated to SBA) for a port hinterland project. Such analysis refers to a project that has a broad impact across society. Presenting an SBA for a port hinterland project allows addressing important issues in port-hinterland relations in a systematic format and dealing with the issue of costs and benefits to the port region or to the hinterland.

This chapter applies the framework to the adequate connection with the hinterland of the port of Zeebruges (Belgium). More specifically, it concentrates on the construction of the A11 motorway between the N49 (Westkapelle, leading to the motorway E34) and the N31 (Bruges–Blauwe Toren, leading to the motorway

E40), which is one of the large projects planned for road traffic to the port of Zeebruges. The project is meant to increase accessibility of the port of Zeebruges to its hinterland considerably. The A11, as part of the Trans-European Transport Network (TEN-T), is about 12.1 km long (see Figure 9.1). Given confidentiality arguments, the projected project cost is for understandable reasons not included in the further course of this chapter. Here, the project cost is equated to 100.

This chapter features three central sub-questions, corresponding to the different elements that compose benefits:

- What is the direct effect on traveling time?
- What are the indirect effects as a result of employment at construction and maintenance?
- Which modification does this project bring with respect to external effects (e.g., the number and severity of road causalities, noise, emissions)?

The first research question is dealt with in full conceptual and calculation detail, as it touches upon seaport activity in its core. The second question is only described conceptually, as it mostly concerns issues that are not core to seaports, but have rather derived economic effects to the local economy. The third question is interesting both for policy-makers and for the concerned operators and users, as it involves the cost impact in case internalisation of externalities related to construction of the A11 motorway gets implemented.

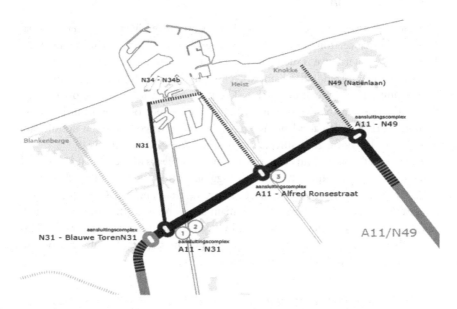

FIGURE 9.1 The A11 trajectory

Source: Agentschap Wegen en Verkeer (2010).

The rest of the chapter is structured as follows. Section 9.2 presents a brief overview of the scope of the research. Section 9.2.1 defines the reference alternative. A concise description of the project alternative is laid out in Section 9.2.2. Section 9.2.3 focuses on the calculation approach, including assumptions and methods used to undertake the research. Section 9.3 deals with the direct impact assessment procedure and required ingredients. Section 9.4 introduces and illustrates the external cost assessment procedure. In Section 9.5, the calculations are summarised and added up. Finally, conclusions are drawn in Section 9.6.

9.2   Scope of the research

In an SBA, the development of the project is compared with the so-called reference alternative, that is, the development without the project. In the project alternative to which the SBA framework is applied, the A11 is in use.

9.2.1   Reference alternative

The reference alternative is the non-implementation of the project. This alternative is contrary to the decision of the Flemish government that classified the A11 motorway initiative as one of the highest priority infrastructure projects in Flanders. In the subsequent sections, the reference alternative is defined.

9.2.1.1   Trajectory

The traffic between the port and the hinterland currently uses smaller regional routes (for instance, from the outer port to the N49, and vice versa, see Figure 9.1). The reference alternative is such a regional route (N376, a 2 × 1 connection) and is characterised by mixed traffic, namely, cyclists and other (slow) regional traffic (e.g., commuting traffic, agriculture trucks). The latter often causes congestion, just like the bridge crossing the Boudewijn Canal when erected. Both result in irritation of (other) drivers, with accidents as a possible consequence. Although not directly designed for heavy freight traffic, the N376 links the N31 and the N49.

At present, the hinterland connection is made by means of the N31 (68% of road traffic uses this route) through the Bruges agglomeration and the N49 (32% of total traffic) from and to Antwerp (Resource Analysis, Technum, IMDC, 2009). During the peak hours, the N31, currently the main access route to the port, is saturated. These approaching routes are also used by cyclists and other regional traffic. This means that without the project, the accessibility of the seaport of Zeebruges by road to and from its hinterland may be considered weak.

9.2.1.2   Some facts and figures

With respect to the type of goods, Table 9.1 shows that the Port of Zeebruges is an important roll-on/roll-off port in Europe. This generates a considerable amount of

TABLE 9.1 Total number of trucks and tonnage for Zeebruges

	2004	2005	2006	2007	2008	2009	2010	2011	2012
Dry bulk	1,596,112	1,718,655	1,956,411	2,011,462	1,953,127	1,598,080	1,693,999	1,652,922	1,622,969
Liquid bulk	4,286,450	4,479,642	6,247,082	5,858,234	6,202,383	7,993,246	7,996,586	8,280,701	7,695,335
Containers	14,012,169	15,604,265	17,985,690	20,323,002	21,202,963	24,894,626	26,403,516	22,742,644	20,317,265
Ro/Ro	11,097,491	11,776,620	12,244,198	12,999,789	11,814,166	9,514,466	12,395,927	13,130,518	12,548,641
General cargo	802,202	1,011,432	1,039,396	884,749	851,779	866,354	1,109,570	1,150,588	1,359,618
Total tonnage	**31,794,424** →	**34,590,614** →	**39,472,777** →	**42,077,236** →	**42,024,418** →	**44,866,772** →	**49,599,598** →	**46,957,373** →	**43,543,828** →
Number of trucks/day	3,749	4,079	4,655	4,962	4,956	5,291	5,849	5,537	5,340

Source: Own compilation based on the Port of Zeebruges Authority.

trucks in transit in Zeebruges. Furthermore, Zeebruges is an important container port. Also in container traffic, the majority of goods are transported by truck. The share of rail amounts to 40%. Improved road connections with the hinterland are necessary. Starting from the fact that 51% of the total transported tonnage is transferred by road, Table 9.1 also gives an indication of the number of trucks (24t truck load – 265 working days) per day.

To be able to assess the benefits of an improvement of the port infrastructure towards the hinterland, a forecast of maritime traffic and hinterland traffic volumes has been produced (Idea Consult, 2002). A future growth in the number of trucks is expected as maritime traffic is expected to increase because the 2008 economic crisis aftermath has disappeared. A first observation concerns the expected annual growth in total maritime traffic of the port of Zeebruges, amounting to 2.9% from 41,601,000 tons in 2007 to 80,071,000 tons in 2030. The share of road transport diminishes from 58.6% in 2007 to 50.2% in 2030. In absolute figures, the number of goods transported by road increases by 2.2% per year. However, over the 2020–2030 period, the share decreases by 1.3 percentage points to 50.2%, while the tonnage increases by 16.7%.

9.2.2 Project alternative

In the project alternative, the actual improvement of road infrastructure linking the port to its hinterland is described. In combination with transforming the N49 to a full-fledged motorway, the redesigning of the N31 and the design of the N44 (Aalter-Maldegem) as primary route type 1, the construction of the A11 allows for a better separation of freight traffic from other traffic streams. The N31 between the Blauwe Toren and Zeebruges as well as the N376 will have lower traffic volumes to cope with. Local roads and local villages (e.g., Lissewege, Westkapelle and the few scattered houses upon the Dudzelestraat) will benefit, and with it, the liveability will improve; the degree to which it will improve will be estimated in the next sections.

Also the Bruges agglomeration will benefit, as the construction of the A11 is expected to shift traffic to the N49/E34. Mainly, economic traffic from, to and between Zeebruges, Ghent, Antwerp and the hinterland is expected to be better spread between the N31 and N49. According to the Strategic Plan Zeebruges, forecast calculations suggest a 50/50 ratio of future geographical distribution between the N31 and N49.

However, given the fact that the Ro/Ro terminals as well as container terminals are located in the outer port of Zeebruges, a significant volume of traffic will remain on the N31 with transit through Lissewege.

9.2.3 Calculation methodology

An SBA is an assessment tool that compares the effects of the project alternative with the reference alternative and, where possible, values these effects in monetary terms.

In order to quantify the impact of the shift from the reference alternative towards the project alternative, it ıs necessary to go more deeply into the time-tied aspects of the analysis. Here, a discount rate of 4% has been chosen, in accordance with other SBA studies.

Successively, the effects of the infrastructure improvement are mapped in a systematic way. The effects can be classified as direct (e.g., time gains related to transport between port and hinterland), indirect (e.g., employment attracted by the improved business climate) and external (e.g., environmental) effects of increased volumes of port-related road transport. Briefly, these effects will be commented on.

The direct effect consists of the benefits of using the project that has been invested in. The benefits of the use stem from the advantages which users experience:

- The improved connection for passenger and truck traffic (i.e., level crossing rail/road traffic will be removed).
- The reduction of tourists' transport time, which adds significant economic value (e.g., increased attractiveness to the city of Bruges, the neighbouring coastal regions), namely, improvement in the value of hotel and catering businesses.
- The travel time advantages and so forth.

Besides these direct effects, indirect effects can also be calculated. Indirect effects are permanent, broader economic effects which turn up elsewhere in the economy. This means that they are not caused directly by the project but are the consequence of direct effects.

An improvement of the transport system coupled to the ongoing improvement of the accessibility promotes, by means of several mechanisms, the economic development of a region:

- The lowering of the transport charges reinforces the competitiveness of companies in the region; the cost of supplying raw materials and transporting of finished products diminishes.
- The region becomes more attractive for the establishment of new companies, for example in the maritime logistics zone; also for tourists, the coastal area becomes more attractive due to the conflict-free motorway A11 and the better quality of the approaching routes.
- The reduction of the transport costs in passenger transport improves the functioning of the labour market; companies can in their search for staff recruit from a larger pool, so that finding suitable staff becomes easier; persons have access to a larger assortment of employers, so that they have more chance to find better suited or paid work.

An important part of this indirect impact is already taken into account in the appreciation of the direct impact. The indirect impact concerns particularly the continued effect of direct time and kilometre cost savings for the users in the economy. Only when it is assumed that important indirect effects have an impact which is not yet incorporated, can it be considered appropriate to take these indirect effects into account.

Within the SBA approach, the following indirect effects can be appreciated:

- The employment impact during the construction.
- The impact on employment, profit and capital turnovers in business commercial matters.
- The impact on coastal tourism.

As mentioned in Section 9.1, as these indirect effects do not directly impact on seaports, they will not be explained in the further calculations in this chapter. The monetary outcomes, however, will be appreciated in the overall impact assessment in Section 9.5.

The external effects are effects which are characterised by lacking a market price. The impact on the environment (e.g., noise, emissions) and safety (e.g., mobility, reduction of road transport victims) are typical external effects. The external effects can be expressed in values to the extent that they are of benefit to users. This impact can, however, still get a price by means of appreciation methods such as using an indicator.

For the A11 project, it concerns mainly modifications in emissions, noise, number and seriousness of traffic victims, and visual quality of the surroundings.

The following sections focus on the appreciation of the project effects.

9.3 Direct effects: impact on the travel time

To identify the impact on the travel time, modelling tools of the Flemish Traffic Centre were used. Section 9.3.1 describes the characteristics and underlying assumptions of those tools. Section 9.3.2 presents the results.

9.3.1 Traffic intensity: modelling

With respect to traffic intensity, the intensity in 2007 (base year) and the forecast intensity in 2020, termed "Business-As-Usual" (hereafter abbreviated as BAU 2020) were assessed. The Traffic Centre simulated the expected traveling time for these two years in two scenarios: a scenario where the A11 is supposed to have materialised and one where the A11 is not built. To do this, the Traffic Centre used the "Provincial Traffic Model Western Flanders v3.5." That model is a static, multimodal, aggregated traffic model.

The "Strategic Freight Model Flanders" is used for determining the impact of hypothetical scenarios on future freight transport. On the basis of this strategic freight model, it is possible to simulate future freight flows by mode (road, rail and inland waterways) and by type of goods, the Standard Goods Classification for Transport Statistics (NST) freight category. An adjusted four-step model is applied.

A distinction between distance-dependent, time-dependent and fixed costs is made. Using employment and population data allows disaggregating the mode matrices to the known zoning ('coarse zones'). One obtains then by mode and by class of goods an origin-destination matrix with the number of tons transported. Analogous to "Provincial Traffic Model Western Flanders," the Strategic Freight Model Flanders also relies upon three types of input data: the observed matrices, employment data and population data.

One sub-scenario of the BAU 2020 assumes that the A11 has not materialised in 2020, but that economic growth, population growth and other infrastructure projects listed in the description do materialise by 2020. This allows isolating the effect of the implementation of the A11. In 2020, the A11 attracts more cars. Between Westkapelle and the node with the eastern port ring road from west to east, there is even an increase of about 50 (2020 without A11) to about 300 (2020 with A11) cars per hour. The number of trucks is rising more rapidly than in the present situation (without A11), while during the evening rush hour, the increase in the number of cars through the implementation of the A11 is less than in the morning. The increase is estimated to be greater in 2020 than in the present situation.

The BAU scenario was established in 2007. Therefore, the effects related to the 2008 economic crisis have also not been taken into account. The benefits are largely attributable to time gained by connecting a higher permitted speed and by mitigating the congestion costs. When only the benefits of the reduction of congestion in the two simulated peak hours are assessed, the benefits of the other times (the peak/off-peak transition hours, weekends and off-peak days) can be neglected. So, it is possible to extrapolate the financial year revenues from the benefits of these two peak hours. The calculation of annual income is based upon the assumption that there are 220 working days. This calculation gives a lower limit of the annual financial benefits, as there is also a gain in the off-peak time other than the mitigation of the congestion cost.

The construction of the A11 will first and foremost provide a faster connection between the N49 and N31. This creates a better link with Zeebruges and might cause a route choice shift, and consequently allows time gains booked elsewhere. In the morning (8–9 AM), the total time savings for passenger cars sum to 430 hours. The time saved has been rounded to five hours. This is the case for all subsequent time savings tables. In the evening (5–6 PM), the total time gain amounts to 445 hours (Table 9.2).

However, even in the off-peak times, time gains can be made. With 220 days, the total time saved on an annual basis is shown in Table 9.2. Here, it is assumed that each off-peak trip will give as much time gain as during rush hour. This assumption is realistic given that there is hardly any congestion on that trajectory.

With respect to the annual total, assuming that each day has two rush hours and every year 220 days, a time gain is obtained of 211,170 hours for cars, 21,950 hours for light trucks and 36,420 hours for heavy trucks (see Table 9.3). However, this is an underestimation: if the benefits are also off-peak, the benefit per year will be greater. If the same factors are applied to convert from rush hour to off-peak

TABLE 9.2 Total time savings per year, expressed in hours, rounded to five hours

	Time gain Morning rush	Time gain Evening rush	Daily time gain	Annual time gain
Passenger car	430	445	5,380	1,183,465
Light truck	50	25	560	122,945
Heavy truck	80	40	855	188,215

Source: Flemish Traffic Centre (Provinciaal verkeersmodel West-Vlaanderen versie 3.5).

TABLE 9.3 Annual time savings and benefits

	Only benefits in rush hours			Equal benefits in rush and off-peak hours		
	Time gain (hours)	Benefits (hours)	Financial benefit (million €)	Time gain (hours)	Benefits	Financial benefit (million €)
Passenger car	192,330	211,170	1.81	1,183,465	1,302,925	11.18
Light truck	17,385	21,950	1.1	122,945	155,265	7.8
Heavy truck	27,340	36,420	1.97	188,215	254,565	13.74
Total	237,055	597,540	4.88	1,494,625	409,830	32.72

Source: Flemish Traffic Centre (Provinciaal verkeersmodel West-Vlaanderen versie 3.5).

hour, a gain is obtained of 5,920 hours for passenger cars, 705 hours for light trucks and 1,155 hours for heavy trucks per day. Assuming 220 working days, an annual benefit is obtained of 1.3 million hours for passenger cars, 155,000 hours for light trucks and 254,000 hours for heavy trucks.

To calculate the financial benefit, total benefits in terms of hours are converted in financial benefits. One uses a value of time of €8.58 per hour for passenger cars, €50.21 per hour for trucks less than 12 tons and €53.97 per hour for trucks greater than 12 tons. When the benefits of the improving traffic flow in the two simulated peak hours are assessed, assuming 220 days, the annual financial benefit amounts to €4.88 million. However, this is a lower limit because even in the off-peak time savings can be made. Otherwise, the annual financial benefit amounts to €32.72 million (see last column in Table 9.3).

9.3.2 Impact on the traveling time: results

Starting from the figures obtained of the Flemish Traffic Centre, an extrapolation of the forecast travelling time is elaborated to get an insight in the evolution in travelling time and traffic intensity after 2020.

With respect to passenger cars, the annual growth rates are supposed to decrease slightly from 1.05% between 2010 and 2020 to 0.70% between 2020 and 2046. In

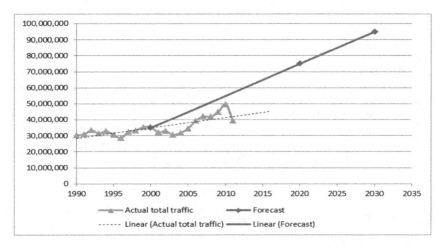

FIGURE 9.2 Effective maritime traffic evolution

Source: Own compilation, based on statistics of Port of Zeebruges Authority (1990–2011), Idea Consult (2002) and Resource Analysis (2008).

combination with the A11, the growth rate of maritime traffic will directly impact on the time gain for light and heavy trucks. This results in developing different scenarios. Idea Consult (2002) reports a forecast of the total traffic for the port of Zeebruges by using elasticities and worked with scenarios of high, medium and low growth. In addition, this study adds a zero growth and a trend growth scenario.

The zero growth scenario (baseline) is used as a reference scenario. The low growth scenario assumes an increase of maritime traffic of 1.7% for the 2017–2021 period, followed by a somewhat lower growth rate of 0.7% (Idea Consult, 2002). Next, the medium growth scenario takes into account a growth of 2.9% for the period 2017–2020 and 2.45% for the period 2021–2046. In the scenario of high maritime growth, it is assumed that the maritime traffic evolves from 2017 to 2020 with an annual growth of 3.4% and of 4.2% thereafter.

Forecasting Ro/Ro and container traffic, the most important business segments in the port of Zeebruges, is more difficult than for the segment "general cargo." For that reason, the effective total traffic figures (dark grey line in Figure 9.2) were projected towards the forecasts published in several studies. Figure 9.2 shows the evolution of the total traffic over the period 1990–2011. A linear trend line was added. Based on these time series, we predict only five periods ahead. In 2015, the total traffic is expected to exceed the level of 44.5 million tons. The trend line clearly differs. The trend growth scenario represents the fifth scenario. Based on the cases outlined previously, it is calculated how the traffic intensity until 2046 will develop.

Based on the cases outlined previously, it is calculated how the traffic intensity until 2046 will develop. The trend growth scenario puts up to the low growth scenario (Figure 9.3).

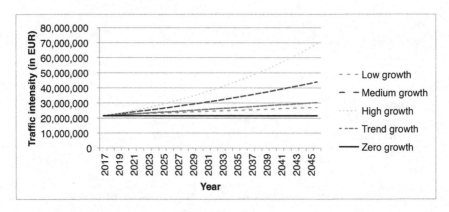

FIGURE 9.3 Evolution of traffic intensity (in €)

Source: Authors' compilation.

TABLE 9.4 Economic benefit for travelling time (indexed values)

	Maritime growth				
	Zero	*Low*	*Medium*	*High*	*Trend*
Annuity	76.9	82.5	95.0	110.1	84.9
Perpetuity	112.0	121.9	145.3	175.0	126.4

A choice can then be made between annuity value[1] and perpetual[2] time benefits. By valuing the time gains annually, the actualised value of the sum of time gains can be calculated for a value in 2011. Subsequently, the outcome is expressed in an index value relative to the project cost of 100. The calculation of the five scenarios yields the following results, as summarised in Table 9.4.

9.4 External effects

Three types of external benefits can be discerned here as a result of investment in the A11 motorway initiative: fewer accidents, less noise and a lower amount of emissions. Visual quality changes are not included here, as it is assumed that sufficient mitigating measures (e.g., visual buffers, sound baffles, silent asphalt, green zone) are taken to avoid this type of externality.

9.4.1 Traffic victims

When the A11 is constructed, traffic will be more divided over the road network. Local traffic will primarily use the current N376, while passage traffic will mainly

TABLE 9.5 Costs per accident in 2011 (in €)

	Personal costs	Costs for the society	Total costs
Fatal victim	1,558,978	155,897	1,714,876
Severely injured	202,667	20,267	222,934
Lightly injured	15,590	1,559	17,149

take the new road A11. This spreading of traffic will lead to a decrease in the risk of accidents and traffic victims.

Next to the value of a statistical life of €1.5 million (Nellthorp et al., 2001; UNITE, 2003), one has to add 10% extra as costs to the rest of the society (column two in Table 9.5). These costs are due to a loss of production, costs of emergency services, and so forth. When the private cost (€1.5 million) and the social cost (10% of €1.5 million) are added up, the total cost is equal to €1.65 million (in 1998).

Subsequently, these cost components have to be converted to the cost at prices in 2011. First, the value will be converted to a value for Belgium. Next, one has to take into account the elasticity value. Nellthorp et al. (2001) use 0.75 as elasticity value in their calculations. So, after this step, the private cost amounts to €1.3 million.

The next adjustment is due to inflation. The private cost taking into account the deflation is consequently equal to €2,057,851.

A last adjustment is the conversion to factor costs: to accomplish the social benefit analysis, a value expressed exclusive of VAT is needed. Finally, the value of the total cost of €1,714,876 is obtained.

Further, the costs of a severe injury and a light injury can be obtained by multiplying the personal cost and costs for the society for fatal victims by 13% and 1% respectively (Nellthorp et al., 2001; UNITE, 2003).

The present value of the annuity to estimate the economic impact of fewer accidents amounts to an index value of 3.98 for the 30-year period, and at 5.75 for the perpetuity value.

9.4.2 Noise

Different relevant studies exist which deal with noise as a source of environmental aspects impacting upon the value of real estate. An important conclusion is that a railway, a motorway or an industrial zone within 500 meters of a house has a negative impact on the property value that can amount to 5%.

In view of the environmental impacts reporting, the number of buildings impacted on by a change in noise hindrance should be listed.

TABLE 9.6 Social benefits from noise (in €)

	Maritime growth		
	Zero/Low	Middle	High
30-year period	12.5	13.1	14.9
Different perpetual value	18.1	18.9	21.5

Table 9.6 summarises the results of the obtained social benefits.

9.4.3 Emissions

The project leads to a change in the quantity and the location of emission of polluting substances in the air in the neighbourhood concerned. In an integral economic analysis, it is also important then to consider this aspect. For the change in economic costs and benefits linked to these emissions, reference values from literature are used.

However, no exact data with respect to emissions for the villages of Dudzele, Westkapelle and Lissewege were available. This does not undermine the results of the chapter, as it is expected that transport companies will invest further in low-emission vehicles.

9.5 Calculation results

Based on the aforementioned estimation and appreciation, Section 9.5 summarises the social benefits of the project A11. For each scenario, the annuity value is given in its full detail in Tables 9.7 and 9.8, expressed as a percentage of the social benefit per scenario.

The result of Table 9.7 corresponds to the assumption that the attractiveness of the coastal area increases by 3%. A sensitivity analysis with 5% was done to give insight in the impact of the assumption, as shown in Table 9.8.

From these tables, it can be seen that the biggest gains are made in the traffic time gains, followed by external effects, more specifically noise reduction and supplementary employment thanks to the project. Similar conclusions stand for both the 3% and 5% approaches.

The perpetual option is additionally calculated, as the road investment is made for a very long period of time. As shown in Table 9.9, with respect to the annuity value, over the period 2011–2046, the social benefit does exceed the project cost regardless of the scenario (zero, low, medium, high and trend growth). The perpetual value leads to a significantly higher benefit than the cost over 30 years, so there is a margin to cover costs after 30 years (i.e., maintenance and replacement costs which in turn result in extra benefits, that is, employment as a result of maintenance and replacement investments) in each scenario.

TABLE 9.7 Overview results annuity 1

		Maritime growth				
		Zero	Low	Medium	High	Trend
Direct effects						
traffic intensity (incl. local and transit traffic)		68.94%	70.12%	71.96%	75.95%	70.96%
Indirect effects						
employment		8.97%	8.50%	7.63%	6.99%	8.37%
business commerce		6.12%	5.80%	5.17%	4.71%	5.70%
catering industry	3%	1.15%	1.09%	0.97%	0.88%	1.07%
External effects						
safety		3.57%	3.38%	3.01%	2.74%	3.33%
noise		11.25%	10.66%	9.92%	10.25%	10.57%
emissions		n/a	n/a	n/a	n/a	n/a

TABLE 9.8 Overview results annuity 2

		Maritime growth				
		Zero	Low	Medium	High	Trend
Direct effects						
traffic intensity (incl. local and transit traffic)		68.42%	69.62%	71.50%	75.51%	70.46%
Indirect effects						
employment		8.90%	8.44%	7.58%	6.95%	8.31%
business commerce		6.07%	5.75%	5.14%	4.68%	5.66%
catering industry	5%	1.90%	1.80%	1.61%	1.46%	1.77%
External effects						
safety		3.54%	3.36%	2.99%	2.73%	3.30%
noise		11.17%	10.58%	9.85%	10.19%	10.50%
emissions		n/a	n/a	n/a	n/a	n/a

TABLE 9.9 Overview social benefit per scenario

		Maritime growth				
		Zero	Low	Medium	High	Trend
3%	Annuity	111.49	117.7	131.97	144.98	119.62
	Perpetuity	158.75	168.65	192.92	225.36	173.37
5%	Annuity	112.34	118.55	132.82	145.83	120.48
	Perpetuity	159.98	169.88	194.16	226.59	174.61

9.6 Conclusions

The aim of the study at hand was to develop a framework for a port hinterland SBA, and to carry out an application to the project A11, a new motorway connection in the hinterland of the port of Zeebruges. The case is specific and requires developing a specific SBA framework, as it links both port, local and leisure traffic. The study answered three questions.

- What is the direct effect on travelling time?
- What are the indirect effects as a result of employment at construction and maintenance?
- Which modification does this project bring with respect to external effects (e.g., the number and severity of road causalities, noise, emission)?

First, the social benefits of the A11 project were calculated under certain assumptions. Clearly, these benefits depend on the evolution of the maritime traffic (proxied in increased traffic intensity) and the evolution of the number of passenger cars. From the results, one can observe that the social benefits are positive regardless of the scenario, and they are largest in the high-growth scenario. However, a comparison with the trend growth scenario suggests that the high-growth scenario is less realistic. It is not surprising that the social benefit of the perpetual value is exceeding that of the annuity value. The low or medium growth scenarios therefore have a higher chance of materialising, with the condition that the other planned investments and policies are implemented, and that no structural economic ruptures occur. The decision not to build the A11 would only aggravate the current traffic problems.

Second, given that the bottlenecks will be eliminated, this will result in a gain of time, both for local and transit traffic.

Third, the underlying road networks will also be improved. Together with the construction of the A11, this will also generate an employment benefit.

Ultimately, because the port-related traffic is expected to no longer use the local roads, the liveability (in terms of noise, emissions, etc.) and safety (e.g., traffic victims) will improve. The latter refers to external effects also taken into account in the calculations.

Some lessons can be learned from developing this framework. First, major investment works should always be based upon a social cost/benefit analysis. Calculating a social cost/benefit analysis is determined on a case-by-case basis using a methodology that has already proven its merits and allows itself to be compared. Second, a social cost-benefit analysis of an investment project plays a key role in determining how to achieve the goals without imposing unnecessary costs on the economy. Therefore, sufficient time for a thorough social cost-benefit analysis should be reserved in the preparation of the project; as well as the incorporation of the results in a final (finance) dossier. Third, to do a more accurate, and thus better, social cost/benefit analysis, exact data collection of costs and benefits is needed. Additional data collection requires both time and money. In the end, the advantages of conducting a social cost-benefit analysis, in terms of avoiding mistakes or choosing a better project, can reach up to 10% of the project costs.

Notes

1 Annualisation is a procedure through which the average benefit per year is worked out. This is simply done by summing up all of the discounted benefits over the appraisal period and by dividing the outcomes by the length of the appraisal period. The present value of a growing annuity formula calculates the actual value of a series of future periodic benefits that grow at a proportionate rate.
2 The perpetuity value formula sums the present value of an infinite amount of future period payments.

References

Agentschap Wegen en Verkeer (2010) *A11 verbinden en verbeteren! Informatiebrochure.* www. zeebruggeopen.be/wp-content/uploads/brochure-a11.pdf.

Aronietis, R., Van de Voorde, E., and Vanelslander, T. (2011) *Competitiveness Determinants of Some European Ports in the Containerized Cargo Market.* Proceedings of the BIVEC-GIBET Transport Research Day 2011, Namur, University Press.

Idea Consult (2002) *Economische positionering van de haven van Brugge-Zeebrugge.* Brussels.

Meersman, H., Van de Voorde, E., and Vanelslander, T. (2010) Port competition revisited. *Review of Business and Economics*, 55(2), 210–232.

Nellthorp, J., Sansom, T., Bickel, P., Doll, C., and Lindberg, G. (2001) *Valuation Conventions for UNITE, Working Funded by 5th Framework RTD Programme.* Institute for Transport Studies, Leeds, University of Leeds.

Resource Analysis (2008) *Haalbaarheidsstudie Seine-Schelde West.* Maatschappelijke kostenbatenanalyse, financiering en macro-economische impact.

Resource Analysis, Technum, Tritel, IMDC (2009) *Haalbaarheidsstudie Seine-Schelde West.* Gevoeligheidsanalyses trafiekprognoses en MKBA.

UNITE (UNIfication of accounts and marginal costs for Transport Efficiency) (2003) *Competitive and Sustainable Growth (Growth) Programme.* Pilot Accounts for Belgium. Work funded by 5th Framework RTD Programme. STRATEC, Brussels, July 2002. Retrieved from www.its.leeds.ac.uk/projects/unite/downloads/D12_Annex1.doc.

Suggestions for further reading

Besides the papers provided as references in the text, we recommend the following texts as suggestions for further reading.

Blauwens, G. (1988) *Welvaartseconomie en kosten-batenanalyse.* MIM, Antwerp.

Blauwens, G., De Baere, P., and Van de Voorde, E. (2016) *Transport Economics.* De Boeck, Antwerp.

Ferrari, C., Parola, F., and Gattorna, E. (2011) *Measuring the Quality of Port Hinterland Accessibility: The Ligurian Case Transport Policy.* Elsevier, Amsterdam.

Finger, M., and Messulam, P. (Eds.). (2015) *Rail Economics, Policy and Regulation in Europe.* Cheltenham, Edward Elgar, 138–170.

Van Hassel, E., Meersman, H., Van de Voorde, E., & Vanelslander, T. (2016) *North–South Container Port Competition in Europe: The Effect of Changing Environmental Policy.* Research in Transportation Business & Management.

Zhang, A. (2008) *The Impact of Hinterland Access Conditions on Rivalry between Ports.* Retrieved from oecd-ilibrary.org.

10

COORDINATION IN HINTERLAND CHAINS

Martijn van der Horst and Peter de Langen

10.1 Introduction

From the first chapters of this book it became clear that ports fulfil multiple functions. They serve mainly as a transport node, but they are also a location for industrial, trade and logistics activities. In addition, ports are part of dynamic global supply chains and part of a network with a foreland and a hinterland. This chapter deals with the hinterland transport chain of ports. Chapter 8 of this book made already clear that ports are increasingly faced with congestion and that hinterland accessibility is a key factor in competition among ports. In that chapter hinterland accessibility or improving the efficiency of port-related transport chains was approached from an infrastructural perspective and the perspective of dynamic traffic management. This chapter approaches hinterland accessibility as an organisational issue. It will show the importance of analysing coordination between actors responsible for an efficient functioning port-related transport chain.

Why is coordination in hinterland chains important? Which coordination problems are relevant and why? Which initiatives are taken to improve bottlenecks in hinterland accessibility? What theoretical perspective(s) can be used to analyse those initiatives? This chapter will give an introduction into these issues, with some empirical illustrations. Although many types of cargo are handled in a port, this chapter is limited to the hinterland transport of containers. Much more is known about port-related transport of containers because efficiency in the hinterland chain is seen as one of the most important determinants in competition between container ports.

Section 10.2 introduces the actors active in port-related transport chains. It examines which coordination problems occur among them, and why. Section 10.3 discusses why efficient hinterland chains are important from a port perspective. Section 10.4 gives a general typology how actors could improve coordination in hinterland chains (so-called coordination arrangements), and how these arrangements are rooted in theory. This section is followed by Section 10.5, where three coordination arrangements are discussed. First, the Extended Gateway concept will be discussed. It shows how a deep-sea terminal operator vertically integrates in the hinterland chain by offering inland waterway transport and railway transport. Next, we will describe a case from the liberalised Dutch railway industry. In this coordination arrangement companies in the port of Rotterdam agreed upon a set of operational rules of the game to improve coordination on the port's railway track. The third example discusses the 'OffPeak program' that was implemented in response to peak loads at the gate of deep-sea terminals in a Los Angeles/Long Beach. Finally, Section 10.6 summarises the most important conclusions.

10.2   Coordination in hinterlands: what is the issue?

Value chains (or networks) can be divided into various components. For example, a transport chain can be divided into maritime transport, port handling activities and inland transportation. At a more detailed level, port handling activities can be further divided in activities such as pilotage, towage, mooring services, unloading the deep-sea vessel, storage of goods, loading hinterland transportation modalities and so on. Between all these separate activities, supply chain coordination is needed to secure alignment between different activities in the chain.

This chapter is concerned with how actors in ports and in port-related transport chains (hinterland chains) try to enhance coordination in order to make the container transport from and to the port more efficient. Coordination problems arise as sufficient coordination between the different parts of the chain (from an overall efficiency perspective) often does not emerge 'spontaneously'. Some examples of coordination problems in port-related transport chains could be:

- A barge owner wants to pick up containers at a deep-sea terminal, but the deep-sea terminal does not know in advance when he will arrive. The result could be long waiting times for the barge.
- A truck driver picks up a container at the terminal, without the required information, leading to additional work at the gate.
- A train brings a number of containers to a particular destination in the hinterland and returns empty, even though there are empty containers that need to return to the port. The train operator does not know about these containers, the shipping line/forwarder in charge of returning the empty containers does not know about the idle capacity on the train.
- Many truckers decide to bring and pick up container at a terminal between 4 and 8 PM. Since they all come at the same time, this results in long waiting hours.

These examples show that synchronising different consecutive activities in a transport chain could lead to increased efficiency. In the preceding examples it seems

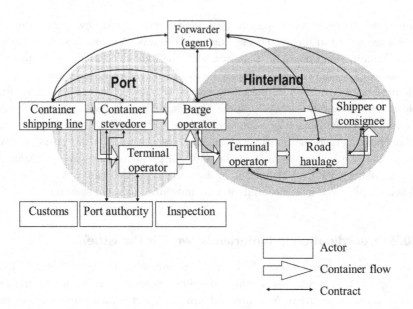

FIGURE 10.1 Actors in the inland shipping hinterland chain

that coordination could be easily arranged. Nevertheless, practice shows that these problems persist. Coordination does not arise spontaneously. The quality of hinterland connections is the result of joint actions by a set of actors with a great deal of so-called operational interdependence. In general, a major complicating factor is the fact that the individual parties perform different activities in the chain and have different interests. Therefore, a way to realise coordination for the chain as a whole is not automatically a good solution for each individual actor.

In the transport chain from deep-sea container vessel towards the shipper (the beneficial cargo owner), many different firms are involved with many interdependencies and contractual relations (see Figure 10.1).

In a typical 'merchant haulage' import chain, where the shipper or its forwarder decides about the hinterland part of the chain, the deep-sea container carrier delivers the container to the deep-sea terminal. The terminal operator unloads and stores the cargo in the terminal area, or (more exceptionally) immediately transfers it to an inland transport mode. The container carrier has a contract with the deep-sea terminal operator and pays for the whole operation at the terminal, that includes unloading the container from the vessel, storage of the container at the terminal and loading the hinterland transportation mode. The inland transport operator, for example, the barge operator, has a contract with the shipper or the forwarder to pick up the container at the terminal and transports it to an inland terminal or to the final destination. The inland transportation company needs to secure a handling slot at the terminal by which the container can be released for further transport (this includes the 'commercial release' and release by customs authorities and potentially other government agencies). The container is also handled at the inland terminal, and for the 'last mile', the shipper or its forwarder has to arrange a trucking company.

This example shows that many actors (at least seven) are concerned with all types of coordination activities to bring one container to its final destination. Sometimes different coordination activities must take place among more than two actors and at the same time. All these actors have different interests. For example, the focus of a transport company is on the efficient utilisation of its assets (barge, locomotives, truck, etc.). The core logic of a deep-sea or an inland terminal is the efficient use of personnel and of the terminal equipment like cranes and straddle carriers. This contrasts with forwarders or logistics service providers, who are in general non-asset-based, and are focused on optimising good and information flows for the shipper.

There are four general reasons why coordination in port-related chains between individual parties with different interests is hard to establish (Van der Horst and De Langen, 2008). First, there is a lack of contracts between firms in the hinterland chain that hinders coordination. For example, there is no contractual relationship between the hinterland transport company and the deep-sea terminal stevedore (see Figure 10.1). Such a contract would give an incentive to both parties to better match the quay planning of the container stevedore with the planning of the transport company. Second, there is a lack of resources or a lack of willingness to invest by at least one firm in the transport chain. Even though all actors in one sector may agree that investments are required to improve coordination, some firms may not be able or willing to take part. This issue is especially relevant for coordination problems involving relatively small firms in a highly competitive environment. In hinterland transport many small companies are active. For example in The Netherlands it is estimated that 50% of the container truck companies are small (one to four trucks), of which 25% are one-truck companies.

In addition to the competitive environment and the small and medium size of companies, risk-averse behaviour and a short-term focus of firms is a third reason. Firms expect that the process of establishing better coordination through cooperation is time-consuming. For many companies the results are uncertain, that is why they are reluctant to put any effort into this process. A fourth reason is the unequal distribution of the costs and benefits of coordination. As stated earlier, well-coordinated hinterland connections are a 'collective result' by a group of different actors with a high interdependence. One actor in the chain has to invest or has to take the initiative in, for example, setting up a joint ICT system, while other actors obtain the benefits. Gain-sharing mechanisms that redistribute the collective benefits may fail because of high transaction costs. In addition, there is a risk of free-rider behaviour.

In hinterland transport of containers, many coordination problems exist. We next summarise the most important ones in inland waterway transport, railway transport and trucking.

10.2.1   Barging: long turnaround of barges in the port area

The long stay of one barge in the port area is a coordination problem because there is insufficient joint planning of the barge operator and the deep-sea terminal. For example, in the port of Rotterdam many barge operators call at a variety of terminals in the port. At the terminals they (un)load a limited numbers of containers, or in other words, the average call size is small. The average rotation time of a barge varies from 21 hours to 36 hours, of which only 7 hours are used for (un)loading. The average call size of the container terminals in the Rotterdam port area is about 30 containers (NextLogic, 2012). Because there is no contractual relationship between the barge operator and the deep-sea terminal operator, there is no direct incentive to improve this coordination problem.

10.2.2   Barging: limited terminal planning

The lack of a contract between the barge operator and deep-sea terminal operator is also (partly) the cause of an insufficient planning of the entire terminal process. Better coordination may lead to an increased utilisation of terminal equipment and barges.

10.2.3   Railway transport: insufficient terminal planning and peak loads at terminal

A major coordination problem between the railway company and the deep-sea terminal operator is peak times at the terminals, and a low punctuality of arriving and departing trains. The peak time at terminals in the port of Rotterdam is usually between 2 and 9 PM, because railway companies drive their container shuttle trains during the night. Also in this case, the terminal operator is limited in taking any action towards the railway company to handle the container trains during other moments of the day. There is no contract between the terminal operator and the railway company. The railway company drives trains on behalf of a freight forwarder and rail operator and not on behalf of the terminal operator.

10.2.4   Railway transport: insufficient coordination between railway company and infrastructure manager

A particular coordination problem in port-related railway chains is between the railway company and the infrastructure manager. If a railway company wants to run a train, he has to ask the infrastructure manager for a so-called train path. A train path is the infrastructure capacity required to run a train between two places over a given time period. Railway companies can request train paths against the payment of a fee based on the length of train path and the type of train (including weight). Often the allocation of train paths by the infrastructure manager takes place on long-term basis (e.g., annually or quarterly). The request by the railway company is only an allocation based on the 'expected' number of container trains. An already requested train path cannot be used by another railway company in case of delays. This system of allocating capacity is quite inflexible, leading to unused track paths.

10.2.5   Trucking: peak loads at the deep-sea terminal

Peak loads in the arrival and departure of trucks are important coordination problems in container trucking. The peaks often take place between 6 and 9 AM and from 4 to 9 PM. On the one hand, it leads to long waiting times for the truckers; on the other hand, a peak load is unattractive for the terminal. It often leads to congestion on the road infrastructure in or just outside of the port area. Besides the already discussed missing contractual relationship between terminal and road haul company, the peak load has two other causes. First, truckers often start their workday with picking up a container at the deep-sea terminal. The shipper of the cargo can handle the container at the beginning of his workday that runs from 8 AM to 6 PM (office hours). Often the trucking companies do not get the possibility from the shipper to bring the container during another moment of the day. Second, many road companies plan the retrieval of a container as the last ride at the end of the day. Peak shaving has been an important issue for a long time in many ports worldwide, but remains difficult to realise.

10.2.6   Across all modes (barging, railway and transport/ trucking): limited exchange of cargo

A relevant coordination problem in container transport for all transport modalities is the limited exchange of cargo between transport companies. The exchange of cargo can allow truck companies to have fewer empty trips. Or for the barge operators it could create the possibilities to operate larger vessels with higher service frequencies and fewer port calls. Also for railway companies, the exchange of containers among each other could generate economies of scale and higher utilisation of trains and wagons. This exchange of cargo, in fact horizontal cooperation, does not develop spontaneously, because transport companies firmly wish to remain independent.

10.2.7   Across all modes (barging, railway and transport/ trucking): delays physical and/or administrative controls

Legislation and regulation on security, anti-terrorism, food quality and the monitoring function of the import, export and transit of goods by the customs authority becomes more complex. This complexity makes good alignment in information exchange between the forwarder, the inspection services, customs authority and transport companies important and difficult at the same time.

10.3   The importance of coordination in hinterland chains

In many ports, containers have become the most important cargo flow with fierce competition between ports. Some of the containers come from or are destined for captive hinterlands in the proximity of the ports. A captive hinterland can be defined as the natural hinterland of the port. For this region the port is geographically well-positioned and the transport costs to the port are substantially lower than to other ports. Nevertheless, most ports not only attract cargoes for the captive hinterland, but also compete fiercely for contestable container cargoes. Container flows can easily be switched between different ports (Slack and Starr, 1993). As a result, ports are eager to enhance the quality of their hinterland transport services to optimise the transport costs. Many studies (e.g., De Langen, 2007) show that having efficient port-related transport chains is an important, or maybe the most important, determinant in competition between container ports nowadays.

Furthermore, there is a lot of potential to cut costs in hinterland transport. There are different estimations about the share of hinterland transport in the overall transport chain. Stopford (2002) indicates that hinterland transportation, including port costs, accounts for about 54% of total chain-related costs. Other costs are those for deep-sea shipping (about 23%), including operating expenses, capital costs and bunker fuel; the cost for the container itself (18%), including leasing and maintenance costs; and about 13% for repositioning the container. Especially in the deep-sea part of the transport chain, cost savings have been realised by deploying larger vessels (up to 21,000 TEU) and horizontal cooperation between container carriers in alliances.

Efficient hinterland connections are not only relevant for port competitiveness but are also an importation location factor. A port is not only a link between transport modes, but because of its central position in global supply chains, it is also an attractive location for activities such as manufacturing and warehousing. Besides proximity to consumers and producers, quality of the labour force, and land price, good hinterland accessibility is also seen as a key location factor for port-related economic activities.

Finally, efficient hinterland chains are important from a sustainability perspective. The containerisation of goods and container transport has many positive characteristics both for the consumers and the society as a whole. Containerisation has accelerated the growth of trade worldwide. At the same time, transport has undesired side effects. The demand for transport leads to congestion and other negative external effects like emissions of greenhouse gases, air pollution, accidents and noise. In almost every seaport, road transport plays a primary role in reaching the origins and destinations in the hinterland. In other words, truck transport has the highest share in the modal split of a port. Although road transport has made some major reductions in the emission of NO_x (nitrogen oxides) and PM (particulate matter) in recent years, it still has more negative external effects than barge or train transport. It is the goal of many port authorities and regional and national governments to reduce these emissions to ensure the licence-to-operate of a port. Better coordination will lead to a reduction of the negative external effects.

10.4 Coordination arrangements in hinterland chains

The previous section showed that coordination problems are not 'solved' spontaneously through the 'invisible hand' of the market economic theories which provide explanations of the existence of economic institutions that enable coordination. Coase (1937) started a huge research stream with the question of why organisations, such as companies, alliances between companies, and governments do exist if the price mechanism is such an efficient coordination mechanism (as was argued in the early twentieth century). In all these cases, Coase argued, *coordination beyond price* emerges. Coase also argued that transaction costs with alternative forms of coordination (e.g., within a corporate hierarchy) can be more efficient than coordination through markets. Coase is seen as one of the founders of Institutional Economics.

Institutional Economics includes a number of theories. For the typology we present next, reference could be made to the work of Williamson (1996). Williamson expanded the work of Coase, introducing behavioural assumptions to the understanding of the best way to organise transaction costs (in a corporate hierarchy vs. through markets). Williamson's transaction costs concept is based on two behavioural assumptions: bounded rationality and opportunistic behaviour. Because of bounded rationality and opportunistic behaviour, transaction costs (e.g., the costs of finding a partner, preparing and concluding a contract, monitoring the execution of the agreement) of contracts can be substantial, especially for complex agreements. In the most efficient governance structure, total production and transaction costs are, in the long run, less than in any other governance structure.

Moreover, our typology of four coordination mechanisms is inspired by the work of Olson (1971) on the conditions under which collective action emerges. Table 10.1 shows four coordination mechanisms, their theoretical foundations and examples.

10.4.1 Introduction of incentives

The first mechanism is the introduction of incentives or changing the incentive structure. The mechanism is related to the concept of property rights. A basic

TABLE 10.1 Four coordination mechanisms and possible coordination arrangements

Coordination mechanism	Main theoretical lens	Possible coordination arrangements
Introduction of incentives	Property rights (Demsetz, 1967)	Bonus, penalty, tariff differentiation, non-financial incentives, tariffs linked with cost drivers
Creation of an interfirm alliance	Transaction costs (Williamson, 1996; Notteboom, 2004)	Long-term contracts with standards on quality and service, formalised procedures, offering a joint product, joint capacity pool, risk sharing agreements for new services
Changing scope	Transaction cost, Transaction value (Williamson, 1996; Notteboom, 2004)	Vertical integration, introduction of an agent, introduction of a chain manager, introduction of an auctioneer, introduction of a new market
Creating collective action	Collective action (Olson, 1971)	Public governance by a government or port authority, public-private cooperation, branch association, ICT system for a sector of industry

Source: Adapted from Van der Horst and De Langen (2008).

principle in an economic model of supply, demand and price determination is that all the cost and benefits of the goods or services traded are reflected in the price. In other words, it is assumed that contracts are complete. However, in the hinterland market it is proven that contracts often are 'incomplete'. In this respect, incomplete means that negative externalities exist, and parties are often not compensated for these effects. In general, incentives can be used to internalise the harmful or beneficial effects (externalities) of a firm's decision with other firms. As an effect they align the interests of individual firms within an efficient overall transport chain. We could distinguish different forms of incentives, as follows:

Bonus–malus systems. A company pays a penalty if it does not meet an earlier made agreement. Or the company gets a bonus when its behaviour is in line with the interest of other links in the chain. For example, the terminal operator gets a penalty if trucks stay longer on the terminal than agreed in advance.

Non-financial incentives. A company that acts in the interest of another party in the chain is rewarded with a non-financial bonus, like, for example, a fixed time slot at a terminal or shorter waiting times.

Contract with clear key performance indicators (KPI). Companies could agree on a contract that clearly describes desired performances for the long run. KPIs in such contracts are linked with penalties and bonuses in case certain performances are not achieved or surpassed.

Differentiated pricing systems. Differentiated pricing systems could include a discount on tariffs for customers that guarantees a fixed certain amount of container slots on a rail shuttle, or a pricing system for peak and off-peak hours.

10.4.2   Creation of an interfirm alliance

The second mechanism for enhancing coordination is the creation of an interfirm alliance between actors in the hinterland chain. Incentives might induce firms to act in the interests of other actors in the chain, but could yield high transaction costs. Alliances are arrangements with more commitment between the companies involved. Different forms of cooperation by a limited number of partners can be regarded as an alliance. These mechanisms may differ in 'intensity' and types. An interfirm alliance with a low intensity is a so-called arm's-length relationship. That could be a long-term contract with standards on quality and service. A larger intensity in the cooperation between companies could be found in a partnership. We can distinguish different types of partnerships like information integration (e.g., exchange schedule information), logistics integration (e.g., effective integration of different business activities) or the integration on the joint development of new transport services. The most intensive form of cooperation is a joint venture. A joint venture is a business agreement in which companies agree to develop, for a certain time, activities by contributing equity. An important characteristic of an interfirm alliance is that the actors involved remain to some extent independent. Alliances include many forms of interfirm cooperation like the following:

Bundling of cargo and equipment. In such a form of horizontal cooperation, bundling will lead to better mutual planning and cost advantages, for example, a joint capacity pool between hinterland transport firms.

Risk sharing agreements for new services. Risk sharing by a number of different companies could help to establish coordination. For example, a large shipper guarantees a base volume of containers for a regional inland terminal. The operator himself must ensure sufficient additional volume for an efficient operation of his terminal.

Dedicated use of facilities or assets by a customer or supplier. In this arrangement, a customer or supplier gets the exclusive right to use a number of facilities (e.g., transport companies may use a storage place or a crane at a terminal). As an effect, he will be more willing to contribute to an efficient use of it.

10.4.3   Changing scope

The third coordination mechanism is changing the scope of an organisation. This mechanism includes hierarchical coordination of the chain (vertical integration). An important feature of changing the scope of a company is that the independence of the parties involved is replaced by a corporate hierarchy. Vertical integration is simply a means to coordinate the different stages in a chain when coordination across markets is not efficient. Like the previous coordination mechanism, this mechanism is rooted in the transaction cost approach. A change of scope will mainly arise when integrating activities leads to lower transaction costs in comparison with performing these activities by two separate companies in the chain. This is the case when it relates to activities or interdependencies which appear frequently, adjusted, involving a high uncertainty, and where specific high investments should be made to achieve good alignment between the parties involved. Besides minimising transaction costs, vertical integration can also lead to increasing transaction value. Apart from vertical integration, there are two other forms that belong to this type of coordination:

Introduction of an intermediary or a 'transport chain director' who is responsible for a proper alignment of different parts of the chain. For example, an independent person could take care for the exchange of information between the transport company and the terminal operator.

Transfer of responsibilities for a particular activity to another party that is better able to achieve good coordination. This is the opposite of vertical integration.

10.4.4 Creating of collective action

The fourth and last mechanism for enhancing coordination is the creation of collective action. Collective action arises when a large group of stakeholders agrees upon a common approach or solution to solve coordination problems. This mechanism is especially relevant when investments have collective rather than individual benefits. Literature on collective action studies the interests, motivations and decision outcomes of these large groups of companies. In a group of actors attempting collective action, individuals will have the incentive to 'free ride' on the efforts of the others. Collecting the joint cost and distribution of the joint efforts often takes place via a public organisation, because it will lead to high transaction cost if this is done by individual market parties. Also industry or sector associations play an important role in establishing collective action. Collective action could take place through:

> **A platform** where various parties from the transport chain are represented. The platform can be used to discuss the joint problems and finding joint solutions.

For example, in the port of Rotterdam, a port community system currently called Portbase was established in 2002 under the name Portinfolink. Portinfolink started as a public–private partnership between the Rotterdam Port Authority and the Port Sector Association Deltalinqs. Such a public–private partnership is efficient given the many and relatively small parties that have to participate to make such an ICT system successful. It also helped in the distribution of the large collective costs and benefits.

10.5 Three examples of coordination arrangements

10.5.1 Extended gate concept: changing the scope of deep-sea terminal operator ECT

The first example of a coordination arrangement is from the port of Rotterdam. It shows the change of scope of deep-sea terminal operator ECT in the hinterland chain (see also Chapter 15). In this coordination arrangement ECT seeks to extend the gate of its deep-sea terminal to inland terminals by offering both container handling and hinterland transport services to both ECT-owned inland terminals and to terminals owned by third parties. ECT owns three inland terminals, in Venlo, near the Dutch-German border; in Willebroek, Belgium, between the port of Antwerp and Brussels; and in Duisburg in the German Ruhr area. In cooperation with logistics service providers, ECT organises container shuttles by rail and barge. In fact, by taking the commercial risk for running container trains and barges, they extended their business into that of rail and barge operators. The hierarchical coordination of the hinterland chain by ECT helps to solve the coordination problem of the long port turnaround time of barges because of too many calls and small call sizes, and the insufficient planning and peak loads of container trains at the deep-sea terminal.

The central idea is to bundle cargo in one firm (ECT) and extend the delivery point from the perspective of the shipper from the deep-sea terminal along a corridor to an inland multimodal terminal (Veenstra et al., 2012). The gate of the deep-sea terminal is more or less placed at an inland location. At the inland terminals that belong to an extended gate network a lot of supplementary or value-added activities could take place. Besides pre- and post-transport by truck, these activities include temporary storage, empty depot or fumigation services (ECT, 2014). Veenstra et al. (2012) stresses that the inland terminal is an extended delivery point of the container. A condition of an extended gate is that the customs regime is also extended to the inland terminal. Under a custom license a container can be brought to the inland terminal. No additional customs documents are needed to bring the container from the port to the extended gate. The custom release will take place at the inland terminal.

We approach the 'Extended Gate Concept' as a solution to improve the coordination between deep-sea terminal and inland transport by rail and barge (Van der Horst and De Langen, 2008). Two additional notes can be made on the existence of the extended gate model of ECT. First, this initiative of ECT can be seen as a reaction to the sustainability policy of the Dutch government and the Rotterdam Port Authority. The port extension project Maasvlakte 2 implies an increase of container throughput in the range of 26 million TEU in 2020 and 33 million TEU in 2035 (Port of Rotterdam, 2011). Because of the environmental performance of transport by inland waterway and rail, a modal shift is foreseen from truck to transport by barge and rail. In 2035, the share of container transport by barge and rail from and to the Maasvlakte should be increase from 52% towards 65%. Second, the extended gate concept of ECT can be regarded as a differentiation strategy. With the port extension project Maasvlakte 2, new terminal operators will enter the market. A change of the business activities by ECT towards the inland transport and inland terminal could lead to a competitive advantage.

10.5.2 Chain Management Port Rail Track: collective action in the Rotterdam railway sector

The second coordination arrangement is the pilot project 'Chain Management Port Rail Track', also in the port of Rotterdam, which started in 2007 and is still active. The purpose of this project is to improve the punctuality of trains by introducing new 'rules of the game' concerning the exchange of information between deep-sea terminal operator, rail terminal operator and rail operator. It concerns information like the estimated time of arrivals, number of containers and real-time reservation of the train tracks.

Primarily, this project can be seen as a form of creating a collective action. A platform was made where the joint problems on the Rotterdam railway track and the joint list with operational rules were discussed. Besides two large terminal operators (ECT and Rail Service Centre Rotterdam) and three railway companies (ERS Railways, Railion and Veolia Cargo), three rail operators (ERS, Hupac and Intercontainer) participated in setting up the list of operational rules. The infrastructure manager Keyrail was the initiator and coordinator of the project, and the Rotterdam Port Authority was also committed to the project.

An interesting feature of this coordination arrangement is that these new operational rules were not made and enforced by the infrastructure manager. The rules were adopted after an informal process of mutual consultation between the parties involved. In one year the parties learned more about each other's behaviour and business processes than in previous years. In the years before, the market parties noted that the operational performance of container trains in the port worsened. A main reason for the bad performance of the railway sector in the port of Rotterdam was due to the fact that in the years from 1995 onwards, the number of market parties that entered the Rotterdam railway market increased. Since 1995, the railway market in The Netherlands started to liberalise, resulting in open access for railway companies. The former state-owned railway company corporatised and several new private firms entered the market. In 2010, 14 railway companies were active in the port-related container transport by rail. Also the number of national and international rail operators increased. A study by Van der Horst and Van der Lugt (2014) shows that managing interdependent activities between so many railway actors with different business interests in a relatively 'new' liberalised market is difficult. The capacity of railway tracks and yards had to be negotiated with, and to be allocated to, about 10 different railway companies, leading to an increase of transaction costs. Furthermore, railway companies lacked incentives to use allocated train paths efficiently. A second coordination problem was on the interface between rail terminals and railway companies. Since liberalisation, coordination between both worsened. In 2007, the average punctuality at terminals was 65% for arriving and 72% for departing container trains (Keyrail, 2013). Instead of one state-owned railway company, the terminal operator had to synchronise its terminal planning with many railway companies.

A first result of the Chain Management Port Rail Track project is that the average punctuality of container trains leaving the rail terminals improved from 65% to 88% in 2010 and 91% in 2011 (Keyrail, 2013). From 2009 onwards, the operational rules became part of the access agreement between infrastructure manager Keyrail and the railway companies. The new rules of the game established by the market parties have now become basically 'institutionalised'. They apply for all railway companies that want to make use of the rail track in the port and the dedicated rail freight track from the port of Rotterdam to the German border (Betuweroute).

10.5.3 OffPeak program: collective action has led to an incentive system to avoid peak loads in Los Angeles/Long Beach

The third coordination arrangement deals with the coordination arrangement to deal with peak loads at the gate of deep-sea terminals and related congestion problems during peak hours. In the port of Los Angeles/Long Beach, different initiatives have been taken to avoid an increase in the amount of trucks arriving during peak hours. A first step was the implementation of different truck appointment systems and the extension of the opening hours of the terminal. Initially, both initiatives were taken in response to legislation in the State of California. Assembly Bill 2650 imposed a fine of $250 for the deep-sea terminal operator for every truck waiting more than 30 minutes. The legislation gave the terminal operators some guidance, but they had the freedom in choosing solutions to reduce the waiting times. It led to dissension among the 13 terminal operators. Some terminal operators chose to implement a truck appointment system. In such a system, trucking companies make appointments to pick up or drop off containers during specific time windows. Some other terminal operators chose to extend the opening times of their terminal. Moreover, the legislation could not prevent that the trucks were waiting inside the terminal, because this could not be fined (Giuliano and O'Brien, 2007).

Around 2004 there was a second development. The first 8,000 TEU vessels called at the ports, leading to large cargo volumes at the terminals from the seaside. Combined with a shortage of workers at the terminals, this caused huge delays for the ships. Because of the weak port performance, container vessels moved to other ports. This was a clear signal to improve the productivity of the terminals, including the truck operations. At the time, the State of California threatened (again) with new legislation to force the terminals to solve the problem of the truck waiting times. As a response, terminal operators collectively came with their own project, including the introduction of incentives. The project was called the OffPeak program. It included a traffic mitigation fee of $40 fee per TEU for loaded containers arriving at the ports by truck during peak hours. The fee is used to help compensate the costs of the off-peak terminal gates. It is adjusted annually to reflect increases in labour costs at the terminals. It is collected by a non-profit company, PierPASS. This organisation was created by the deep-sea terminal operators; it provides five off-peak terminal gates at all 13 terminals within the ports of Los Angeles and Long Beach. The projects led to an increase of more than 40% for the off-peak gate usage (Port of Los Angeles, 2014).

10.6 Conclusion

This chapter has dealt with coordination in hinterland chains. Coordination problems arise between different actors with a great deal of operational interdependence. The interdependent activities are not effectively coordinated, with poor efficiency of the whole transport chain as a result. For container ports, efficient hinterland chains are seen as one of the most important determinants in the competition between container ports. Having good hinterland access is also an importation location factor of other port-related businesses likes manufacturing and logistics services. This chapter has shown many coordination problems, such as the long stays of barges, trains and trucks in the port or at the terminal; limited call sizes and integrated planning at the deep-sea terminal; unused or overused rail and road infrastructure; and limited exchange of cargo. These problems may severely reduce the efficiency of hinterland chains. Coordination problems in hinterland chains arise because of the involvement of many different individual parties with different business logics, a lack of contracts between some firms in the chain, insufficient resources or willingness to invest and an unequal distribution of the costs and benefits of coordination. Insights from institutional economics were used to make a distinction between four coordination mechanisms, the introduction of incentives, the creation of interfirm alliances, changing the scope, and the creation of collective action.

The chapter concluded with three coordination arrangements from practice. The Extended Gateway concept showed how Rotterdam-based deep-sea terminal operator ECT has changed its scope from deep-sea terminal operators towards hinterland transport by barge and train. Besides a solution to improve coordination between terminal and inland transport, the extended gateway concept can be seen as a strategy to promote the use of barge and railway transport and to differentiate itself from other terminals. The coordination arrangement Chain Management Port Rail Track showed a collective action in the Rotterdam railway sector. In an informal setting, and with support from the infrastructure manager and the port authority, the railway sector in the port of Rotterdam agreed upon a set of operational rules of the game to improve coordination of the port's railway track. The new operational rules are now part of the general access agreement between the infrastructure manager and all of the railway companies. Third, the OffPeak program in the port of Los Angeles/Long Beach was discussed. It showed how 13 terminal operators collectively agreed on the introduction of incentive peak loads at the terminal gates.

These examples show the relevance of understanding coordination problems as well as the factors that influence the effectiveness of the different coordination mechanisms in improving coordination. Nevertheless, the challenge of how to improve coordination in ports and hinterland chains remains a key issue for the port community. Further academic research to assist companies in the port as well as the port authority is needed. Such research can extend and deepen the knowledge from an institutional perspective, as used in this chapter, but can also focus on developing gain-sharing mechanisms and can focus on the strategic and behavioural aspects of cooperative projects.

References

Coase, R. H. (1937) The nature of the firm. *Economica*, 4, 386–405.

De Langen, P. W. (2007) Port competition and selection in contestable hinterlands; the case of Austria. *European Journal of Transport and Infrastructure Research*, 7(1), 1–14.

Demsetz (1967) Towards a theory of property rights. *American Economic Review*, 57, 347–359.

ECT (2014) *European Gateway Service*. Retrieved from www.extendedgatewayservices.com, accessed 21 November 2014.

Giuliano, G., and O'Brien, T. (2007) Reducing port-related truck emissions: The terminal gate appointment system at the Ports of Los Angeles and Long Beach. *Transportation Research Part D: Transport and Environment*, 12(7), 460–473.

Keyrail (2013) *Project Ketenregie* (Project Chain Management). Retrieved from www.keyrail.nl/ketenregie, accessed 3 November 2013.

NextLogic (2012) *Performance Measurement: Results Baseline Measurement*. Retrieved from www.nextlogic.nl/ketenoptimalisatie-containerbinnenvaart/performance-meting

Notteboom, B. (2004) *Inter-firm Collaboration, Learning and Networks: An Integrated Approach*. London, Routledge.

Olson, M. (1971) *The Logic of Collective Action: Public Goods and the Theory of Groups*. Cambridge, MA, Harvard University Press.

Port of Los Angeles (2014) *Extended Terminal Gate*. Retrieved from www.portoflosangeles.com, accessed 21 November 2014.

Port of Rotterdam (2011) *Port Vision 2030: Port Compass*. Strategy Document, Port of Rotterdam Authority.

Slack, B., and Starr, J. (1994) Containerization and the load centre concept. *Maritime Policy and Management*, 21(3), 185.

Stopford, M. (2002) *Is the Drive For Ever Bigger Containerships Irresistible?* Lloyds List Shipping Forecasting Conference, 26th April 2002.

Van der Horst, M. R., and De Langen, P. W. (2008) Coordination in hinterland transport chains: A major challenge for the seaport community. *Journal of Maritime Economics & Logistics*, 10(1), 108–129.

Van der Horst, M. R., and Van der Lugt, F. (2014) An institutional analysis of coordination in liberalized port-related railway chains: An application to the port of Rotterdam. *Transport Reviews*, 34(1), 68–85.

Veenstra, A., Zuidwijk, R., and Van Asperen, E. (2012) The extended gate concept for container terminals: Expanding the notion of dry ports. *Maritime Economics and Logistics*, 14, 14–32.

Williamson, O. E. (1996) *The Mechanisms of Governance*. New York, Oxford University Press.

Suggestions for further reading

Besides the papers provided as references in the text, we recommend the following texts as suggestions for further reading.

Bergqvist, R., Macharis, C., Meers, D., and Woxenius, J. (2015) Making hinterland transport more sustainable a multi actor multi criteria analysis. *Research in Transportation Business & Management*, 14, 80–89.

Van den Berg, R., and De Langen, P. W. (2011) Hinterland strategies of port authorities: A case study of the port of Barcelona. *Research in Transportation Economics*, 33(1), 6–14.

Van der Horst, M. R. (2016) *Coordination in Hinterland Chains. An Institutional Analysis of Port-related Transport*. TRAIL Thesis Series no T2016/19, The Netherlands Research School TRAIL.

PART 2

Ports and networks

Operations

11

OPTIMISATION IN CONTAINER LINER SHIPPING

Judith Mulder and Rommert Dekker

11.1 Introduction

Seaborne shipping is the most important mode of transport in international trade. In comparison to other modes of freight transport, like truck, aircraft, train and pipeline, ships are preferred for moving large amounts of cargo over long distances, because shipping is more cost-efficient and environmentally friendly. Reviews of maritime transport provided by the United Nations Conference on Trade and Development (UNCTAD) show that about 80% of international trade is transported (at least partly) over sea. In 2013, containerised cargo is with a total of 1.5 billion tons responsible for over 15% of all seaborne trade, which resulted in a world container port throughput of more than 650 million TEU (twenty-foot equivalent units).

In the shipping market, three types of operations are distinguished: tramp shipping, industrial shipping and liner shipping (Lawrence, 1972). Tramp shipping is used to follow the demand of available cargoes to be delivered. Hence, tramp ships do not have a fixed schedule, but ensure an immediate delivery for the most profitable freight available, resulting in irregular activities. The behaviour of tramp ships is thus comparable to taxi services. In industrial shipping the cargo owner also controls the ships used to transport the freight. The objective of industrial operators is to minimise the cost of shipping the owned cargoes. Liner ships follow a fixed route within a fixed time schedule and serve many smaller customers. The schedules are usually published online and demand depends on the operated schedules. Hence, liner shipping services are comparable to train and bus services.

In this chapter, we will focus on container liner shipping and related terminal operations. In Section 11.2, we discuss decision problems in the container liner shipping market. Section 11.3 introduces the decision problems in terminal operations that are related to the liner shipping operations and Section 11.4 discusses geographical bottlenecks. Section 11.5 provides a case study of Indonesia to explain some of the liner shipping problems in more detail. Finally, Sections 11.6 and 11.7 conclude this chapter and discuss expected future developments.

11.2 Container liner shipping

We will focus on the liner shipping operations concerned with the transport of containers. Liner shipping operators face a wide variety of decision problems in operating a liner shipping network. First, at the strategic planning level, they need to decide on the composition of the fleet with regard to the fleet size and mix problem and on which trade routes to serve in the market and trade selection problem. At the tactical planning level the network needs to be designed, prices need to be set and empty containers have to be repositioned. Finally, at the operational level, operators need to determine the cargo routing through the network and how to deal with disruptions. Furthermore, they can make adjustments to the earlier set prices and need to determine a plan to store all of the containers on the ship during the loading process. These problems are considered in the cargo routing, disruption management, revenue management and stowage planning problems, respectively.

Some problems have to be considered at both the tactical and the operational planning level, such as setting the sailing speed and optimising the bunkering decision and designing a (robust) schedule. In this section, we will introduce all these decision problems. In these problems we will make use of the following terminology. Liner shipping operators will also be referred to as liner shipping companies, liner companies or liners. Liner ships follow fixed routes, which are sequences of port calls to be made by the ship. Route networks consist of a set of services, which are routes to which a ship is allocated. The exact arrival and departure days at each port of call are also published by liner companies. When we refer to the route together with the arrival and departure days, we will talk about a schedule. Finally, a round tour refers to one traversal of a route and a (sea) leg refers to the sailing between two consecutive ports.

11.2.1 Strategic planning level

The strategic planning level consists of long-term decision problems. Generally, these problems are only solved at most once a year. Examples of long-term decision problems in container liner shipping are the fleet size and mix problem and the market and trade selection problem.

11.2.1.1 Fleet size and mix

In the fleet size and mix problem, a liner company decides on how many ships of each type to keep in its fleet. Container ship sizes have increased substantially because of the growth in container trade and because of competitive reasons. For example, the Emma Maersk (introduced in 2006) has an estimated capacity of more than 14,500 TEU. Before the introduction of the Emma Maersk, the capacity of the largest container ship in the world was less than 10,000 TEU. In 2013, Maersk introduced a series of ships belonging to the Triple E class with capacities of over 18,000 TEU, while both MSC and CSCL introduced container ships with a capacity of more than 19,000 TEU in 2015. Ships benefit from economies of scale when they are sailing at sea, but they might suffer diseconomies of scale when berthing in ports. However, the effect of the economies of scale at sea is much larger than the effect of the diseconomies of scale in ports (Cullinane and Khanna, 1999). Hence, economies of scale in larger container ships can lead to substantial savings if the capacity of the ship is adequately used. However, if the demand decreases and the liner company is not able to fill these large ships any more, higher operational costs are incurred by these large ships. Therefore, fleet size and mix problems are used to balance the possible benefits of economies of scale with the risk of not being able to use the full capacity of the ships. Because building a new container ship may take about one year and ships usually have life expectancies of between 25–30 years, future demand and availability of ships play an important role in the fleet size and mix problem. Pantuso et al. (2014) present an overview of research conducted on the fleet size and mix problem. Most of these works incorporate ship routing and/or deployment decisions in order to ensure feasibility of demand satisfaction and capacity constraints.

11.2.1.2   Market and trade selection

Before a liner container shipping company starts building a network and operating the routes, it has to decide which trade lanes to participate in. The Asia-Europe trade lane is an example of a popular trade lane. Clearly, the selected trade lanes influence the type and number of ships required. For example, trade lanes serving the US will usually not use vessels from the Maersk Triple E class, since they cannot sail through the Panama Canal and most ports in the US are not capable of handling these large vessels. Furthermore, the type and amount of cargoes that have to be transported and the required sailing frequency may influence the ship types used on the trade lane.

11.2.2   Tactical planning level

Medium-term decision problems belong to the tactical planning level. Liner companies usually change their service networks every 6–12 months, but more often in case of worldwide disruptions. Problems that have to be solved again each time the service network is adjusted are considered to belong to the tactical planning level. The examples that will be discussed next are the network design problem, the pricing problem and the empty container repositioning problem.

11.2.2.1   Network design

The network design problem in liner shipping can be split into two sub-problems. The first sub-problem is the routing and scheduling problem, which is concerned with determining which ports will be visited on each route, in which order the ports will be called at and what the arrival and departure times at each port will be. Many studies only consider the routing decisions in the network design problem and do not address the scheduling problem of determining the actual arrival and departure time.

The second sub-problem considers the fleet deployment and frequency. Here, the liner company determines which ships will be used to sail each route and with what frequency the ships will call at the ports along the route. In general, a weekly frequency is imposed, which facilitates planning by shippers, but this can be relaxed to a biweekly frequency for low-demand routes or multiple port calls per week for high-demand routes.

Sometimes sailing speed optimisation is considered as a third sub-problem of the network design problem, but in most studies the sailing speed is either assumed to be fixed and known, or will follow directly from the imposed frequency. Usually, the cargo routing problem is already included (using expected demand as input) in the network design problem in order to evaluate the profitability of a service network. The cargo routing problem will be discussed in more detail with the operational planning level problems.

The structure of the routes in a network can be divided into several types, like non-stop services, hub-and-spoke systems, hub-and-feeder systems, circular routes, butterfly routes, pendulum routes and non-simple routes. A non-stop service provides a direct connection between two ports: a ship sails from one port to the other and immediately back to the first port; sometimes this is called a shuttle service, although that also requires a high frequency. In a hub-and-spoke system, usually one port is identified as the main or hub port. All other ports (also called feeder ports) are served using direct services from the hub port. However, it is also possible that multiple hubs are applied, which are connected with each other and used as trans-shipment ports to satisfy demand between different feeder ports, in which case they might also be referred to as main ports. In the hub-and-feeder system, feeder ports might also be visited on routes with multiple port calls. Circular routes are cyclic and visit each port exactly once, while butterfly routes allow for multiple stops at the same port in one cycle. Pendulum routes visit the same ports in both directions, only in reverse order. Finally, ports can be visited multiple times on non-simple routes. Examples of some of these route types are provided in the case study in Section 11.5.

The liner shipping network design problem has attained quite some attention in the literature. An important contribution to the literature has been made by Brouer et al. (2014). They provide both a base mixed integer programming formulation for the network design problem and benchmark data instances, introducing the opportunity to compare networks.

11.2.2.2 Pricing

The goal of liner companies is to maximise profit by transporting containers from one port to another. The revenue of the company is determined by the amount of containers that are transported and the price that will be charged for each container. The pricing problem is concerned with which price to charge for each possible demand pair. Factors that influence the price are, for example, distance, trade direction, expected demand and expected capacity. The pricing problem is more a marketing, micro-economic problem than an operations research problem. Although it is an interesting problem, it has hardly been touched. Two approaches exist: cost-plus and what the market can pay. Yet, even determining the cost is a difficult allocation problem.

11.2.2.3 Empty container repositioning

Containers delivering import products in a region can be reused to transport export goods to another region. However, most regions face an imbalance between import and export containers. This trade imbalance results in an excess of empty containers in regions with more import than export and a shortage of containers for high export regions. The empty container repositioning problem tries to reallocate the empty containers in order to solve the imbalance, where costs are associated with transporting a container from one region to another. The repositioning of empty containers is considered to be very costly, because there is no clear revenue associated with it.

11.2.3 Operational planning level

The operational planning level captures the problems that occur during the execution of the routes in the service network. In order to solve operational level problems, reliable information about the actual situation is needed. Hence, operational problems usually need to be solved relatively shortly before the solutions have to be implemented. Next we will discuss the cargo routing problem, the disruption management problem, the revenue management problem and the stowage planning problem.

11.2.3.1 Cargo routing

The cargo routing problem takes the liner shipping network and container demand as an input. The goal of this problem is to find a cargo flow over the network, satisfying the capacity constraints imposed by the allocated container ships, which maximises the profit of transporting the containers. Costs are associated to (un) loading and transshipment operations. A transshipment occurs when a container has to be unloaded from a ship and loaded to another ship in order to arrive at its destination. Additionally, penalties can be imposed for demand that is not met. It is also possible to include transit time constraints to guarantee that containers will arrive at their destination.

Formulations of the cargo routing problem can be distinguished in before origin/destination-based link flow formulations, segment-based flow formulations and path-based formulations (Meng et al., 2014). All flow formulations consider the amount of flow at a link or segment of the route as decision variables in the model.

Flow balance constraints ensure that all flow starts at the origin port and arrives at the destination port, but the exact route followed by a container might not be immediately clear from the model. In the origin/destination-based link flow formulation, both the origin and the destination of the container are stored for each link in the network, while the origin/destination-based link flow formulations only store the origin or destination of the container. In this way, the number of decision variables can be reduced significantly. In a segment-based flow formulation, consecutive links of a route are already combined into segments before building the model. Segment-based flow formulations reduce the number of decision variables even more, but limit the possibility of transshipment operations to the ports at the beginning and end of the predefined segments.

Finally, in path-based formulations complete container paths from origin to destination are generated beforehand and used as a decision variable in the model. These paths might also include transshipment operations. The disadvantage of path-based formulations is that the number of paths might explode, such that more complex methods, like column generation, are needed to solve the problem. However, path-based formulations can usually be solved faster than flow-based formulations. Furthermore, transit time constraints are easily incorporated in path-based formulations, while this is generally much more troublesome in flow-based formulations. Little research is performed on the separate cargo routing problem: usually it is considered a sub-problem of the network design problem. Recently, Karsten et al. (2015) considered the cargo routing problem with transit time constraints. They propose a path-based formulation exploiting the ease to include transit time in this type of model. Their findings show that including transit time constraints in the cargo routing model is essential to find practically acceptable container paths and does not necessarily increase computational times.

11.2.3.2 Disruption management

During the execution of the route schedules, ships may encounter delays. The disruption management problem focuses on which actions should be taken in order to get back on schedule after a disruption has occurred. Examples of actions that might be performed are changing the sailing speed, swapping port calls, cut and go (leave the port before all containers are [un]loaded) or skipping a port. Usually, the goal of disruption management is to find a sequence of actions with minimum cost such that the ship will be back on schedule at a predetermined time. Brouer et al. (2013) propose a mixed integer programming formulation to solve this problem and prove that the problem is NP-hard. However, experimental results show that the model is able to solve standard disruption scenarios within 10 seconds to optimality.

11.2.3.3 Revenue management

At the operational level, more information about the demand and available capacity of a ship is available. Therefore, it might be profitable for liner companies to vary their prices based on the available capacity between a port pair. Liners will probably charge higher prices related to low capacity pairs, while they might reduce the prices on legs where the available capacity is high.

11.2.3.4 Stowage planning

The stowage planning problem determines at which location containers are stored on the ship during the loading process. The stowage planning is a very complicated process with many constraints. Essential constraints are, for example, the stability of the ship both during the next sea leg and during the (un)loading process. Furthermore, containers may have to be stored at specific locations on the ship, like reefer containers. However, the storage of the containers also influences the (un)loading process in the next ports. Ideally, all containers with destination in the next port are stored on top of the stack, but this may take too many movements in the current port. Hence, a trade-off between the number of moves required to store and to discharge a container has to be made. Tierney et al. (2014) prove that the container stowage planning problem is a NP-complete problem.

11.2.4 Both tactical and operational level

Finally, some problems can either be considered at two different planning levels or have to be considered at two planning levels at the same time. For example, sailing speed optimisation and bunkering optimisation can be considered at the moment a new service network is designed, but the solutions to these problems can be reconsidered during the execution of the routes. Furthermore, robust schedule design is an example of a decision problem that combines decisions to be taken at the tactical and at the operational level.

11.2.4.1 Sailing speed and bunkering optimisation

Both sailing speed optimisation and bunkering optimisation are typical problems that can be considered at two different planning levels. At the tactical planning level, the environmental aspects of sailing speed are usually considered. At the operational level, sailing speed is mostly used as an instrument to reduce delays incurred by the ship. The bunkering optimisation problem is concerned with deciding at which ports ships are going to be refuelled. Initially, a bunker refuelling plan is made given estimates of the bunker price at the moment the ship will be at a bunkering station. Shipping lines also regularly make bunkering contracts, containing the ports where bunker can be purchased, the amount to be purchased, the price to be paid and the validity duration of the contract (Pedrielli et al., 2015). However, due to fluctuations in prices or fuel consumption, this initial plan might have to be adjusted at the operational planning level. At this stage, more accurate information about the fuel prices and availability at the ports and the bunker level of the ship is available.

Sailing speed plays an important role in the bunker fuel consumption of ships, and hence sailing speed optimisation is often included in the bunkering optimisation problem (Yao et al., 2012).

11.2.4.2 Robust schedule design

Robust schedule design can be seen as a combination of the scheduling problem at the tactical level and disruption management or sailing speed optimisation at the operational level. The order in which ports are visited is considered to be an input of this problem. The goal is to jointly determine the planned arrival and departure times in each port and the actions that will be performed during the execution of the route when delays are incurred. The difficulty of this problem is that the tactical and operational planning level problems cannot be solved separately, but have to be considered simultaneously.

11.3 Terminal operations

Liner shipping operations are closely related to terminal operations, and decisions about ships cannot be taken while disregarding their effects on terminals. In fact, terminals are the largest bottleneck for shipping. It is important to have the right berth slots and to be loaded and unloaded quickly and in a predictable manner. There are many ports all over the world with a large number of ships waiting in front of the harbour to be allowed to berth. Accordingly, we will discuss those terminal operations aspects which directly affect shipping, namely, berth scheduling, crane allocation and container stacking.

11.3.1 Berth scheduling

Both at a tactical and an operational level, liner shipping schedules are made while taking berth availability into account. On a tactical level, when designing the liner shipping routes, agreements are made with terminals on berth availability and productivity (how many cranes and crane teams will be employed and how many container moves will be done per hour). This enables the shipping line to calculate the port time of its ships and to complete the vessel scheduling. Naturally buffer times are incorporated in the schedule and in the berth schedule in order to take care of schedule deviations. Quite often agreements are made on demurrage charges (penalties related to delayed cargoes) if terminals need more time or if the shipping line arrives too late at the terminal. At the operational level the berth schedule is adjusted according to actual information. Quite often liner ships are too late. In 2015 Drewry shipping reports that ships are on average one day late. So the berth schedule is updated at a relatively short term (2 weeks) to take care of changing circumstances, while taking the tactical berth planning as a start.

11.3.2 *Crane allocation*

One level deeper than berth scheduling is the crane allocation. As container ships typically visit many ports, the cargo destined for a particular port will be distributed over many holds in the ship. After unloading, a ship will load cargo for several destinations which all have to be put in different ship holds. As a result the scheduling of the cranes is a difficult stochastic problem (handling times of containers are quite variable). The last crane to finish determines the moment when the ship can leave and hence the port time. A good balancing of the workload between the cranes is therefore necessary, but also very difficult to achieve. Another complication comes from the fact that port workers often work in shifts with fixed starting and end times, and a terminal will have to accommodate these restrictions.

11.3.3 *Container stacking*

A final aspect we like to mention is the stacking of the containers in the yard. The issue is not only that containers are stacked on top of each other, which complicates the retrieval of a bottom container, it is also the location on the yard of the containers to be loaded. If a ship moors right before the place where its (to be loaded) containers are located, then travel distances to the quay cranes are short and no bottlenecks are likely to occur. However, if a ship (due to delay or congestion at the berths) berths somewhere else, or if containers are spread out over the yard, then the terminal has to transport the containers over longer distances by which the loading could potentially be delayed. Container stacking is closely related to stowage planning, as the latter determines the order in which containers are to be loaded. In a perfect world one can take the order in which containers are stacked into account while making the load planning, but that creates a very complex problem, which also suffers from the variations in the loading. Hence costly reshuffles, where top containers are placed somewhere else to retrieve bottom containers are needed in large quantities.

11.4 Geographical bottlenecks

Canal restrictions form the main geographical bottlenecks for container ships. The Suez Canal and the Panama Canal are two well-known canals imposing restrictions on container ships. The type of restrictions may differ between different canals. For example, the main restriction imposed by the Suez Canal is the compulsory convoy passage through the canal. This results in long waiting times if a container ship misses the planned convoy. The Panama Canal, on the other hand, imposes limits on the size of ships that want to sail through the canal.

Two other examples of geographical bottlenecks are the Strait of Malacca and the Gulf of Aden. These waterways are narrow, but are strategically important locations for world trade, making them vulnerable to piracy.

Finally, ports may also impose geographical bottlenecks. Large ships might not be able to access certain ports, because the access ways have tight draft restrictions.

11.5 Case study: Indonesia

Shipping is an important mode of transport in Indonesia because the country consists of many islands. We consider six main ports in Indonesia. The six ports are located on five islands of Indonesia; hence transport over land is only possible between Jakarta and Surabaya. Transportation between all other combinations of these cities is only possible by sea or air.

We will use the Indonesian case to illustrate some of the decision problems introduced in Section 11.2. Thereto, we will assume that Table 11.1 gives the expected weekly demand in TEUs between the Indonesian ports. The last row and column give the row and column sums, denoting the total supply from and demand to a port, respectively. The supply and demand values denote the number of containers leaving and arriving in the port, respectively. The difference between demand and supply indicates how many empty containers have to be repositioned from or to the port. For the six ports in Indonesia, the empty container repositioning problem is of limited importance, since there are no large differences between supply and demand.

Table 11.1 shows that Jakarta and Surabaya are the two ports with the largest container throughput, while trade with Sorong is relatively small. In this specific case, this might lead to problems, since Sorong is also located relatively far away from the other ports. Liner shipping companies prefer to offer services calling at the ports of Jakarta and Surabaya and consider it too costly to call at Sorong. By charging higher prices for containers that have to be transported from or to the port of Sorong, liners can make stops at Sorong more attractive. Hence, the liner company may use the pricing strategy to ensure that services calling at Sorong will also be beneficial. However, to determine exactly which prices they have to charge in order to maximise their profit, the liner company needs more details on the cost structure of the network they will provide.

TABLE 11.1 Expected weekly demand in TEU between the Indonesian ports

	Belawan	Jakarta	Surabaya	Banjarmasin	Makassar	Sorong	Supply
Belawan	–	6,500	1,000	100	75	25	**7,700**
Jakarta	6,750	–	2,000	4,000	2,800	450	**16,000**
Surabaya	1,000	2,500	–	3,750	4,800	2,150	**14,200**
Banjarmasin	100	3,600	3,500	–	10	0	**7,210**
Makassar	100	3,500	4,000	75	–	0	**7,675**
Sorong	50	650	2,000	0	0	–	**2,700**
Demand	**8,000**	**16,750**	**12,500**	**7,925**	**7,685**	**2,625**	**55,485**

Figure 11.1 shows examples of a hub-and-feeder system, a circular route, a butterfly route and a pendulum route calling at the six Indonesian ports. In the hub-and-feeder system of Figure 11.1a the port of Surabaya is the hub port, while Belawan, Jakarta, Banjarmasin, Makassar and Sorong are feeder ports. The route Surabaya-Jakarta-Belawan-Surabaya is referred to as F1. F2 is a direct feeder route between Surabaya and Sorong. The third feeder route, F3, calls at Surabaya, Banjarmasin and Makassar, after which it returns to Surabaya. The circular route in Figure 11.1b has as characteristic that each port is called at exactly once during the round tour. Figure 11.1c shows the butterfly route Belawan-Surabaya-Banjarmasin-Makassar-Sorong-Surabaya-Jakarta-Belawan on which Surabaya is visited twice. Finally, in the pendulum route of Figure 11.1d all ports are visited twice, only the second time in reverse order.

Table 11.2 shows the distances in nautical miles (nmi) between the six Indonesian ports and Table 11.3 provides some characteristics of five ship types. Types 1, 2, 3 and 5 are obtained from Brouer et al. (2014), while type 4 is suggested by the Indonesian government and costs are obtained using interpolation. Note that the

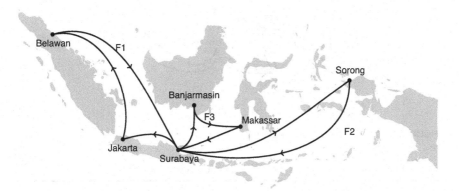

FIGURE 11.1A Example of a hub-and-feeder system

FIGURE 11.1B Example of a circular route

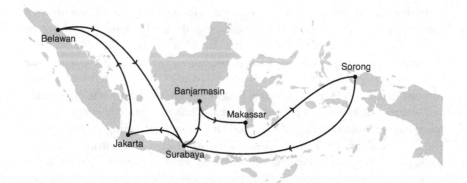

FIGURE 11.1C Example of a butterfly route

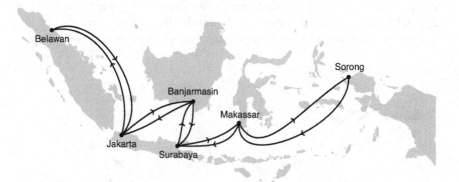

FIGURE 11.1D Example of a pendulum route

TABLE 11.2 Distances between the Indonesian ports in nautical miles

	Belawan	*Jakarta*	*Surabaya*	*Banjarmasin*	*Makassar*	*Sorong*
Belawan	–	1,064	1,488	1,430	1,708	2,807
Jakarta	1,064	–	438	614	806	2,102
Surabaya	1,488	438	–	328	520	1,816
Banjarmasin	1,430	614	328	–	353	1,577
Makassar	1,708	806	520	353	–	1,375
Sorong	2,807	2,102	1,816	1,577	1,375	–

Source: Based on data from www.ports.com/sea-route/.

fuel usage in ton/day of type 4 is larger than the usage of type 5, because type 4 has a higher design speed than type 5. These data can be used to get some insight in the route cost using different ship types and network structures. In the calculations we use a simplified version of the fuel cost function as provided in Brouer et al. (2014):

FORMULA 11.1   Simplified version of the fuel cost function as provided in Brouer et al. (2014)

$$F_s(v) = 600 \left(\frac{v}{v_s^*} \right)^3 fs.$$

Here, $F_s(v)$ denotes the fuel cost in USD per day for a ship of type s sailing at a speed of v knots (nmi/hour); v_s^* and f_s are the design speed and fuel consumption in tons per day of a ship of type s sailing at design speed and can be found in Table 11.3. Note that the bunker cost varies over time, but is assumed to be constant and equal to 600 USD per ton in this study (Brouer et al., 2014). Table 11.4 shows the route distance in nautical miles, the duration in weeks, the frequency, the number of ships required to obtain the frequency and the sailing speed in knots for each route. Distances can be found by adding the distances of the individual sea legs, while the duration and frequency are manually fixed in this example.

In liner shipping it is common to use weekly port calls at a route. Route durations are typically a round number of weeks such that a round number of ships is needed to

TABLE 11.3 Data of the ship characteristics

Ship type	Capacity (TEU)	Charter cost (USD/day)	Draft (m)	Min speed (knots)	Design speed (knots)	Max speed (knots)	Fuel usage (tons/day)
1	900	5,000	8	10	12	14	18.8
2	1,600	8,000	9.5	10	14	17	23.7
3	2,400	11,000	12	12	18	19	52.5
4	3,500	16,000	12	12	18	20	60.0
5	4,800	21,000	11	12	16	22	57.4

Source: Based on Brouer et al. (2014).

TABLE 11.4 Route characteristics for the different ships

Route	Distance (nmi)	Duration (weeks)	Frequency (per week)	Required ships	Speed (knots)
F1	2,990	2	1	2	11.33*
F2	3,632	2	1	2	12.61
F3	1,201	1	1	1	12.51
Circular	6,476	4	1	4	12.27
Butterfly	6,862	4	1	4	13.62
Pendulum	7,802	5	1	5	13.55

Note: An asterisk indicates that the speed is outside the feasible range for some ship types.

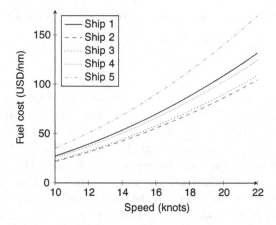

FIGURE 11.2 Fuel cost in USD per nautical mile

sail this route. For example, a route with duration three weeks and which is sailed by three ships ensures a weekly frequency. Given the duration and frequency, the number of required ships can be found by taking the product of these two values.

Figure 11.2 shows the fuel price in USD per nautical mile at different speeds for the five ship types. The fuel price is a convex function, meaning that when the speed is doubled, the fuel cost per nautical mile is more than doubled. Hence, a constant sailing speed during the route will minimise the fuel cost. The speed is calculated under the assumption that every port call takes 24 hours and the durations as given in Table 11.4. The following formula can then be used to determine the speed on each route:

FORMULA 11.2 Determination of the speed on each route

$$v = \frac{\delta}{168t - 24n},$$

where δ is the route distance in nautical miles, t the route duration in weeks and n the number of port calls on the route. An asterisk in the column denoting the speed of Table 11.4 indicates that the speed is outside the feasible speed range for some ship types. The frequency is chosen in such a way that it is feasible for each ship type when sailing at maximum speed. Hence, the necessary speed can only be lower than the minimum speed of the ship type, in which case the ship will sail at minimum speed and will wait in one of the ports to obtain a weekly frequency.

Table 11.5 shows the weekly cost in USD for each of the routes given the frequency and duration as given in Table 11.4. The route costs consist of three components: the fixed ship costs, the port call costs and the fuel costs. When a liner

TABLE 11.5 Route cost per week for the duration and frequency as given in Table 11.4

Route/Ship	Cost in USD/week				
	Type 1	Type 2	Type 3	Type 4	Type 5
F1	176,268	196,765	252,848	336,691	446,793
F2	228,411	328,026	285,297	373,868	497,669
F3	88,076	98,537	121,253	162,296	214,803
Circular	408,876	438,257	531,147	702,468	933,207
Butterfly	490,525	503,210	598,818	779,713	1,038,189
Pendulum	571,488	596,234	714,297	935,318	1,243,643

company needs three ships to satisfy the required route duration and frequency, it bears weekly the fixed ship costs of all these three ships. Hence, the fixed ship cost is given by $7Sc_s^f$, with S the number of required ships and c_s^f the daily fixed ship cost of type s, which can be found in Table 11.3. The port call cost is the sum of the port fees of the ports visited on the route. If we assume that all port fees are the same, the port call cost is given by $F_p nq$, where F_p is the port fee per port visit and q the route frequency. In this example, we assume that $F_p = 650$ USD. The fuel cost is given by the product of the frequency, the number of days that a ship needs to sail one round tour and the fuel cost per day: $q\frac{\delta}{24v}F_s(v)$, where $F_s(v)$ is the fuel cost in USD per day when sailing at speed v as given by equation (11.1). Consider a liner route with a duration of two weeks to which four ships are allocated. Each port on the route will then be called twice a week, resulting in a frequency of twice a week. Each ship needs two full weeks to sail a round tour, so in one week it will sail half of the route. Since there are four ships allocated to the route, in total two full round tours are made during a week (since the frequency is two). This explains the multiplication with the frequency in the fuel and port call cost. The total route cost in USD per week is now given by:

FORMULA 11.3 Total route cost in USD per week

$$c_s' = 7Sc_s^f + q\frac{\delta}{24v}F_s(v) + F_p nq,$$

where c_s' is the route cost in USD per week for a ship of type s. It can easily be seen that doubling the capacity of a ship will not result in a doubling of the weekly route cost. This illustrates the effect of economies of scale: larger ships will in general have higher total costs, but lower costs per TEU, which is also exemplified in Table 11.6 by showing the weekly route cost per TEU under the assumption that the ship is fully utilised. The table also shows that the effect of economies of scale can differ quite a lot between ship types.

TABLE 11.6 Economies of scale in ship size at full utilisation

Route/Ship	Cost per TEU in USD/week				
	Type 1	Type 2	Type 3	Type 4	Type 5
F1	196	123	105	96	93
F2	254	149	119	107	104
F3	98	62	51	46	45
Circular	454	274	221	201	194
Butterfly	545	315	250	223	216
Pendulum	635	373	298	267	259

The disadvantage of the circular route is that the capacity cannot be utilised efficiently. When containers from, for example, Surabaya to Jakarta are transported using the circular route, they will be on board of the ship on all sea legs except the leg from Jakarta to Surabaya. Butterfly routes are better able to utilise the available capacity, since some ports are visited twice on a round tour. In the butterfly route, the ports of Surabaya and Jakarta are visited directly after each other, such that the containers are only on board during one sea leg of the route. The pendulum route visits all ports twice, hence it needs the lowest capacity. In a hub-and-feeder network, usually many ports are connected by only one or a few sea legs. This ensures that hub-and-feeder networks are able to utilise the available capacity very efficiently.

Figure 11.3 shows the utilised capacity at each sea leg in the four different route networks under the assumption that all demand has to be satisfied using only the given network. For the butterfly route, we assumed that the containers that have to be transported from Makassar to Banjarmasin will stay on board of the ship during the route segment Surabaya-Jakarta-Belawan-Jakarta-Surabaya. Alternatively, these containers can be unloaded during the first call at Surabaya and loaded again during

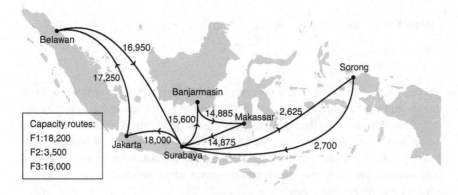

FIGURE 11.3A Utilised capacities in TEU for the hub-and-feeder system

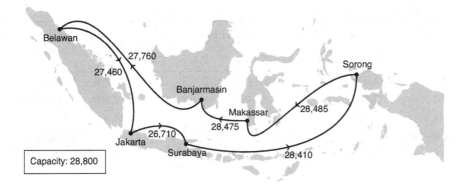

FIGURE 11.3B   Utilised capacities in TEU for the circular route

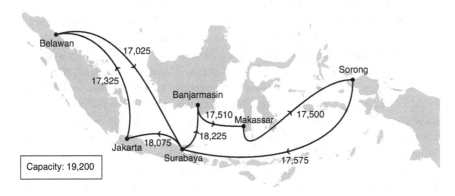

FIGURE 11.3C   Utilised capacities in TEU for the butterfly route

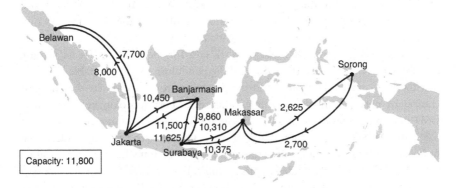

FIGURE 11.3D   Utilised capacities in TEU for the pendulum route

TABLE 11.7 Network cost per week when shipping all demand

Route	Req. cap. (TEU)	Port calls per week					Av. cap. (TEU)	Cost (USD/week)
		Type 1	Type 2	Type 3	Type 4	Type 5		
F1	18,000	0	1	0	2	2	18,200	1,763,733
F2	2,700	0	0	0	1	0	3,500	373,868
F3	15,600	0	1	0	0	3	16,000	742,948
HF–Total								**2,880,549**
Circular	28,485	0	0	0	0	6	28,800	**5,599,239**
Butterfly	18,225	0	0	0	0	4	19,200	**4,152,755**
Pendulum	11,625	0	0	0	2	1	11,800	**3,114,280**

the second port call at Surabaya, in which case transshipment costs at Surabaya are incurred. The utilised capacities are found by adding all container flows that need to traverse the given sea leg in order to arrive at their destination. Table 11.7 shows the required capacity in TEU for each route, the number of port calls per week for each ship type in order to have enough capacity to satisfy all demand, the available capacity in TEU using these ship types, and the total route costs in USD per week. The required capacity is found by taking the maximum utilised capacity of the route. Next, we make a combination of ship types such that enough capacity is available at each route. Given these ship allocations, the total route cost can be found by multiplying the weekly route cost for a ship type by the number of port calls per week divided by the route frequency. Note that the type and number of ships needed vary a lot between the three different route structures. For the hub-and-feeder system, $2 \times 1 + 1 \times 1 = 3$ ships of type 2 (since feeder route 1 has a duration of two weeks and feeder route 3 has a duration of one week), $2 \times 2 + 2 \times 1 = 6$ ships of type 4 and $2 \times 2 + 1 \times 3 = 7$ ships of type 5 are needed. The circular route uses $4 \times 6 = 24$ ships of type 5, while the butterfly route uses $4 \times 4 = 16$ ships of type 5. Finally, the pendulum route uses $5 \times 2 = 10$ ships of type 4 and $5 \times 1 = 5$ ships of type 5. Hence, the optimal solution to the fleet size and mix problem is highly dependent on the network structure.

Table 11.7 also shows the total network cost for the hub–and–feeder system, the circular route, the butterfly route and the pendulum route. The table indicates that the hub–and–feeder system and the pendulum route are by far the cheapest choices of networks in this example. They both cost approximately 3 million USD per week, while the circular and butterfly routes cost respectively about 5.5 million USD and 4 million USD per week. One remark has to be made: in the hub–and–feeder system, a lot of containers need to be transshipped, adding additional costs that are not included in this example. In total 15,450 containers have to be transshipped per week in the hub–and–feeder system. If a transshipment costs, for example, 100 USD per container, the total cost of the hub–and–feeder system will rise to almost 4.5 million USD per week. Hence, the hub-and-feeder system will then have higher costs than the butterfly and pendulum routes. Of course, one could also make route networks with combinations of these routes, which might be more cost-efficient.

The good performance of the hub-and-feeder system and pendulum route is (partly) caused because of the better utilisation of capacity in the hub-and-feeder system. Another advantage of hub-and-feeder systems is that liners can allocate different ship types to the different types of routes. Feeder ports usually have less demand than hub ports, hence it makes sense to allocate smaller ships to the feeder routes than to the main routes. If all ports are visited on similar routes, like circular, butterfly and pendulum routes, all these ports are visited by the same ship type. Hence, large ships might visit very low demand ports if these ships are able to berth in the smaller ports (smaller ports might have stricter draft restrictions than hub ports). Otherwise many small ships are needed in order to satisfy the demand of the large ports. However, a disadvantage of hub-and-feeder networks is that usually many transshipments are needed in order to satisfy the demand, which increases both the transportation price and transit time. In airline passenger transport, hub-and-feeder systems are very popular; an important reason for this is that transshipments are made by passengers at no apparent cost.

Finally, we determine the profit and efficiency of the networks when we assume that each container will generate a revenue of 200 USD if it is transported, (un)loading and transshipment costs are all 40 USD per container. Recently, the problem has also been studied on request of the Indonesian government, resulting in a single pendulum route to be sailed. This pendulum route is also known under the name Pendulum Nusantara. We use the mixed integer programming model of Mulder and Dekker (2016) to determine the optimal route network given an initial set of routes. Routes are constructed in the following way using the ordering of ports used for the pendulum route. Ports may be visited at most twice during a route: once on the eastbound trip and once on the westbound trip. All feasible routes are generated and given as input to the mixed integer programming problem, which makes use of a path formulation to solve the cargo allocation.

Table 11.8 shows the profits of these networks and Figure 11.4 shows the optimal route network. We see that the pendulum route performs indeed better than

TABLE 11.8 Efficiency and profit of the different networks

Network	Shipped distance (nmi/TEU)	Profit (USD)
Hub-and-feeder	1,428.06	3,159,651
Circular	3,269.79	1,058,961
Butterfly	2,227.57	2,284,445
Pendulum	996.80	3,642,916
Optimal	852.66	4,897,109

FIGURE 11.4A Optimal route network: route 1

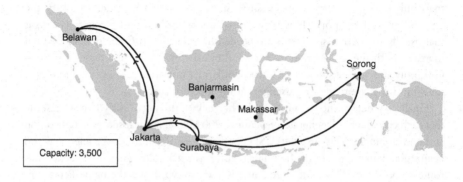

FIGURE 11.4B Optimal route network: route 2

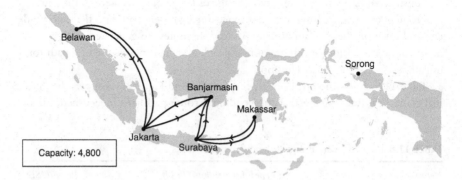

FIGURE 11.4C Optimal route network: route 3

the hub-and-feeder network and the butterfly and circular routes. The efficiency of the networks is measured by the shipped distance in nmi per TEU. The shipped distance for direct shipping is equal to 836.57 nmi/TEU. Table 11.8 shows that the pendulum and optimal networks are both efficient networks with respect to shipped distance. Furthermore, the hub-and-feeder network is much more efficient than the circular and butterfly routes as expected.

The optimal route network as shown in Figure 11.4 consists of two pendulum routes (routes 2 and 3) and a non-stop service (route 1), which is a special type of pendulum route with only two port calls. The pendulum route structure ensures efficient transportation between all demand pairs. All routes have a frequency of once a week, reducing the number of required ships.

11.6   Conclusion

Maritime transportation is very important in the world economy. The types of operations in the shipping market are distinguished in tramp, industrial and liner shipping. This chapter considers decision problems that occur in the operations of container liner shipping companies. The decision problems can be distinguished in three different planning levels: strategic, tactical and operational. The strategic planning level consists of long-term decision problems, while the tactical and operational planning levels contain medium and short-term problems, respectively.

This chapter discusses the fleet size and mix and market trade selection problem on the strategic level; network design, pricing and empty container repositioning on the tactical level; and cargo routing, disruption management, revenue management and stowage planning on the operational planning level. Furthermore, sailing speed and bunkering optimisation and robust schedule design are covered. These decision problems are related to both the tactical and the operational problem. The chapter also introduces three decision problems in the terminal operations that are related to container liner shipping: berth scheduling, crane allocation and container stacking.

A case study is used to explain the concepts of the problems in more detail. The case study is based on six main ports in Indonesia. Four networks with different structures (hub-and-feeder system, circular, pendulum route and butterfly route) are proposed. Calculations show that the costs of the hub-and-feeder system and the pendulum route are much smaller than the cost of the other two networks when transshipment costs are not considered. Explanations for this result are that capacity is better utilised in hub-and-feeder systems compared to the other network structures and ships can be chosen more freely, since more shorter routes are used. When transshipments are charged at 100 USD per container, the hub-and-feeder system performs comparable to the butterfly route and still considerably better than the circular route.

11.7   Expected future developments

The shipping industry is lagging behind compared to the airline and road industry with respect to the implementation of operations research methods. The promising results from the airline and road industry will probably lead to the introduction of more decision support systems for optimisation in the shipping industry. Also, optimisation is required to solve the large inefficiencies in transport caused by imbalances and variability in demand. In the last two decades, a lot of progress has been made in developing operations research methods to solve important decision problems in the shipping industry, and this is expected to continue in the future.

References

Brouer, B. D., Álvarez, J. F., Plum, C.E.M., Pisinger, D., and Sigurd, M. M. (2014) A base integer programming model and benchmark suite for liner-shipping network design. *Transportation Science*, 48(2), 281–312.

Brouer, B. D., Dirksen, J., Pisinger, D., Plum, C.E.M., and Vaaben, B. (2013) The Vessel Schedule Recovery Problem (VSRP) – A MIP model for handling disruptions in liner shipping. *European Journal of Operational Research*, 224(2), 362–374.

Cullinane, K., and Khanna, M. (1999) Economies of scale in large container ships. *Journal of Transport Economics and Policy*, 33(2), 185–207.

Karsten, C. V., Pisinger, D., Ropke, S., and Brouer, B. D. (2015) The time constrained multi-commodity network flow problem and its application to liner shipping network design. *Transportation Research Part E*, 76, 122–138.

Lawrence, S. A. (1972) *International Sea Transport: The Years Ahead*. Lexington, MA, Lexington Books, ISBN 0-66-984319-9.

Meng, Q., Wang, S., Andersson, H., and Thun, K. (2014) Containership routing and scheduling in liner shipping: Overview and future research directions. *Transportation Science*, 48(2), 265–280.

Mulder, J., and Dekker, R. (2016) *Will Liner Ships Make Fewer Port Calls Per Route?* Econometric Institute Report EI2016-04, Erasmus University Rotterdam.

Pantuso, G., Fagerholt, K., and Hvattum, L. M. (2014) A survey on maritime fleet size and mix problems. *European Journal of Operational Research*, 235(2), 341–349.

Pedrielli, G., Lee, L. H., and Ng, S. H. (2015) Optimal bunkering contract in a buyer-seller supply chain under price and consumption uncertainty. *Transportation Research Part E*, 77, 77–94.

Tierney, K., Pacino, D., and Jensen, R. M. (2014) On the complexity of container stowage planning problems. *Discrete Applied Mathematics*, 169, 225–230.

Yao, Z., Ng, S. H., and Lee, L. H. (2012) A study on bunker fuel management for the shipping liner services. *Computers & Operations Research*, 39(5), 1160–1172.

Suggestions for further reading

In 1983, Ronen provided the first overview paper on the contribution of operations research methods in ship routing and scheduling. Since this first paper, every decade a follow-up overview paper appeared, reviewing new research conducted in that decade (Ronen, 1993, Christiansen et al., 2004, 2013). Initially, the reviews were mainly focused on the ship routing and scheduling problem, but more and more other shipping problems are included in these reviews. Furthermore, Christiansen et al. (2007) provide an extensive overview of maritime transportation problems. Finally, Meng et al. (2014) give an overview of research related to container routing and scheduling in the liner shipping industry in the last 30 years.

Christiansen, M., Fagerholt, K., Nygreen, B., and Ronen, D. (2007) Maritime transportation. In: Barnhart, C. and Laporte, G. (eds.) *Handbook in OR & MS*, vol. 14. Amsterdam, Elsevier, 189–284.

Christiansen, M., Fagerholt, K. Nygreen, B., and Ronen, D. (2013) Ship routing and scheduling in the new millennium. *European Journal of Operational Research*, 228(3), 467–483.

Christiansen, M., Fagerholt, K., and Ronen, D. (2004) Ship routing and scheduling: Status and perspectives. *Transportation Science*, 38(1), 1–18.

Ronen, D. (1983) Cargo ships routing and scheduling: Survey of models and problems. *European Journal of Operational Research*, 12(2), 119–126.

Ronen, D. (1993) Ship scheduling: The last decade. *European Journal of Operational Research*, 71(3), 325–333.

12

REVENUES AND COSTS OF MARITIME SHIPPING

Albert Veenstra

12.1 Introduction

The international ocean transport industry carries the majority of international trade. The reason for this is that, per ton of cargo, shipping is by far the cheapest transport option. Nevertheless, also in ocean transport, there still is considerable pressure on cost levels, and different parts of the shipping industry have seen a move towards lower costs per ton or unit of cargo. This was often accomplished by building larger and larger ships.

A ship is an expensive asset. While secondhand vessels may be procured for perhaps $10 million, large, advanced ships can easily cost $300 million or more. The economic life of vessels is also long, 20–25 years, which makes buying a vessel a relatively long-term decision. A further complication is that the environment in which ships operate is global, and highly uncertain. The economic situation in the shipping industry fluctuates much more wildly than the world economy. This means that the value of ships can also fluctuate. As a result, there are two main ways of making money with a ship: by operating it, or by buying and selling it when the price is right. This also creates pressure on costs, since this uncertainty can often only be dealt with through significant proactive savings on costs.

In markets where larger ships are not or are no longer an option, the pressure on costs found other avenues to achieve a reduced reduction in the cost per ton of cargo. This is the topic of this chapter. In this chapter, we will discuss the structure of costs in the shipping industry, and the various ways in which the shipping industry has been able to attain cost savings. The global nature of ocean shipping plays an important role in this. The global environment in which shipping operates is characterised by a multinational legal framework that provides many opportunities to cherry-pick regulations around the world and save costs as a result.

Given the importance of the management of costs in shipping, there are also negative consequences. These are the result of extreme cost savings, resulting in ships that are poorly maintained, crewed with poorly educated people, and paying hardly any tax. Apart from the obvious risks of these ships breaking up and causing pollution, they are often also vulnerable to criminal abuse. These consequences will also be discussed in this chapter.

In this chapter, we will discuss the importance of costs in shipping. First we will introduce the cost structures in shipping. We will then elaborate on revenue in shipping, different contract types and the way costs are distributed in these contracts. We then discuss the role of shipping in international trade, and the international regulatory framework for shipping. We will then provide some insight into the way the shipping community uses its international footprint to optimise costs. This behaviour has major repercussions for the way shipping functions today.

12.2   Cost structures

The structure of costs in shipping is basic material in any introductory book in maritime economics. We refer the reader to Stopford (2009), chapters 2 and 6 for an extensive discussion.

The analysis of costs starts from the assumption that the ship is the main object of study. This idea goes back to Zannetos (1966): "the ship is the firm." It means in essence that we will discuss the cost structure of ships, and not of shipping companies.

It is generally accepted that a ship operation has three important cost categories:

1   Capital costs
2   Operating costs
3   Voyage costs.

There is another important cost component in maritime operations, cargo handling costs, but the shipowner usually does not pay for these costs. They will therefore be ignored in the remainder of this chapter.

12.2.1   Capital costs

Capital costs are the result of the purchase of the ship. In most cases, ships are bought with a bank loan of some sort, and a limited amount of capital from the ship owner. The most common type of bank loan in shipping is a ship mortgage. As a result, the ship owner pays interest on the loan, and these costs are capital costs.

The main determinants for capital costs are obviously the size of the loan (which depends on the capital position of the owner, the value and earnings potential of the ship and market conditions), the duration of the loan, the currency of the loan, the interest rate and several other elements that are related to the specific form of the mortgage. For instance, it is often the case that the shipowner can defer the loan repayments until the time he resells the ship. This is called a balloon payment.

There are also other financing schemes, such as corporate loans or investment arrangements where parties take part in the capital of the shipping company. In some of these cases, other types of capital costs such as dividends may become applicable.

12.2.2 Operating costs

Operating costs are the costs resulting from the operational readiness of the ship. These costs include:

- Crew costs
- Stores
- Spare parts and maintenance
- Insurance
- Administration.

Crew costs are all the costs related to the personnel on board. These costs are directly determined by the composition and number of crew, and the salary level. These conditions, both the prescribed crew composition and the salary level, can differ from country to country. There is an international standard for shipping personnel, a safe level of crewing, minimum wages and so on. It depends strongly on the flag state if these rules are enforced.

Stores, spare parts and maintenance costs determine to a large extent the quality of the shipping operation. The stores refer to food, water and other requirements of the crew on board. Spare parts and maintenance costs refer to the technical state of the ship. Attempts to save on these costs will quickly result in a deterioration of the ship and the performance of its crew.

Finally, insurance and administration costs refer to the business side of a shipping operation. Proper record keeping and a sufficient level of insurance is required for ships to be engaged in transport contracts and sail in certain waters. A ship has to keep records of its crew's nationality, of waste storage and tank cleaning activities at sea, of maintenance activities performed as a result of inspections, and so on. Ship insurance consists of two parts: hull and machinery insurance against damages, and so-called property and indemnity (P&I) insurance. The latter is a legal liability insurance for the owner of the ship. In the maritime world, this type of insurance is usually offered by cooperatives of shipowners, the so-called P&I clubs.

12.2.3 Voyage costs

If capital costs and operating costs are covered, the ship still has not sailed a single sea-mile. The costs related to actual sailing are called voyage costs. These costs are mainly fuel costs, together with several other items: port-related costs and canal dues.

Fuel costs are almost completely determined by the speed of the ship. There is some differentiation between speeds of different ships, but on the whole, the range of speeds in shipping is limited. The fuel consumption is related to speed as follows:

FORMULA 12.1   Fuel consumption related to speed

$$consumption \propto speed^3$$

Given this third power in the equation, changes in speed lead to a disproportional change in fuel consumption. This means that slow steaming of ships can result in substantial savings in the shipowner's fuel bill. In times of crisis, this is a favoured option for shipping companies to save costs and remain solvent.

Port costs for ships consist of the charges for nautical maritime services, such as pilotage, towage and mooring. These are costs that the shipowner usually pays himself. In addition, while sailing, the ship may have to traverse a canal, such as the Suez Canal, the Panama Canal or the Kieler Canal. The authorities of these canals usually charge fees that are based on the type of ship and the size of the ship. Both the Suez and the Panama authorities have their own certified measurement procedure for the size of the ship that is the basis of their charging system.

12.3   Revenue in shipping

Shipowners rent out their ships to parties that have cargo to transport. This process is called chartering. The party who needs transport is called a charterer. The contract that is eventually drawn up is called a charter party (see, for instance, Gorton et al., 2009).

In practice, different types of charter arrangements have developed over the years. These charter arrangements differ in terms of the division of responsibilities of charterer and shipowner, and consequently in terms of the costs that are paid by both parties. The different type of charter contracts are stated in Table 12.1.

TABLE 12.1 Overview of charter contract types and division of costs

	Capital costs	Operating costs	Voyage costs	Price
	Paid by the shipowner			
Voyage charter	X	X	X	$/ton
Time charter	X	X		$/day
Trip time charter	X	X	X	$/day
Bareboat charter	X			$/day

Source: Stopford (2009: 182).

A voyage contract is a contract for a specific trip, from port A to port B, and for a specific volume of cargo. The price is paid in dollars per ton of cargo, and the shipowner pays generally for all the costs. A variant of this type of contract is a trip time charter, which is also a contract for a specific trip, but paid in dollars per day. This could be a choice if the sailing time is uncertain.

A time charter is a contract for a specific period of time, in which the charter can have the ship make several consecutive trips. Usually the contract specifies the general sailing area, and/or the area in which the ship will be at the end of the contract. The price is also charged in dollars per day, and the shipowner does not pay for the voyage costs, since he does not know how far the ship will sail. The contract will usually specify the sailing speed and the fuel consumption of the ship, as a guarantee for the charterer. Charter contracts may also contain complicated clauses to calculate the compensation for the charterer if the ship does not perform according to the specified speed and fuel consumption. Veenstra and Van Dalen (2011) show that this leads to strategic behaviour of shipowners underquoting speed and over-quoting fuel consumption in time charter contracts.

A bareboat charter contract is a long term contract, in which the owner relinquishes control of the ship, including its crew, to a charterer. The shipowner then only pays the capital costs. The charterer may even choose to reflag the ship.

12.4 Chartering practice

In many shipping companies, the fleet of ships is operated under a mix of these contract types. Clever chartering includes the best choice for a contract type given the market circumstances and the current position of the ship. Rising or declining fuel costs, as well as market prospects several months ahead, play an important role in the choice for a voyage charter or a time charter contract.

The basis for these decisions is the average daily running costs of the ship and average daily earnings. Calculating the potential profit on the activities of a ship is called the voyage cash flow or simply voyage calculation:

FORMULA 12.2 Cash flow

Total cash flow = freight earnings − voyage costs − operating costs

FORMULA 12.3 Average daily income

Average daily income = total cash flow/sailing days

Freight earnings are either the freight rate in $/ton times cargo tons, or $/day times the duration of the contract. The average daily income is the coverage the shipowner receives for capital costs and repayments of the loan.

A related calculation shipowners make is the comparison of a voyage charter rate with a time charter rate. This calculation is performed as follows:

FORMULA 12.4 Comparison of a voyage charter rate with a time charter rate

Time charter equivalent rate = (voyage charter rate × cargo volume − voyage costs)/ duration of the contract

If the current time charter rate in the market is higher than the time charter equivalent value of the voyage charter rate, it is better to go for a time charter contract. There is considerable research on the relationship between voyage charter rates and time charter rates (see, for instance, Veenstra (1999) and Kavussanos and Alizadeh (2002)). Much of this research tries to find evidence for a long run relationship between voyage and time charter rates. The evidence is weakly positive that this relationship, called the term structure of charter rates, exists.

12.5   The role of shipping in international trade

Ocean shipping is the main carrier of international trade. The International Chamber of Shipping, for instance, mentions a figure of 90% of the total merchant trade (in tonnes) being carried by ship. The part of trade that is carried by sea is referred to as seaborne trade. An overview of the development of seaborne trade, spread over its main categories, is depicted in Table 12.2.

Note that the total volume of seaborne trade is about 9,000 million tons, roughly evenly spread over oil/gas, main bulks and other dry cargo. The *main bulks* are a specific category of bulk commodities that includes iron ore, coal, grain, bauxite and phosphate rock. All other dry cargo includes as main category containerised cargo.

TABLE 12.2 Seaborne trade 1970–2012 (million tons)

Year	Oil and Gas	Main Bulks	Other Dry Cargo	Total
1970	1,440	448	717	2,605
1980	1,871	608	1,225	3,704
1990	1,755	988	1,265	4,008
2000	2,163	1,295	2,526	5,984
2005	2,422	1,709	2,978	7,109
2006	2,698	1,814	3,188	7,700
2007	2,747	1,953	3,334	8,034
2008	2,472	2,065	3,422	8,229
2009	2,642	2,085	3,131	7,858
2010	2,772	2,335	3,302	8,409
2011	2,794	2,486	3,505	8,784
2012	2,836	2,665	3,664	9,165

Source: Based on UNCTAD Maritime Transport Review 2013.

The relationship between shipping and trade has sprouted a separate part of maritime economics that focuses on the availability of the relevant infrastructure for countries to take part in trade. This infrastructure is very maritime oriented: ports, access channels, deep water, and, for a while at least, it also includes shipping companies and ships. The now forgotten UNCTAD Code of Conduct for Liner Conferences from 1975 was an agreement that enabled developing countries to maintain a shipping industry for the carriage of their own trade.

In a world in which ports often already exist, but have tended to be unsuitable for the demands of modern international trade, this same area of maritime economics reverted to the reform of ports and maritime regulatory framework in countries to facilitate the introduction of foreign, corporate investors and commercial shipping and ports operations. These corporate investors also came with an international clientele that required improvements in operational performance: quicker turnaround times for ships and other modes of transport and lower costs of handling cargo. All this resulted in an intense pressure on costs of shipping.

12.6 The regulatory framework for shipping

Ships sail the seven seas, and they have done so since time immemorial. In the fifteenth and sixteenth centuries, when the entire earth came to be divided into countries by explorers, eventually the question of the ownership of the sea became inevitable. A major innovation, introduced by Dutchman Hugo de Groot, was the concept of Mare Liberum. In other words: the high seas are not owned by anyone. Parts of the sea, which directly border a country can be claimed as national territory, but beyond a small distance of several nautical miles, ships are in the high seas.

This decision had two important repercussions. One was that ships still need a link with a country, to signal ownership, to signal nationality when the ship arrives in a port or at a beach, to fall back on protection by a country's navy and to establish some rule of law on the ship (see also Stopford (2009) for a discussion on this topic).

The second consequence was that countries started to claim parts of the seas as their national territory. The rules were established in a United Nations convention called the UN Convention on the Law of the Sea (UNCLOS). In this convention, the boundaries for national territory at sea were established: the 12-mile zone, and the exclusive economic zones of countries. This does not mean that there are no disputes about national territory at sea. There are many, for instance around the Spratley Islands in the Chinese Sea, and the island groups between Japan, Korea and China.

In this chapter, we will focus on the solution for establishing the nationality of a ship. Since the sixteenth century, this has been done by putting a national flag on the ship, and more recently, by entering the ship in a national registration of ships. Many countries have these so-called ship registers, and often they are associated with a maritime tradition and national pride.

The registration of the ship means that a ship is made part of the national territory of that country, and that all relevant laws and regulations apply to that ship, wherever it is. This holds for tax laws, laws for the crew of the ship, and laws pertaining to the adoption of international conventions that were ratified and entered into national law by that country and the right to national protection (Stopford, 2009: 666). In other words, the flag establishes a legal framework that applies to everyone and everything on that ship.

There are two other components that make up the legal framework for shipping. The first one is the set of technical rules for the ship. These rules are governed by independent bodies, called classification societies. Examples include the Lloyd's Register, Bureau Veritas and the American Bureau of Shipping. They play an important role in establishing the type of ship and its formal size; it includes the cargo carrying capacity, and other technical descriptions of the ship that are entered in a worldwide shipping register, maintained by Lloyd's Register (and published by IHS, formerly Lloyd's Register-Fairplay). The type and size of the ship needs to be independently established because they are the basis for many charging mechanisms in ports, waterways and canals.

The second is the regulation of so-called port states. While ships are governed by their own nationality through registration and the flag, they often sail in areas that belong to other countries. The countries that border an ocean or sea found that they sometimes had difficulty disciplining ships that sailed into their ports. Ships that were found to cause damage or pollution, or that were simply unseaworthy could not be detained; they might simply sail away, and sail into the next port without any repair or repercussion. Port state control awarded a coastal state the right to board ships with a foreign flag, and to verify the level of compliance of that ship to international standards and conventions. In addition, neighbouring port states entered into agreements to share information on these inspections with each other, so that ships who sailed away could be detained at the next port and could be inspected again to see whether the necessary corrective action(s) had taken place. See for a discussion, for instance, Hare (1996). The ports around the Atlantic Ocean are part of the so-called Paris MOU (www.parismou.org) for port state control.

This international regulatory framework is elaborate and complicated, and, at the same time, affords shipping companies possibilities to look for opportunities to minimise the regulations that apply to their ships. One of the most commonly used possibility is the choice of a so-called flag of convenience.

12.7 Flags of convenience

Given the continuous pressure of the costs of shipping, and the international nature of shipping, in the 1950s shipowners started to realise that other flag states existed that could perhaps impose fewer rules and regulations, and consequently offer possibilities to save costs. See the autobiography of Norwegian shipowner Erling Naess (1977) for an account of the debate that ensued when "flagging out" ships became an established practice. The main alternative flag states used in those early days were Panama, Liberia and Honduras. The first two are still important flag states today.

TABLE 12.3 Overview of flag state performance

Flag states	SOLAS	MARPOL	LL66	STCW78	ILO MLC	CLC/fund92
Antigua/ Barbuda	√	√	√	√	√	√
Bahamas	√		√	√	√	√
Cambodia	√		√	√		√
Honduras	√		√	√		
Liberia	√	√	√	√	√	√
Malta	√	√	√	√	√	√
Mongolia	√	√	√	√		
Panama	√	√	√	√	√	√
Sri Lanka				√		√
Thailand				√		

Source: ICS Flag State Performance Table 2013, selected countries.

Note: √ means the flag state ratified convention.

SOLAS = safety of life at sea; MARPOL = Maritime Pollution Convention, including annexes; LL66 = Load Lines conventions; ILO MLC = International Labour Convention; CLC/Fund92 = the International Convention for Civil Liability for Oil Pollution.

The impact of choosing the "right" flag can be shown with the ICS[1] flag state performance table, that records annually which conventions the flag states in the world ratified. A sample of that table is reproduced in Table 12.3.

Note that some unexpected countries appear in Table 12.3: Mongolia, for instance, is a landlocked country. Many of these countries have a register of ships, because it brings in foreign currency, not because they are so involved in ocean shipping or have a maritime tradition. On the other hand, the classic flags of convenience, such as Panama and Malta, have ratified all the important conventions nowadays. This still does not mean that they impose the same rules as classic maritime nations.

Table 12.4 lists some maritime nations, and the number of ships under national and foreign flags. Note that in most cases, the share of foreign flags is higher.

The US Maritime Administration (MARAD) regularly looks into the cost differences between having a US flag and a foreign flag. The United States is decidedly one of the most protectionist maritime countries, with its 1920 Jones Act and its follow-up, the US Code). This study shows the cost differences for several standard ship types as follows (Table 12.5).

Observe that costs can be reduced by more than half if an American shipowner should choose a foreign flag. The Jones Act prevented this in many cases, since the ship would not be allowed to operate within American coastal waters anymore. For many other shipowners, who operate in cross trades anyway, this is not a limitation, and they choose to register their ships in foreign registers.

TABLE 12.4 Selected maritime nations and their flags

Maritime country	Number of vessels			Vessels deadweight (tons)			
	National flag	Foreign flag	Total	National flag	Foreign flag	Total	Foreign flag as a % of total
Greece	738	2,583	3,321	64,921,486	159,130,395	224,051,881	71.02
Japan	717	3,243	3,960	20,452,832	197,210,070	217,662,902	90.60
Germany	422	3,567	3,989	17,296,198	108,330,510	125,626,708	86.23
China	2,060	1,569	3,629	51,716,318	72,285,422	124,001,740	58.29
Korea, Republic of	740	496	1,236	17,102,300	39,083,270	56,185,570	69.56
United States	741	1,314	2,055	7,162,685	47,460,048	54,622,733	86.89
China, Hong Kong SAR	470	383	853	28,884,470	16,601,518	45,485,988	36.50
Norway	851	1,141	1,992	15,772,288	27,327,579	43,099,867	63.41
Singapore	712	398	1,110	22,082,648	16,480,079	38,562,727	42.74
Bermuda	17	251	268	2,297,441	27,698,605	29,996,046	92.34
Russian Federation	1,336	451	1,787	5,410,608	14,957,599	20,368,207	73.44
United Kingdom	230	480	710	2,034,570	16,395,185	18,429,755	88.96
Netherlands	576	386	962	4,901,301	6,799,943	11,701,244	58.11
Indonesia	951	91	1,042	9,300,711	2,292,255	11,592,966	19.77
Switzerland	39	142	181	1,189,376	3,700,886	4,890,262	75.68

Source: Adapted from UNCTAD Review of Maritime Transport, 2013.

TABLE 12.5 Cost differentials

Ship type	Percentage difference foreign against US flag
Container ship	45%
Ro/Ro	31%
Bulk carrier	33%
Average	37%

Source: Cost differentials based on average daily operating costs, foreign and US flag (MARAD, 2011: 4).

Another interesting observation made in the MARAD (2011: 5) study is the change in cost structure as a result of flagging out. Under a US flag, crewing accounts for 68%, while under a foreign flag, this is only 35%. This means that the importance of the other cost components will increase more or less proportionally.

In The Netherlands, in the early 1990s, a similar analysis was made for the Dutch ship register. One of the main issues of many shipowners at that time was that the Shipping Inspectorate enforced crewing guidelines on ships under a Dutch flag that led to larger crews than under some other flags. This obviously had a positive effect on the operating costs, and destroyed the competitiveness of Dutch ships in the international transport market (see the brochure Maritime Turning Point of the Royal Dutch Shipowners' Association, 1993).

12.8 Slow steaming

While choosing the right flag saves operating costs, slow steaming can save considerably on voyage costs. Slow steaming is exactly what is says: sailing at a slow pace. This has been a proven tactic for shipowners for many years when the market slowed down. Vessels would slow down, and take longer to carry out their contracts. This would mean that they bought time in the market instead of having to wait for a new contract empty. It was so common that the magazine *Lloyd's Shipping Economist* would report the part of the fleet that was slow steaming for a number of ship types.

The reason slow steaming is so effective has already been shown in equation (12.1). As fuel costs are proportional to the third power of speed, slow steaming disproportionally lowers fuel consumption and therefore fuel costs.

While slow steaming was a trick used in tanker and dry bulk shipping, in recent years it has also been adopted in container shipping. In container shipping it is also much more effective, since the sailing speeds are markedly higher than in bulk shipping. Marine Insight (2012) offers a discussion on the impact of slow steaming. First of all, they note that slow steaming is not beneficial for the ships. Ships are designed for an efficient speed. Sailing below that speed may cause damage to the engine.

On the other hand, Maersk (2009) states that by slow steaming from 24 knots (Kn) to 20 Kn, they saved 14% fuel and CO_2 per vessel. Later, Maersk performed tests to run engines at engine loads as low as 10%, without damaging them. They refer to this as super slow steaming. The resulting room to vary speed of all their ships offered a fuel and CO_2 saving of 30%.

While slow steaming offers considerable cost savings for shipowners, for shippers it is not such a positive development. Streng (2012) shows shippers may benefit from greater scheduling reliability if ships slow steam only slightly. If ships slow down too much, shippers' losses are higher than the shipowners' gains (see also Hailey [2013]).

12.9 Optimising ownership structures

We have established that choosing the "right" flag in shipping can make a large difference in the operating cost structures. This process of choosing the nationality of the ship has developed into a larger mechanism, where shipowners are also looking for the "right" crew, the "right" technical or commercial management company, the "right" leasing company or even the "right" charterer for the ship. The international environment in which shipping operates provides many choices and opportunities to save costs.

While this is beneficial for shipowners, it also has a negative side. International ocean shipping is regularly associated, in the popular press, with calamities, such as groundings of ships, ships being lost at sea, oil spills, and, more recently, increasingly with piracy. Less well known is that there is also a grey shipping industry, with ships being used for illegal trades,[2] ghost ships that steal cargo by posing as legitimate operators and then vanish, and ships that run blockades to countries that are sanctioned, such as Iran.[3] In many cases, tricks pulled by these ships involve quick changes of ship names, owners and flags, as well as masking the identity of the true owners.

OECD (2003) have identified the possibilities of how shipowners can hide or mask their identify. They find that it is very easy to do so, and that the mechanisms used are mostly based on common company registration rules, where directors, owners and shareholders can be represented by nominees or intermediaries.

In this section, we will elaborate a bit more on the way shipowners construct the ownership and control structure of their ships. Veenstra and Bergantino (2000) suggest that there should be a distinction in ownership, management and registration of a ship. Each of these elements could be either home based (in their example Dutch) or foreign. They use this ownership structure definition to identify dynamics in the Dutch fleet as a result of the policy changes around the mid-1990s in The Netherlands. But in fact ownership structures of ships can be much more complicated than this. IHS Fairplay[4] maintains seven roles related to vessel ownership and management: registered owner, ship manager, technical manager, commercial operator, bareboat/demise charter, group beneficial owner and document of compliance company. The first four are fairly straightforward. A bareboat/demise charter is the charterer under a bareboat or demise charter contract. The group beneficial owner is the parent company of the registered owner. This field can also be used for the bank (the so-called disponent owner) if the bank has foreclosed on the mortgage of the ship. Finally, the document of compliance company is a company that takes on all technical responsibilities under the International Safety Management (ISM) code.

A typical ownership structure of a shipping company is demonstrated in the left-hand side of Figure 12.1. A ship is commonly owned by a so-called single-ship company. This is sometimes a requirement of the register: a ship should be owned by a company in the country of registration. Another reason is the detention right of parties against the owner of the ship. These parties may detain any ship of the same owner. It is then safer to let the direct owner just own one ship. A number of single-ship companies are then owned by a holding company that may be owned by the eventual beneficial owner. This last party is the actual owner of the ship, and the party that controls the operations of the ship. Tracing beneficial ownership in shipping is not easy. Commercial data vendors in shipping provide ownership and control data for all ships above 100 GT,[5] but if the owner does not want to be

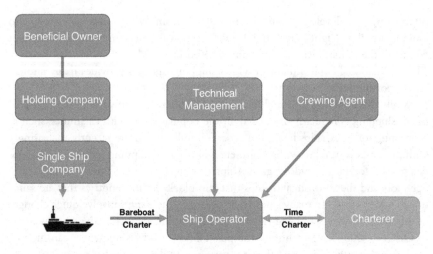

FIGURE 12.1 Ownership and control structure

known, it is easy to construct an ownership tree that does not fit in the data structures of these data vendors.

Note that the three companies can all have different nationalities. The beneficial owner may be Greek, while his ship holding company is based in the UK, and the single-ship company in Panama, because the ship is registered there.

In practice, the ship owner may have bareboat chartered the ship to another company, who operates the ship under a time charter contract for a charterer. The technical operations may be taken care of by a technical management company, while a crewing agent provides a continuous supply of able seamen. In this control structure, as displayed at the right-hand side of Figure 12.1, the bareboat charter may decide to choose another flag than the original ship owner. This is a possibility under a bareboat charter. So the ship now has two registrations and two flags. The technical management may be based, say, in Germany, and the crewing agent in the Philippines.

For a charterer, or a port state, the questions now become: Who controls this ship? Who decides on the quality of the crew? Who is responsible for maintenance of the ship? Who signs charter parties? Who is liable for damage, pollution, or loss of cargo caused by this ship's operation? If this is unclear, there are also opportunities for parties to cut costs and save money on all of these items. These ships can easily become a danger for the port states they visit, because low levels of maintenance, poorly educated crews and negligent ship management will result in higher probabilities of negative external effects and casualties caused by or related to the ship (see for instance Knapp and Franses, 2007).

12.10 Concluding remarks

In this chapter we have discussed cost structures in shipping. Apart from a description of cost items, we have also introduced the international environment in which ships and their owners operate. This international environment creates pressure on costs, but also offers opportunities to choose jurisdictions for the ship that facilitate the saving of costs. If left unchecked, these results in substandard shipping operations have a negative impact on the environment, and on the image of shipping as the efficient carrier of international trade.

Fortunately, there are mechanisms in place that provide checks and balances. Port state control, classification societies and flag state regulations, as well as several international conventions offer the right set of rules for international shipping. However, the complex ownership and control structures that may be set up around ships still offer parties that want to, the means to create anonymous shipping operations that can avoid the regulatory net, at least for a while. Ensuring quality shipping around the world still requires vigilance from all relevant authorities.

Notes

1 International Chamber of Shipping.
2 See, for instance, the case of the MV A1 Hufoof.
3 See the case of the Ukrainian ship Faina.
4 http://www.ihsfairplay.com/About/Definitions/definitions.html, visited 2 December 2014.
5 Gross tonnage, or the non-linear measure of a ship's overall internal volume.

References

Gorton, L., Ihre, R., Hillenius, P., and Sandevärn, A. (2009) *Shipbroking and Chartering Practice*. London, Informa.

Hailey, R. (2013) *Shippers 'Lose Out in Slow Steaming'*. Lloyd's List, 7 January 2013.

Hare, J. (1996) Port state control: Strong medicine to cure a sick industry. *Georgia Journal of International and Comparative Law*, 26, 571.

ICS (2013) *Flag State Performance Table*. http://www.ics-shipping.org/docs/flag-state-performance-table

Kavussanos, M.G., and Alizadeh-M, A.H. (2002) The expectations hypothesis of the term structure and risk premiums in dry bulk shipping freight markets. *Journal of Transport Economics and Policy*, 36(2), 267–304.

Knapp, S., and Franses, P.H. (2007) Econometric analysis on the effect of port state control inspections on the probability of casualty: Can targeting of substandard ships for inspections be improved? *Marine Policy*, 31(4), 550–563.

Maersk (2009) *Trade Flows in an Uncertain Economy*. Presentation by Soren Andersen at Transportøkonomisk Forening, 2009.

MARAD (2011) *Comparison of U.S. and Foreign-Flag Operating Costs*. U.S. Department of Transportation, Maritime Administration, Report, September 2011.

Marine Insight (2012) *The Guide to Slow Steaming on Ships*. Brochure. Retrieved from www.marineinsight.com/wp-content/uploads/2013/01/The-guide-to-slow-steaming-on-ships.pdf

Naess, E. D. (1977) *Autobiography of a Shipping Man*. London, Seatrade Publications.

OECD (2003) *Ownership and Control of Ships*. Report, Maritime Transport Committee, March 2003.

Stopford, M. (2009) *Maritime Economics* (3rd ed.). Abingdon, Routledge.

Streng, M. (2012) *Slow Steaming: An Economic Assessment of Lowering Sailing Speeds on a Supply Chain Level.* Master thesis, Erasmus School of Economics, Rotterdam.

Veenstra, A. W. (1999) The term structure of ocean freight rates. *Maritime Policy & Management,* 26(3), 279–293.

Veenstra, A.W., and Bergantino, A.S. (2000) Changing ownership structures in the Dutch fleet. *Maritime Policy & Management,* 27(2), 175–189.

Veenstra, A.W., and Van Dalen, J. (2011) Ship speed and fuel consumption quotation in ocean shipping time charter contracts. *Journal of Transport Economics and Policy (JTEP),* 45(1), 41–61.

Zannetos, Z.S. (1966). *The Theory of Oil Tankship Rates: An Economic Analysis of Tankship Operations (No. 4).* Cambridge, MA, MIT Press.

Suggestions for further reading

Besides the papers provided as references in the text, we recommend the following texts as suggestions for further reading.

Karakitsos, E., and Varnavides, L. (2014) *Maritime Economics – A Macroeconomic Approach.* Basingstoke, Palgrave Macmillan.

Notteboom, T., and Cariou, P. (2011) Are Bunker Adjustment factors aimed at revenue-making or cost recovery? Empirical evidence on the pricing strategies of shipping lines. In: Cullinane, K. (ed.) *International Handbook of Maritime Economics.* Cheltenham, Edward Elgar.

Stopford, M. (2009) *Maritime Economics* (3rd ed.). Abingdon, Routledge, chs. 6, 7.

Talley, W.K. (ed.). (2012) *The Blackwell Companion to Maritime Economics.* Chichester, Blackwell Publishing Ltd., chs. 5, 10, 11, 12.

Talley, W. K., Agarwal, V. B., and Breakfield, J. W. (1986) Economics of density of ocean tanker ships. *Journal of Transport Economics and Policy,* 20(1), 91–99.

13

CONTAINER TERMINAL OPERATIONS

An overview

Iris F. A. Vis, Héctor J. Carlo and Kees Jan Roodbergen

13.1 Introduction

Container vessels are scheduled to visit a set of container terminals along their route. Import containers will be unloaded from the vessel while export containers will be loaded at each terminal. Berthing times at terminals are considered as non-value activities and for that reason need to be minimised. Consequently, terminal operations need to be organised in such a way that unloading and loading operations can be performed efficiently. Those operations take place at five different areas, namely at the berth, quay, transportation, yard and gate. We define the berth and quay areas as the seaside of the terminal. The landside of the terminal includes the storage yard and gate area. Those two areas are connected by a transportation area where transfer vehicles transport containers from the vessel to the yard and vice versa.

We can describe the processes at a container terminal as follows. Arriving vessels will moor at one of the berths. Quay cranes can start the unloading process as soon as the vessel has moored. The quay cranes retrieve containers from the vessels' hold and deck according to an unloading plan and transfer them to vehicles upon further transportation to the yard. In the yard the import containers are temporarily stored by equipment such as yard cranes. At a certain point in time the containers are retrieved and transferred to other modes of transportation (e.g., vessel, trucks or trains) and leave the terminal gate for the next trip towards their final destination. Export containers arrive via the gate at the terminal. Typically, those containers are first stored at the yard to wait for their vessel. The loading process is basically the reverse of the unloading process and traditionally happens after the unloading process has finished. A loading plan determines the order in which containers need to be loaded on the vessel. Vis and De Koster (2003) and Steenken et al. (2004) provide a detailed overview of all operations in a container terminal.

TABLE 13.1 Number of scientific peer-reviewed papers per topic in the period 2004–2012

Topic	Number of papers
Seaside operations	80
Transportation operations	55
Yard operations	90

Carlo et al. (2014a, 2014b, 2015) present a trilogy of overview papers on methods available to design, plan, and control container terminal operations. Each of the papers discusses seaside, transportation and yardside operations, respectively. In this chapter we provide an overview of the insights presented in the trilogy. Table 13.1 shows the number of peer-reviewed papers published in scientific journals on each of the topics studied. We will not discuss the contents of individual papers, but focus on explaining decision problems, performance measures and modelling issues. In Section 13.2 we discuss seaside operations in more detail. Transportation operations and yard operations are elaborated in Sections 13.3 and 13.4, respectively. Section 13.5 presents the conclusions.

13.2 Seaside operations

In this section, we provide a short overview of the types of seaside operations that can be distinguished. Next to that we discuss relevant decision problems and solution approaches that are available in the literature to tackle each of them. For an extensive overview of the literature in the area of seaside operations, the reader is referred to Carlo et al. (2015).

We define seaside operations as the operations performed at the berth and quay. Arriving vessels at a container terminal will be assigned to a berth (refer to Section 13.2.2). Two main decision problems in the design phase of a container terminal consider the number of berths and the type of berth used. In Section 13.2.1 we will discuss the various types of berths in more detail. Commonly, rail-mounted non-automated cranes positioned on the quay (i.e., quay cranes) are used to unload containers. A set of quay cranes (QCs) are assigned to a vessel to retrieve containers from bays, which are found on sections of the deck and the hold of the vessel. A spreader is attached to the QC to perform the picking process. Trolleys enable the transport of the container from the ship to the quay and vice versa. More advanced QCs have multiple spreaders and double trolleys enabling pickup and transfer of multiple containers at the same time. Upon unloading, the QC will position the container on an internal transportation vehicle for further transport in the terminal. Scheduling policies will determine the order in which operations at specific holds/decks will be performed and by which QC (refer to Section 13.2.3).

Important information for this part of the process concerns the unloading plan. An unloading plan provides a list of containers to be retrieved and their specific locations aboard the vessel. The QC operator has some degree of flexibility

TABLE 13.2 Seaside operations: decision problems

Decision problem	Known in literature as
Number of berths available	
Number and types of QCs to be purchased	
Types of berth(s) used	
Allocating vessels to berths	Berth allocation problem (BAP)
Number of QCs to be assigned to each vessel	
Assigning specific QCs to vessels	Quay crane assignment problem (QCAP)
Sequencing and assigning unloading and loading tasks to QCs	Quay crane scheduling problem (QCSP)
Define which containers to unload and where they are located	Unloading plan
Define which containers to load and at what location	Stowage plan

to determine the exact order in which containers are unloaded from a specific hold. Traditionally, quay cranes first unload all containers assigned before loading containers (i.e., single-cycle mode operations). Consequently, in the single-cycle mode, approximately half of the QC movements performed are empty movements. Double-cycle operations reduce the amount of empty QC movements by combining unloading and loading operations.

A stowage plan determines the storage location of each container to be loaded on the vessel. In the loading process hardly any flexibility is allowed and the order in which containers will be loaded in each of the holds is determined up front. An efficient vessel loading process requires coordination between seaside, transportation, and yard side operations.

Table 13.2 summarises the decision problems related to seaside operations, and if applicable, gives the commonly used problem name. We note that in the literature those problems are typically studied in isolation. Complex mathematical modelling is required to integrate the different problems mentioned.

13.2.1 Types of berths

Most container terminals have berths that allow (un)loading operations from a single side of the vessel. This type of berth is typically called a traditional or a marginal berth. Indented berths are an alternative to traditional berths that allow QCs to operate on a vessel from more than one side. Three quays surround indented berths such that yards can be positioned at three sides of the vessel. The Amsterdam Container Terminals in The Netherlands (formerly Ceres Paragon terminals) implemented the first indented berth (Vis and Van Anholt, 2010). Benefits from indented berths can be found in a larger productivity given that other terminal processes are designed well. At the same time several new challenges can be noted. First, repositioning of QCs becomes more challenging due to the need for both horizontal and vertical movement. Second, in QC scheduling policies both neighbouring and opposite cranes have to be taken into consideration. Third, decision making with regard to transportation and storage of containers gets an additional complicating factor, namely at what side of the berth will operations be performed. Fourth, it has to be decided how many vessels can moor at the same time at an indented berth. A direct consequence of allowing multiple vessels to moor in the same indented berth is that the first ship that enters the berth cannot leave until the last vessel that entered the berth has left (Imai et al., 2007).

13.2.2 Berth allocation problem

The main decision in the berth allocation problem (BAP) is to assign vessels to positions at the berth. Vessel characteristics (such as dimensions and draft) and berth characteristics (such as length and depth) are taken into account for the BAP. Other drivers in the BAP are the total time required to (un)load a vessel, distances to be travelled to and from the specific storage locations used for the containers designated for the vessel, and the turnaround time agreed on in the contract. In the past years, three overview papers have been published discussing berth allocation problems (Theofanis et al., 2009; Bierwirth and Meisel, 2010; Carlo et al., 2015). A schematic overview of the classification of decision problems (Bierwirth and Meisel, 2010 and Carlo et al., 2015) is given in Figure 13.1.

In short, a BAP problem can be described by a set of attributes. First, a definition of the type of berthing positions used should be given. Discrete berth positions are predefined in such a way that only a single vessel can be assigned to a single berth position. If more than a single vessel can moor in a berth space, continuous berth positions will be used. Hybrid berth positions denote that a vessel can occupy multiple discrete neighbouring berth positions. Second, it has to be defined if arrival times of a vessel will be considered. Static arrival times indicate that vessels are available at the scheduled time of operation. Dynamic arrival times consider expected arrival times within a planning horizon. Third, a description has to be given as to how the time required for all operations will be defined. In most papers one of the three options as presented in Figure 13.1 are selected, namely a handling time measured in relation to the number of containers to be handled, the number of QCs assigned to the vessel or the distances to be travelled from the berthing position to the storage locations of containers. Any combination of attributes can be modelled. Performance measurement is part of the objective of those models. A large variety of deterministic and stochastic performance measures can be seen in literature which are related to vessel operation times, resource utilisation, distances to be travelled and deviation between actual and desired berthing positions.

A large set of solution approaches is available whereby it is possible to study each of the combinations of attributes as shown in Figure 13.1 (refer to Carlo et al. (2015)). Most papers start with a mathematical formulation of the problem and

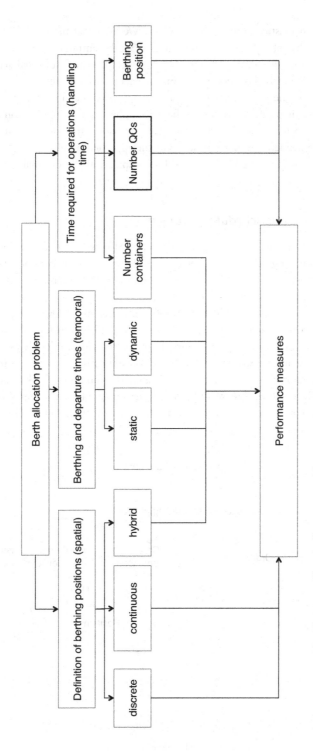

FIGURE 13.1 Classification of berth allocation problems

Source: Based on Bierwirth and Meisel (2010) and Carlo et al. (2015).

present a set of heuristics to solve the problem. We note that most of the existing solution approaches classify as static approaches. So, the authors studied an offline version of the berth allocation problem which is solved before the actual arrival of a vessel. Dynamic berth allocation problems take actual arrival times of vessels into account in solving the problems. Static solution approaches can serve as a base for solving the dynamic problems as long as they are applied in a rolling horizon framework with the option to alter previously made decisions. In that case, numerical experiments are needed to show robustness of the approach that has been designed. New methods that are designed are typically compared to the optimal version of the static BAP.

13.2.3 Quay crane scheduling problems

The goal of quay crane scheduling problems (QCSPs) is to assign and sequence unloading and loading of tasks to a set of QCs. Figure 13.2 presents an overview of the attributes used in the classification scheme of Bierwirth and Meisel (2010), which was slightly altered by Carlo et al. (2015).

The attribute task represents the type of job considered in the QCSP. A task typically varies between a bay of a vessel, a specific area of a bay, a group of containers stacked on top of each other, and a group of containers with similar characteristics or individual containers. Specific precedence constraints between tasks might be specified. Constraints might be needed to indicate if pre-emption of tasks is allowed. Models typically make one to several QC assumptions. The attributes shown in Figure 13.2 represent assumptions on the availability of cranes, their readiness, and

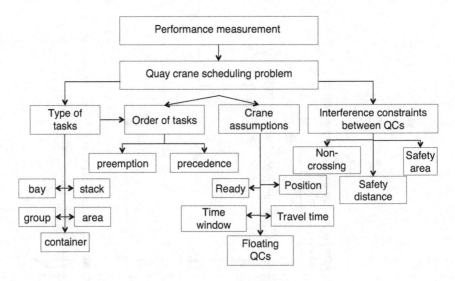

FIGURE 13.2   Classification of quay crane scheduling problems

Source: Based on Bierwirth and Meisel (2010) and Carlo et al. (2015).

their initial and final positions. Furthermore, travel times between bays might be considered. Another attribute related to QC assumptions specifies if floating QCs are used. The last category of attributes indicates the type of interference constraints used for the QCs assigned to the vessel. One type of constraints specifies the avoidance of QC-crossing. The other type of constraints is related to respecting safety. Safety constraints can be defined as the distance between cranes on the same side or in areas between opposite cranes or floating platforms.

The objective of the QCSP is a function of one or more performance measure attributes that will be minimised when scheduling the cranes. Specific performance measures being studied are task completion times, utilisation rates, throughput of cranes and travel times of QCs. Some papers study the relationship between the (un) loading process with other types of material handling equipment at the terminal. Consequently, transport vehicle and yardside equipment utilisation might also be addressed in the objective.

13.3   Transport operations

The berth area and yard are connected via a transport area. Transfer vehicles take care of the transport of containers to and from the vessel. The transportation process typically will be designed in such a way that operations at the seaside and landside are streamlined to avoid bottleneck operations. For an overview of literature in the field of transport operations we refer to Carlo et al. (2014a).

Various decision problems need to be tackled in performing transport operations. At a strategic level a terminal needs to decide on the type of vehicles to be used. Upon arrival of a vessel it has to be decided how many vehicles will be assigned to perform all transport operations related to that vessel. Finally, at an operational level routing and dispatching decisions have to be made to connect tasks to vehicles to enable transport via specific routes from origin to destination. Avoidance of collisions, congestion, and deadlocks are key attributes in routing decisions.

Typically, in deciding on the type of vehicle used, a trade-off will be made between investments, operational costs and performance. Figure 13.3 presents a classification of types of vehicles commonly used in container terminals. Self-lifting vehicles have a stacking capability and can lift a container themselves. Non-self-lifting vehicles need another type of material handling equipment to receive or drop off a container. Both automated and non-automated types of vehicles are available. Automated vehicles do not need a human operator and are typically controlled by a central computer system. Automated guided vehicles (AGVs) and automated lifting vehicles (ALVs) are available. AGVs typically travel on fixed paths in the terminal marked by, for example, wires in the ground. Free-travel AGVs use GPS control to move around resulting in more flexibility and the need for more complex traffic management tools. For an overview of research in the area of AGVs we refer to Vis (2006). Straddle carriers, non-automated self-lifting vehicles with the capability to stack containers up to several levels high, might be used for both transport and stacking operations. Yard trucks carry containers on a chassis.

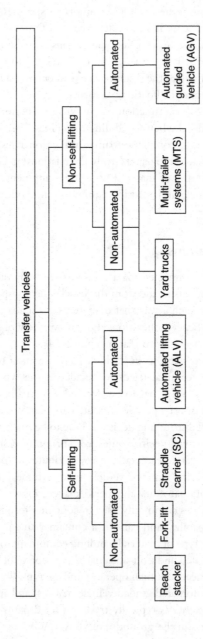

FIGURE 13.3 Different types of transfer vehicles

Source: Based on Carlo et al. (2014a).

Twin-load technology has recently been introduced for transfer vehicles. This kind of technology allows vehicles to transport two containers at the same time. A direct benefit is a decrease in the number of trips required to handle all transport requests. However, an increased complexity can be noted in the planning of trips due to the need to visit multiple blocks in the yard to pick up containers that according to the loading plan have to be loaded sequentially. Simulation typically is used in the literature to compare different types of vehicles (Carlo et al., 2014a).

13.3.1 Control policies

Carlo et al. (2014a) present a classification scheme for literature in the field of transport operations. Main decision problems as mentioned in the introduction of this section are depicted in Figure 13.4. Papers can be grouped in respectively discussing only unloading or loading operations or a combination of the two. Some authors also take inter-terminal transport operations into account. Typically, solutions for control problems are provided for a specific type of vehicle (refer to Figure 13.3) and a known number of vehicles. Transportation requests typically can be described by arrival times and if applicable, due times. Both deterministic and stochastic settings are considered within the literature. The main performance measures in modelling transport operations can be classified as vehicle costs, distances travelled, task completion times, and vessel operation times. Although transport operations are on the interface of seaside and landside operations, most papers study specific control problems in isolation. Figure 13.4 illustrates the close link between the (un) loading process at the vessel and the storage process at the yard.

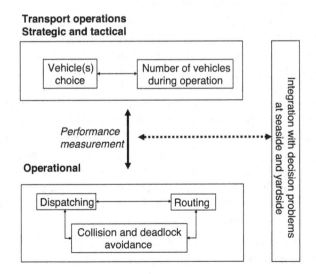

FIGURE 13.4 Decision problems in transport operations

Source: Based on Carlo et al. (2014a).

Vehicle dispatching can be defined as assigning a vehicle to a transport request. Most of the existing literature focuses on real-time vehicle dispatching. Two categories of dispatching rules can be distinguished: vehicle-initiated and load-initiated dispatching (Egbelu and Tanchoco, 1984). Vehicle-initiated dispatching can be defined as assigning an idle vehicle to a request. Load-initiated dispatching means that an available request is assigned to a vehicle. Usually, container terminals apply vehicle-initiated dispatching given that there are far more requests to be transported than vehicles are available.

Vehicle routing includes decisions on the number of trips to be performed and the order in which requests are performed along what paths. Most authors in the literature study the static version of the problem and test the robustness of the solution approaches designed in a stochastic environment. The problem can be formulated as an integer programming model. A large variety of heuristic-based solution approaches have been defined to delineate vehicle routes.

Deadlock and collision avoidance strategies can be formulated separately or as part of vehicle routing and dispatching strategies. A specific stream of the literature focuses on defining methods for deadlock and collision avoidance (Carlo et al., 2014a). In summary, the following solutions are defined. First, the area used by AGVs can be divided in blocks each to be used by a single vehicle. Next, constraints in AGV movement can be formulated to be used in routing policies. Finally, online routing algorithms taking time-dependent AGV behaviour into account are formulated (Gawrilow et al., 2012).

13.4   Storage yard operations

Containers are temporarily stored in the yard waiting for further transport via a vessel or other mode of transportation. Carlo et al. (2014b) presents a classification scheme of decision problems and shows what solution approaches are available to tackle each of them by means of a literature review. At a strategic level, container terminal managers decide on the layout of the yard and the type of material handling equipment to be used. At an operational level arriving containers will first be assigned a storage location. Second, the container will be dispatched to a crane which travels via a selected route to the location to perform the storage operation. Upon leaving the yard, a similar path is followed to retrieve the container. Consequently, control policies for decisions with regard to storage space assignment, crane dispatching and routing, and reshuffling of containers needs to be in place. Each of the decisions typically influences each other. However, most papers discuss decision problems in isolation. In Section 13.4.1 we discuss yard design in more detail. Section 13.4.2 provides an overview of control policies.

13.4.1   Yard design

In defining the yard design, two decision problems need to be tackled including the layout of the yard and the selection of material handling equipment used. Multiple rectangular blocks together form a yard where containers can be stored. Storage and retrieval operations are performed by material handling equipment assigned to one or multiple blocks. A block can be specified in the number of rows, where each consist of a number of ground slots where containers can be stored (i.e., bays) and a stacking height denoted by the number of tiers. The type of equipment selected typically influences the exact layout of the yard.

In general, two types of yard layouts can be distinguished which differ in the location of the transfer points and the relative positioning of the blocks to the quay. Transfer points (i.e., I/O-points) are used to transfer containers from the transfer vehicle (refer to Section 13.3) and the yard cranes. Blocks can be positioned parallel to the quay in which one or multiple rows are used as truck lanes enabling the transfer of containers at each bay (Asian layout). Blocks can also be positioned perpendicular to the quay where transfer points are located at the respective ends of each block to handle operations at the landside and seaside of the yard (European layout).

The differences between both types of layouts can be described in terms of investment costs, operational costs, storage capacity and handling times. Parallel layouts typically have lower investment costs, but higher operational costs. In perpendicular layouts, cranes can travel faster but have to travel over a longer distance to store and retrieve containers. As a direct consequence the travel distances for transport vehicles are shorter. As mentioned, parallel layouts have truck lanes which result in a lower storage capacity. The distance between rows in a block depends on the type of equipment used. A third type of layout is the one used by straddle carriers, which need space between rows to travel from one end to the other. If gantry cranes, spanning multiple rows, are used, a shorter distance between rows is required.

Consequently, main decisions in yard layout design are (1) the required number of blocks and their configuration and (2) the relative positioning of the blocks to the quay. Research studies typically analyse the layout of the yard in relation to the overall container terminal performance by means of simulation studies. General conclusions are hard to draw, where the outcomes of each study highly depend on assumptions made and the terminal characteristics (e.g., number and type of requests, type of equipment used and control policies applied).

Different types of material handling equipment can be used in the storage yard. Figure 13.5 shows an overview of different configurations.

Material handling equipment is either assigned to a single block or to multiple blocks. In the latter case, freely moving straddle carriers or gantry cranes can be used. The main difference between straddle carriers and gantry cranes is that a straddle carrier spans a single row of containers while a gantry crane spans multiple rows. Two different types of gantry cranes can be used: rail-mounted gantry cranes (RMGCs) and rubber-tyred gantry cranes (RTGCs). The latter type typically has an operator and can travel freely within and between blocks. RMGCs exist in automated (then typically called automated stacking cranes) and non-automated versions. The movement of RMGCs is defined via the rails it travels on. Wiese et al.

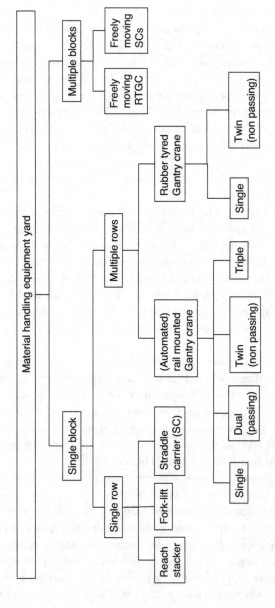

FIGURE 13.5 Different configurations material handling equipment yard

Source: Based on Carlo et al. (2014b).

(2009) provide an overview of yard equipment and show the usage of the different types in terminals worldwide.

Different configurations of gantry cranes can be implemented. Dual gantry crane configurations are defined by one large and one small gantry crane that can pass each other. Passing constraints require a lot of precision and as a result currently only dual automated stacking cranes have been implemented. Twin gantry crane configurations allow two identical cranes to operate in a single block. Safety distance constraints and an exchange for transferring loads from one crane to another are enforced. Both RMGCs and RTGCs can operate in this configuration. The triple crane configuration can be seen as a combination of a dual and twin configuration. One large passing crane typically works in the middle of a block to pre-position containers for the smaller cranes operating at the landside and seaside of the block, respectively. In the existing literature, comparative analyses are performed by means of simulation and travel time estimates to provide insights on different types of equipment. Second, guidelines are derived for specific conditions when what type of equipment is preferred over others.

13.4.2   Control policies

Equipment in the yard stores and retrieves containers. At an operational level we distinguish three main categories of decision problems, namely storage assignment decisions, routing and dispatching of equipment, and reshuffling policies. Reshuffling operations are performed to enable access to a desired container that needs to be retrieved but has other containers placed on top of it.

In the literature we note that those decision problems are studied for the different types of yard layout and equipment (see Figure 13.6) as introduced in Section 13.4.1. Carlo et al. (2014b) presents a classification of literature on yard operations. Basically, each paper can be linked to a decision problem, a type of layout and equipment used, stochastic or deterministic ready and due times and a specific (combination of) performance measures. Typical performance measures are task completion times, distances travelled and utilisation of yard space or cranes.

Housekeeping operations in yards are performed to balance workload over yard blocks. Equipment in a yard can work in single or double-cycling policies. Double-cycling can be defined as serving a storage request followed by a retrieval request to avoid empty travel movements. Typically double cycling is used in perpendicular layouts. In parallel layouts typically operations are performed in dedicated block areas where containers with similar characteristics (e.g., vessel, type, destination) are grouped (Saanen and Dekker, 2011).

In the dispatching problem a container will be assigned to a crane to perform the required storage or retrieval operation. In the routing problem, the order in which a crane performs the assigned containers will be determined. Routing problems for yard cranes can be classified as vehicle routing problems. The performance measure used depends on the type of transfer vehicle used. If non-lifting vehicles are deployed, the yard and transportation processes are coupled. In that situation,

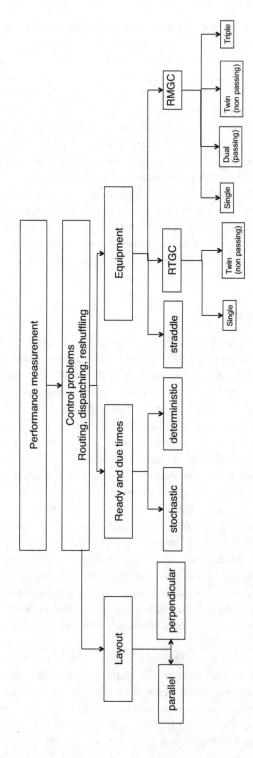

FIGURE 13.6 Classification decision problems at the operational level

Source: Based on Carlo et al. (2014b).

typically the objective is to minimise the total waiting time of transfer vehicles at the yard to receive or drop off a container. If lifting vehicles are used and as a direct result both processes are de-coupled, the main performance measures are to maximise throughput or minimise completion times. Both categories of performance measures are directly related to the overall performance measure in the terminal, namely, to minimise vessel operation times. The type of crane used in the yard determines to a large extent the complexity. If multiple non-passing or passing cranes are used, additional decisions on which crane performs what job along what route and additional constraints with regard to exchanging loads and passing safety concerns need to be added.

In a yard block, containers are typically stored on top of each other to minimise space used. A direct consequence is that the containers might not directly be available for retrieval if needed. In that case reshuffling operations are performed to remove the containers positioned on top of the requested container. Either those operations are performed before the operations for a specific vessel start or during the actual operation. In the literature four types of studies can be distinguished related to the decision problem of reshuffling operations. The first three types of methods linked to reshuffling operations are performed before the actual start of the unloading and loading of a vessel. First, storage assignment methods are presented that aim to prevent or minimise the expected number of reshuffling operations based on forecasts of retrieval sequences. Second, tools for pre-marshalling operations are introduced to reorganise the yard block such that during the actual operation no reshuffles are required. Third, remarshalling methods are designed to remove containers from their storage location to be pre-positioned in a separate block dedicated to the designated vessel. For both pre-marshalling and remarshalling it is assumed that the order in which the containers will be retrieved is known beforehand. For usage during actual operation, methods are designed that simultaneously determine the order in which retrieval operations and reshuffle operations can be performed.

In general, we note that most authors present a mathematical model while assuming deterministic ready and if applicable due times. Mainly meta-heuristic methods are introduced to solve the various control problems. Robustness and performance of the methods designed is typically tested by means of simulation.

13.5 Conclusions

In this chapter we have described the various container terminal operations in more detail. The operations can be categorised as seaside, transportation, or landside operations. For each of those categories we have provided an overview of design issues (e.g., layout and equipment used) and control policies. We have shown how the literature in each of those areas can be classified and have provided a sketch of the challenges encountered in modelling and solving the problems presented. For a full in-depth overview of literature in each of the fields, including material handling equipment, operational paradigms, and new avenues for academic research based on current trends and developments in the container terminal industry, the reader is referred to Carlo et al. (2014a, 2014b, 2015).

References

Bierwirth, C., and Meisel, F. (2010) A survey of berth allocation and quay crane scheduling problems in container terminals. *European Journal of Operational Research*, 202, 615–627.

Carlo, H. J., Vis, I.F.A., and Roodbergen, K. J. (2014a) Transport operations in container terminals: Literature overview, trends, and research directions. *European Journal of Operational Research*, 236, 1–13.

Carlo, H. J., Vis, I.F.A., and Roodbergen, K. J. (2014b) Storage yard operations in container terminals: Literature overview, trends, and research directions. *European Journal of Operational Research*, 235, 412–430.

Carlo, H. J., Vis, I.F.A., and Roodbergen, K. J. (2015) Seaside operations in container terminals: Literature overview, trends, and research directions. *Flexible Services and Manufacturing Journal*, 27, 224–262.

Egbelu, P. J., and Tanchoco, J.M.A. (1984) Characterization of automatic guided vehicle dispatching rules. *International Journal of Production Research*, 22(3), 359–374.

Gawrilow, E., Klimm, M., Möhring, R. H., and Stenzel, B. (2012) Conflict-free vehicle routing: Load balancing and deadlock prevention. *European Journal of Transport Logistics*, 1, 87–111.

Imai, A., Nishimura, E., Hattori, M., and Papadimitriou, S. (2007) Berth allocation at indented berths for mega-containerships. *European Journal of Operational Research*, 179, 579–593.

Saanen, Y. A., and Dekker, R. (2011) Intelligent stacking as way out of congested yards: Part 1. *Port Technology International*, 31, 87–92.

Steenken, D., Voß, S., and Stahlbock, R. (2004) Container terminal operation and operations research – A classification and literature review. *OR Spectrum*, 26, 3–49.

Theofanis, S., Boile, M., and Golias, M. M. (2009) Container terminal berth planning: Critical review of research approaches and practical challenges. *Transportation Research Record*, 22–28.

Vis, I.F.A. (2006) Survey of research in the design and control of automated guided vehicle systems. *European Journal of Operational Research*, 170(3), 677–709.

Vis, I.F.A., and De Koster, R. (2003) Transshipment of containers at a container terminal: An overview. *European Journal of Operational Research*, 147, 1–16.

Vis, I.F.A., and Van Anholt, R.G. (2010) Performance analysis of berth configurations at container terminals. *OR Spectrum*, 32(3), 453–476.

Wiese, J., Kliewer, N., and Suhl, L. (2009) *A Survey of Container Terminal Characteristics and Equipment Types*. Working paper, University of Paderborn. Retrieved from http://wiwi.uni-paderborn.de/fileadmin/lehrstuehle/department-3/wiwi-dep-3-ls-5/Forschung/Publikationen/wiese_et_al_a_survey_of_container_terminal_characteristics_and_equipment_types_2009.pdf.

Suggestions for further reading

Besides the papers provided as references in the text, we recommend the following texts as suggestions for further reading.

Carlo, H. J., Vis, I.F.A., and Roodbergen, K. J. (2014a) Transport operations in container terminals: Literature overview, trends, and research directions. *European Journal of Operational Research*, 236, 1–13.

Carlo, H. J., Vis, I.F.A., and Roodbergen, K. J. (2014b) Storage yard operations in container terminals: Literature overview, trends, and research directions. *European Journal of Operational Research*, 235, 412–430.

Carlo, H. J., Vis, I.F.A., and Roodbergen, K. J. (2015) Seaside operations in container terminals: Literature overview, trends, and research directions. *Flexible Services and Manufacturing Journal*, 27, 224–262.

Steenken, D., Voß, S., and Stahlbock, R. (2004) Container terminal operation and operations research – A classification and literature review. *OR Spectrum*, 26, 3–49.

Theofanis, S., Boile, M., and Golias, M. M. (2009) Container terminal berth planning: Critical review of research approaches and practical challenges. *Transportation Research Record*, 22–28.

14

PORT-CENTRIC INFORMATION MANAGEMENT IN SMART PORTS

A framework and categorisation

Leonard Heilig and Stefan Voß

14.1 Introduction

Soon after the container shipping industry emerged in the 1960s, major ports began to adopt information technologies, standards, and information systems for supporting an efficient planning, information sharing, and management of port operations and procedures. The ongoing adoption clearly indicates that digital innovation facilitates re-engineering of port processes and triggers modernisation of ports (Heilig et al., 2017).

Ports are complex environments consisting of several stakeholders and inter-organisational port processes. For performing activities in individual port processes, the use of information systems has turned out to be indispensable. However, in such complex environments, it becomes increasingly important to further improve the overall visibility of port processes and to tie loose ends in order to increase the efficiency, safety, and sustainability in the overall ecosystem of the port. Particular ports with space problems or those surrounded by urban environments face severe traffic and environmental problems due to increasing trade volumes and larger vessels (Giuliano and O'Brien, 2007; Zhao and Goodchild, 2010). An advanced management and utilisation of information may help to improve collaboration and to proactively avoid inefficient cargo flows and mission-critical situations. The role of information technologies (IT) and information systems in achieving a competitive edge has been pointed out in several works (see, e.g., Robinson [2002] and Notteboom [2004]). Growing national and international regulatory requirements furthermore encourage ports to implement single windows for the collection, dissemination, and integration of trade-related information at a single entry point to fulfil all import, export, and transit-related regulatory requirements (Tsen, 2011). Thus, a secure and efficient data exchange between stakeholders is not only important for increasing the productivity and efficiency of port operations, but also for meeting safety and security concerns (UNECE, 2005).

Recently, several large digital transformation projects have been started in several ports, like the smartPORT logistics (SPL) project in the port of Hamburg (HPA, 2014b) and the Algeciras BrainPort 2020 (ABP2020) project in the port of Algeciras Bay (Pernia and De los Santos, 2016), striving to implement the idea of a data-driven smart port aiming to exploit various integrated sources of information and real-time analytics to improve the efficiency, flexibility, collaboration, and safety in port processes, taking into account changing environmental circumstances and implications in terms of economic and ecological benefits (Heilig et al., 2017).

Over the last decades, a potpourri of different information systems has emerged in port environments, supporting specific port operations on the seaside, landside, and in transshipment areas of the port, predominantly in container terminals. Although port-related information systems play an essential role in port operations, we identify a lack of a comprehensive overview of IT and information systems applied in port operations. As the integration of these systems builds an essential foundation for implementing the concept of smart ports, we develop a framework for port-centric information management in smart ports. This framework demonstrates the role of information systems and information integration for both individual and overall port processes from an institutional and functional information management perspective. It is further used to categorise existing and future information systems, standards, and technologies. Given the framework, we explain the implementation of smart ports with an illustrative example from a functional information management perspective. Consequently, the framework may help to better understand the current and future role of information systems for solving operational issues and for triggering modernisation in ports.

The remainder of this chapter is structured as follows. Section 14.2 presents the overall framework and explains its relation to different operating areas and processes in the port community. This further includes a brief overview on important IT infrastructure systems. In Sections 14.3–14.5, we provide an overview of relevant information systems, used to implement both local and port-centric information management. In Section 14.6, we apply the framework to explain the implementation of smart ports based on a sample application and discuss important information management aspects. Finally, some conclusions are given in Section 14.7.

14.2   Framework for port-centric information management

Information management can be defined as the economic (efficient) planning, acquisition, processing, allocation, and distribution of information as a resource for decision support (institutional information management), as well as the design and implementation of respective infrastructure (functional information management) (Voß and Gutenschwager, 2011). Thus, a key element is to efficiently utilise information in business environments for gaining business value. This implies the need for achieving and sustaining business-IT alignment (see, e.g., Luftman [2004]) and might therefore involve several stages of IT-enabled business transformation (see, e.g., Heilig et al. [2017] and Venkatraman [1994]), also referred to as digital transformation.

In the past, these transformations predominately took place on a local port level. Especially container terminal operators designed and implemented huge IT environments to efficiently manage container terminal operations. Those adoptions helped to streamline and automate individual port processes and thus were an essential driver for increasing the efficiency in port activities. To plan and handle the inter-organisational handover between different port processes, information platforms in the form of port community systems (PCS) were developed in the mid-1980s. The development of common electronic data interchange (EDI) document standards made a significant contribution in harmonising information exchange in ports (Heilig et al., 2017). Moreover, a central PCS builds a foundation for implementing a single window. Given these three resulting roles of information management in ports, we see that digital transformation triggers the modernisation of ports, covers important regulatory requirements, and builds a basis for achieving competitive advantages.

The rather sparse coordination and information exchange between port community actors during port operations, however, frequently leads to both individual process inefficiencies and general issues on the overall port level, especially in the case of shared resources and resulting capacity bottlenecks (e.g., road infrastructure, terminal gate) or when cargo needs to be taken over from one process to another process handled by another actor. The latter often results from a lack of responsiveness and coordination, which is strongly affected by changing environmental conditions and interdependencies. Without realising a more active information exchange, port actors are often not capable of adapting quickly to the new situation. Also the control of physical port infrastructure has a huge impact on the efficiency and safety of port operations in general. Considering the growing cargo volumes, a port-centric inter-organisational information management bridging the information gaps between port actors becomes increasingly important.

In this section, we explain the different components that together build a novel port-centric perspective on information management in ports. The framework, which takes into account the different operation areas (light grey rectangles) and related activities in ports, is depicted in Figure 14.1. For an extensive overview on related port and terminal operations, the reader is referred to detailed surveys, such as given in Steenken et al. (2004), Stahlbock and Voß (2008) and Notteboom (2004). We further differentiate between IT infrastructure systems and information systems (dark grey rectangles). Information systems are exploited on a local port level to support certain operation areas, and on an overall port level where local and external information sources need to be purposefully integrated and managed. While former activities require individual port actors to implement information management practices, information for cross-process activities and value-added port information services (e.g., decision analytics, location-based services) need to be handled by an inter-organisational port-centric information management, for instance, established by a syndicate of port stakeholders, a third-party, or by the port authority. In the following sections, we adopt an institutional information management perspective

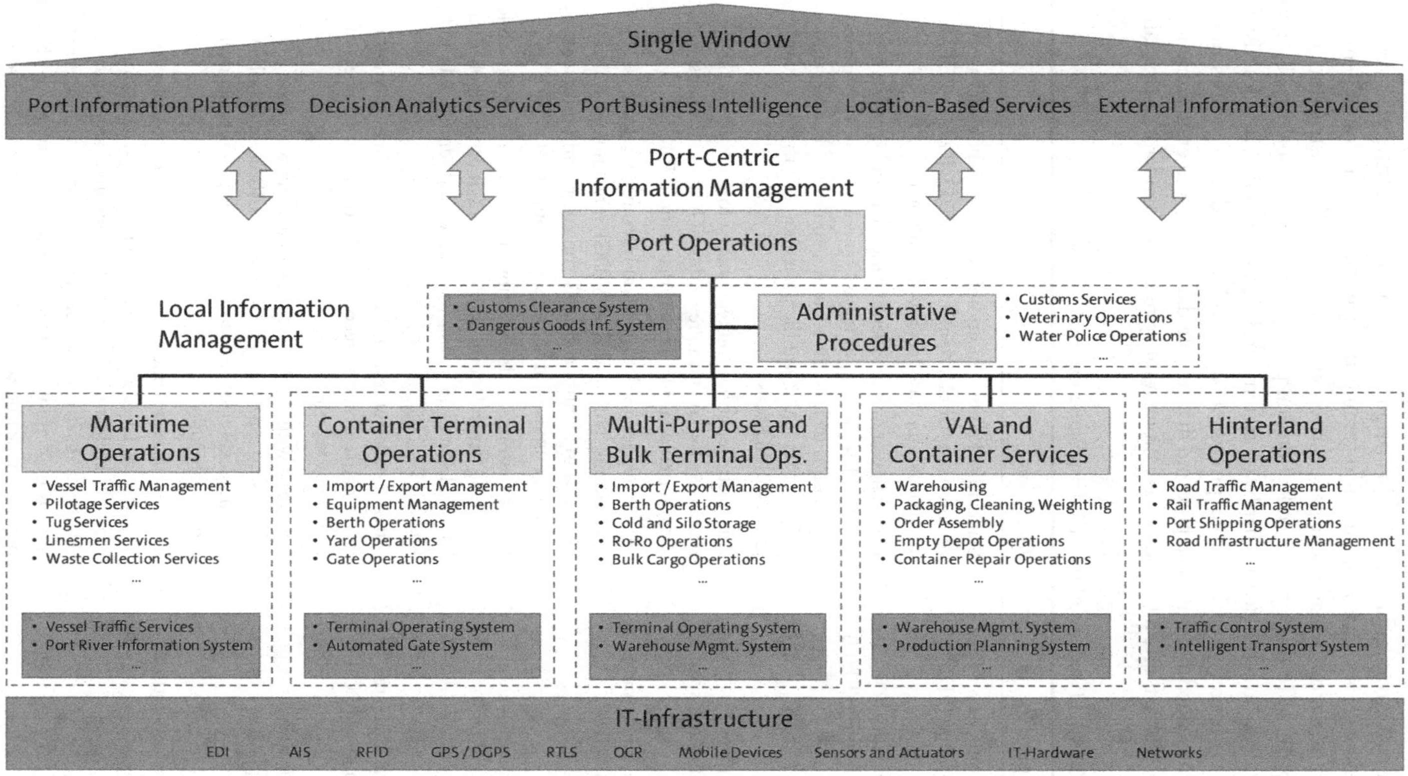

Single Window

Port Information Platforms · Decision Analytics Services · Port Business Intelligence · Location-Based Services · External Information Services

Port-Centric Information Management

Port Operations

Administrative Procedures
- Customs Services
- Veterinary Operations
- Water Police Operations

Port Operations
- Customs Clearance System
- Dangerous Goods Inf. System

Local Information Management

Maritime Operations
- Vessel Traffic Management
- Pilotage Services
- Tug Services
- Linesmen Services
- Waste Collection Services
- ...

- Vessel Traffic Services
- Port River Information System

Container Terminal Operations
- Import / Export Management
- Equipment Management
- Berth Operations
- Yard Operations
- Gate Operations
- ...

- Terminal Operating System
- Automated Gate System

Multi-Purpose and Bulk Terminal Ops.
- Import / Export Management
- Berth Operations
- Cold and Silo Storage
- Ro-Ro Operations
- Bulk Cargo Operations
- ...

- Terminal Operating System
- Warehouse Mgmt. System

VAL and Container Services
- Warehousing
- Packaging, Cleaning, Weighting
- Order Assembly
- Empty Depot Operations
- Container Repair Operations

- Warehouse Mgmt. System
- Production Planning System

Hinterland Operations
- Road Traffic Management
- Rail Traffic Management
- Port Shipping Operations
- Road Infrastructure Management
- ...

- Traffic Control System
- Intelligent Transport System

IT-Infrastructure

EDI · AIS · RFID · GPS / DGPS · RTLS · OCR · Mobile Devices · Sensors and Actuators · IT-Hardware · Networks

FIGURE 14.1 Port-centric information management

with a focus on port logistics aspects, before we explain the implementation of smart ports from a functional information management perspective.

Before going into more detail regarding the specific information systems and the information management perspectives, we extend the IT infrastructure part of Figure 14.1. To this end, the collection and processing of operational data is highly dependent on advanced IT hardware systems and standards. This includes distributed hardware systems to store, process, analyse, and access data as well as mobile and embedded systems to acquire and disseminate data in physical environments. In Table 14.1, we give a brief overview on relevant systems and standards. As explained

TABLE 14.1 IT systems of the IT infrastructure layer

IT system	*Description*
Electronic data interchange (EDI)	Standardised communication between stakeholders based on international EDI formats like UN/EDIFACT (EDI for administration, commerce, and transport).
Radio frequency identification (RFID)	Contactless automatic identification technology that enables identification of tagged objects. The most common applications include the identification of trucks and cargo (Wang et al., 2006) and the protection of cargo with tamper-resistant RFID seals.
Automatic identification system (AIS)	Automatic identification technology to track vessels on waterways for collision prevention. Their use is imposed by the International Maritime Organization (IMO).
Global positioning system (GPS)	Mostly applied for location and inventory tracking applications. Differential GPS (DGPS) technology is used in terminals to identify and track container yard positions as well as to steer unmanned vehicles and equipment.
Real-time location system (RTLS)	Enables automatic identification and position tracking while not being dependent on satellites. Thus, it can be applied in confined spaces including warehouses and tunnels (Ma and Liu, 2011).
Optical character recognition (OCR)	Often installed at terminal gates and yard cranes in order to detect damages and to recognise identification numbers of containers or vehicles.
Mobile devices (e.g., smartphones, tablets)	Equipped with powerful computing, communication, and sensing capabilities including GPS, RFID, and mobile communication services to receive and transmit data over mobile networks.
Sensors and actuators	Used to monitor conditions and control physical objects, respectively. Laser technologies and camera systems, for example, build an essential basis for terminal automation (Heilig et al., 2017). Nowadays, opportunities for Internet of Things (IoT) implementations, forming a network of physical objects interacting with each other, are intensively discussed.

in the following sections, those systems provide key functionality to oversee and control port operations.

14.3   Local information management

14.3.1   Maritime operations

A prime objective of maritime operations is the safety of navigation for any vessel in the port area. Besides, several other processes need to go hand in hand, such as pilotage, tug, and linesman services. A vessel traffic service (VTS), also referred to as vessel traffic information system (VTIS), supports the collection, analysis, and dissemination of information to safely manage the navigation of vessels. Predominately, the implementation of a VTS shall reduce the risk of hazardous vessel accidents, which increasingly occur in port areas with an increased vessel traffic density. For this purpose, the information system utilises several data sources and various technologies to track the location, movements of vessels, and to communicate to the vessel operator including vessel movement reporting systems (VMRS), AIS, radar systems, radio communication systems, signals and traffic lights, and video surveillance systems. The port of London, UK, for example, introduced Thames AIS in 2007 to improve the situational awareness of vessels on the river Thames and to extend the collection of safety critical information of the existing VTS (PLA, 2014).

Besides, the availability of vessel traffic information is essential for improving vessel scheduling and limiting idle times. Pre-planned arrivals and delays are determined on the basis of current vessel movements. Prior to arrival, other port operations can be adjusted accordingly in order to increase the port's productivity and to reduce vessel turnaround times and emissions produced by a vessel. For the navigation of vessels, tidal windows and turnaround manoeuvres need to be taken into account among other aspects. In tidal-dependent seaports, tidal windows are used to determine whether a vessel with a certain draught and speed can reach its destination; otherwise, the vessel must wait until the water level is high enough to approach a port. Vice versa, a favourable tidal window might be reasonable so that vessels leave the port earlier, even though there are still containers to be loaded onto the vessel, also referred to as the "cut and run" principle (Notteboom, 2006). For restricted waters, dedicated information systems are essential to control maritime traffic and to avoid traffic congestion (Hu et al., 2010), while considering tidal windows and vessel priorities. More specifically, larger container vessels might demand one-way traffic and a coordination of incoming and outgoing maritime traffic flows. In river port areas, the sailing route of vessels may include locks that could be another capacity bottleneck due to a lack of real-time coordination.

An extension of VTS are port river information systems. To cope with specific geographic and tidal requirements, these information systems collect and process tide gauge and meteorological data from tide sensors and weather stations in real time to ensure a safe entry and exit of vessels. Real-time data on tide conditions can be further used to advance waterway scheduling and berth allocation. In the port of

Hamburg, Germany, PRISE (Port River Information System Elbe) has been developed, extending the existing PCS to enhance the information exchange between terminal operators, pilots, shipping companies and shipping agents, tugs, and mooring staff in order to improve handling processes for large vessels that need to be carried out within narrow time windows. PRISE gets water level forecasts from the German Federal Maritime and Hydrographic Agency (HHLA, 2014).

14.3.2 Terminal operations

Terminal operations represent the core of port operations, managing the transshipment and flow of cargo between the waterside and hinterland of a port. This mainly involves operations in container terminals, multipurpose terminals, and bulk terminals. Generally, terminal operating systems (TOS) support terminal-related planning activities and operations (e.g., berth and yard operations). Consequently, a TOS can be defined as an integrated information system that acquires, analyses, and disseminates information from different data sources to ensure an efficient handling of cargo as well as a cost-effective allocation of handling equipment and personnel. A TOS commonly utilises different technologies to monitor and handle the flow of cargo, such as OCR, GPS, RTLS, and RFID. TOS are often implemented as or integrated with an enterprise resource planning (ERP) system to facilitate back-office operations in an integrated fashion. This might also include other business applications, such as those related to booking, production planning, and warehouse management. Those business applications are also required for offering value-added logistics (VAL) services, such as packaging, warehousing, and order assembly, as well as for managing containers in empty depots or container repair stations. To further support decision making, TOS are often extended by simulation tools and advanced planning and scheduling (APS) modules.

Navis SPARCS N4 (Navis, 2014) is currently regarded as the leading TOS. It provides extensive means to customise the TOS functionality. Planning modules contain domain knowledge important to enhance port operations. The solution further enables real-time monitoring and process automation at the vessel, quay, and in the yard (Navis, 2014). Moreover, extensions for other types of cargo (e.g., break bulk) are available. Another example is CITOS (Computer Integrated Terminal Operations System), which is an advanced ERP system that specifically supports the integration of port operations. Integrated expert subsystems process information to facilitate berth allocation, stowage planning, and resource allocation. A communication link to Singapore's PORTNET (see Section 14.4.1) is enabled (Gordon et al., 2005). Moreover, certain activities in terminals require dedicated information systems that are integrated with the TOS.

Gate appointment systems, for example, are used to better balance truck arrivals in order to reduce gate congestion. Several terminal operators have implemented gate appointment systems enabling the arrangement of appointment via web applications (Giuliano and O'Brien, 2007). In addition, gate operations need to ensure that all information on the cargo is recorded correctly, including information on damages and hazard classifications, and that the truck driver is authorised to enter or leave the terminal. Terminal operators commonly utilise OCR and/or RFID technologies to automatically identify vehicle and container numbers and check respective records provided by the TOS. Often, the gate system integrates information of multiple subsystems into one user interface to efficiently handle a transaction including TOS information and surveillance camera images. If relevant information is missing, it can be manually recorded either by the truck driver or gate personnel. Some ports, such as the port of Hamburg, have introduced self-service stations that enable drivers to register and check in standard containers by using container data and a valid trucker smart card. In particular in peak hours, self-service systems are an effective means to reduce the gate workload.

In container yards, an information system is essential to plan and register new containers as well as to track their position within the container yard. Different kinds of manned and unmanned gantry cranes are available to load and unload containers to and from different transport vehicles, respectively. In particular, automated transfer cranes (ATC) rely on the availability and accuracy of job and container information to autonomously handle respective yard moves. In other types of terminals, for instance in bulk terminals and empty container depots, effective means to handle and keep track of the materials and empty containers, respectively, must be installed, such as systems to weigh, transfer, store, and track cargo according to individual requirements.

14.3.3 Hinterland operations

Besides an efficient transshipment of cargo in the port, efficient hinterland connections must exist. While a large portion of cargo is transported via rail, the majority of cargo is transported on the road via trucks in many international ports. Growing international trade volumes have resulted in significantly increased volumes of freight traffic leading to severe issues including traffic congestion and increased vehicle emissions in ports (Giuliano and O'Brien, 2007, Zhao and Goodchild, 2010). In order to overcome those daily problems, ports have started implementing traffic control systems tracking the current traffic situation of roads within the port based on stationary measuring points (video/infrared/laser vehicle detection, induction loops, etc.). The traffic control systems are mainly used to manage public and port-related traffic flows within the port area by using special traffic signs and signals. Moreover, the analysis of real-time traffic measurement data enables a more accurate prediction of travelling times and traffic congestions.

Advanced applications for different modes of transport involve intelligent transport systems (ITS) using automatic vehicle identification (AVI), floating car data (FCD), and wireless vehicular ad hoc network (VANET) technologies. Instead of equipping roads and highways with expensive stationary measuring instruments, as currently seen in several ports, GPS data is acquired from trucks to analyse positions, velocity, travel time and routes in order to calculate the travel time for each link of the respective road network (Schäfer et al., 2002). As the data is processed in real time based on hundreds of observations including historic data records, the system is able to predict arrival times and congestions more accurately. Traffic control systems further provide the basis for truck acceleration programs in the port area.

Besides traffic information, weather forecasts based on meteorological measurements may be important to announce or react to certain weather conditions. For instance, dense fog or extreme winds increase the risk of severe accidents causing not only material losses – as seen, for instance, on the Donghai Bridge, China, which connects the mainland of Shanghai and the Yangshan offshore deep-water port. Given current data on weather conditions, traffic control information systems are used to warn and route road users or even to close road sections, if necessary. Along with an increased traffic density, the demand for an efficient service area and parking space management is growing. Service areas and parking spaces serve not only as a place to give truck drivers the possibility to rest instead of waiting in a traffic jam, but also allow to avoid or reduce congestion.

As rail transport represents another essential guarantor for hinterland accessibility, rail logistics information systems are important to coordinate trains, track train occupation, support train loading processes, and support information exchange among involved stakeholders, in particular among railway undertakings (RU).

14.4   Port-centric information management

The overview on operation area–focused information management in the previous section demonstrates the importance of information utilisation and decision making for specific port activities. In particular regarding the information management in maritime and hinterland operations, we can already identify forms of inter-organisational port-centric information management where information systems are used to manage and coordinate actions of different stakeholders in certain operation areas. In this section, we focus on information systems for establishing a port-centric information management aiming to integrate and utilise information from different port operation areas.

14.4.1   Port information platforms

Port information platforms refer to inter-organisational information systems focusing on the paperless information exchange between port stakeholders. A PCS is the most important system in this category aiming to improve general administrative and logistics processes by providing a common information platform used to exchange relevant documents between port actors and authorities, such as related to customs handling, import and export declarations, transport orders, and so forth (Keceli et al., 2008). The value of a PCS depends on the number of actors using the system, known as network effect, as well as on the information quality and associated benefits for all involved actors. Thus, a fundamental challenge for the success of a PCS is the adoption of a single information system among port community members and the willingness to share information (Van Baalen et al., 2009). EDI standards are generally supported by PCS. The range of functions and characteristics of PCS, however, greatly varies among seaports (for an extensive overview of existing PCS, the reader is referred to Posti et al. [2011]). According to the International Port Community Systems Association (IPCSA; IPCSA, 2014), key features include an easy, fast, and efficient information exchange (e.g., based on EDI standards), customs declaration, electronic handling of all information regarding import and export of containerised, general and bulk cargo, status information and control, tracking and tracing through the whole logistics chain, processing of dangerous goods, and processing of maritime and other statistics. Examples include DAKOSY PCS (Hamburg), Portbase (Amsterdam and Rotterdam, The Netherlands), eModal (applied by several ports in the USA), and PORTNET (Singapore; Seattle, USA).

However, common PCS are limited to a static, asynchronous document exchange based on EDI services for planning and organising the handover of cargo between actors as well as regulatory procedures. As outlined in the introduction, information platforms supporting a dynamic, real-time information exchange and decision support play an essential role for achieving competitive advantages. In the port of Hamburg, for instance, the Hamburg Port Authority (HPA) started the SPL project in 2010 with the objective to improve traffic and cargo flows within the port area by investing in a modern information systems and port infrastructure (HPA, 2014b). The main idea is to integrate different traffic control centres (road, sea, railway) into a main port traffic centre that supports decision making and an ongoing interaction with actors being actively involved in transportation activities based on real-time data (Heilig et al., 2014). This also includes an integration of traffic and infrastructure management based on a network of various sensors and actuators, which enables the routing of traffic flows dependent on the current traffic situation in the port. The sensors are further used to measure the conditions with respect to infrastructure and environmental impacts. A cloud-based information platform facilitates the integration of information from various subsystems (for an extensive overview on cloud computing, the reader is referred to Heilig and Voß (2014)). This includes the interaction with drayage truck drivers through a mobile application providing real-time location-based information and driver assistance functionality.

14.4.2 External information services

External information services refer to third-party information systems that are not managed by the port, but provide relevant sources of information to be included in value-added port information and decision support services. For example, this may include real-time data on traffic, parking, and weather conditions.

14.4.3 Decision analytics and business intelligence

Port decision analytics and business intelligence services implement methods and techniques to fully utilise and analyse information for supporting decisions and for discovering useful business insights, respectively. Therefore, these services rely on integrated sources of information, for instance, provided by port information platforms. Possible applications could be solutions for inter-organisational routing of inter-terminal operations (see Section 14.6) and more precise travel time forecasting. The growing demand for data-driven real-time analytics leads to big data problems. That is, efficient means to handle growing volumes of data in various formats at an increasing speed need to be implemented. With respect to optimisation problems, this usually involves the implementation of heuristics. Regarding data analysis, most big data solutions rely on the divide and conquer principle; that is, computations are largely distributed and parallelised (for an overview on big data, the reader is referred to Chen et al. [2014]).

14.4.4 Location-based services

Location-based services (LBS) integrate the location or position information of a mobile device with other information to provide added value information services to a user (Spiekermann, 2004). Therefore, LBS can be applied in port operations for conveying both general and individual information and decision support, taking into account the current position and the context-related conditions of individual port actors. Consequently, a purposeful integration with port information platforms, external information systems, and related components for decision analytics and business intelligence is necessary.

14.5 Single window

Main objectives of single window implementations are the streamlining, harmonisation, and coordination of reporting formalities, processes, and procedures for fulfilling regulatory requirements, mainly by electronic means (UNECE, 2005). Common PCS provide a foundation for implementing a single window. The next step would be to integrate various sector-specific information systems for developing national and transnational electronic commerce and logistics platforms as well as to enhance the access to specific information for government authorities. Due to empirical findings of Lai et al. (2008), electronic integration is positively associated with logistics cost performance. For delivering superior logistics services, ports need to consider customer requirements in the implementation process in order to create added value and competitive advantage (Lai et al., 2008).

14.6 Implementation of smart ports

After focusing on the institutional information management perspective taking into account several current IT and information systems, we explain the idea and role of smart port applications based on an illustrative example. This example tackles the problem of inter-terminal transportation (ITT; Heilig and Voß, 2016) between container terminals and other auxiliary facilities of the port (e.g., dedicated rail terminals, empty depots, and repair stations). It considers a real-time platform information exchange, decision support, and location-based services based on an inter-organisational port-centric mobile cloud information platform.

In a nutshell, the implemented mobile cloud information platform, referred to as *port-IO* (Heilig et al., 2017), allows the coordination of truck movements within the port area based on the current positions of trucks and traffic situation. Transport orders can also be outsourced to subcontractors with available capacities in the port area, if considered as economically reasonable. Truck drivers with smartphones are able to receive order information and are assisted when approaching the port area. The mobile cloud platform contains all functionality to enable real-time information exchange and planning. For the latter, it implements different methods to optimise the related inter-terminal truck routing problem (ITTRP), aiming to reduce the transport costs while reducing empty trips. By coordinating transports, the number of empty trips can be reduced considerably. Consequently, *port-IO* is an example of a smart port application used to improve the real-time coordination and decision support among port actors by taking into account environmental conditions.

Given the proposed framework, we are able to better categorise and explain the implemented components of *port-IO*. Moreover, it is possible to identify potentials for further functionality, for instance, by integrating other information and analytics services. In the following, important information management aspects of the framework application are discussed.

- *Business scenario*: In the beginning of smart port projects, it is essential to understand the needs of the port community and thereby derive high-level requirements of the port community. In this regard, one of the main questions is how an envisioned smart port application is able to sustainably address requirements with respect to the efficiency, flexibility, collaboration, and safety in overall port processes. As indicated earlier, the main idea of *port-IO* is to reduce traffic volumes by efficiently coordinating container transports in order to gain both economic and ecological benefits by reducing transport costs and vehicle emissions, respectively.
- *Sources of information*: One of the most important steps is to identify important sources of information. In the case of *port-IO*, we identify several sources of (real-time) information that need to be collected for supporting inter-organisational decision making, such as driver-specific information (e.g., current position), order information, distance, and traffic information. This information may stem from several local and external information systems (e.g., order management systems, traffic control systems). In *port-IO*, a cloud information platform is implemented to gather, integrate, store, and process data. A mobile application is installed on the truck drayage drivers' mobile device to gather their current position and status within the port area.
- *Utilisation of information*: Once required data is collected and stored in a central information platform, it can be further processed to support decisions and

gain valuable business insights. In *port-IO*, we develop a decision analytics service in form of a cloud-based vehicle routing optimisation component. The implemented heuristic determines coordinated transport routes by taking into account real-time data on transport orders, actual positions of available trucks, the current traffic situation, and resulting travel times. Moreover, data analysis services could be implemented, for example, to more accurately forecast travel times or service times in respective terminals and port facilities. The underlying cloud infrastructure further allows to scale data-intensive computations and use on-demand computing resources to handle big data problems. Regarding overall port operations, the increased visibility based on an efficient utilisation of information might further help to (automatically) control port infrastructure elements (e.g., lifting bridges) with the aim to optimise cargo flows within the whole port.

- *Dissemination and allocation of information*: As the flexibility and responsiveness in smart port operations mainly results from a real-time information exchange, it is important to convey information to the right stakeholder at the right time and through an appropriate information channel according to the individual port processes and stakeholder requirements. In this regard, the individual context and information gap of stakeholders as well as information access rights need to be considered precisely. *Port-IO* mainly provides three channels to disseminate information. The driver receives planned route schedules and is able to interact with the dispatcher via the mobile application. The mobile application implements driver assistance functionality, for instance, by providing navigation based on the given tour plan. Moreover, a web application in the form of a cloud-based software-as-a-service (SaaS) is implemented to centrally manage and plan inter-terminal operations as well as to oversee the positions of available trucks in the port area. Standardised web application programming interfaces (APIs) can be used to integrate information services of *port-IO* with other information systems, for instance, managed by individual port actors. In addition, LBS could be integrated in order to provide location-based information for truck drivers like parking capacities or, building upon the idea of social networks, the position of other close truck drivers from the driver's buddy list.
- In general, it is important to demonstrate the incentives of using a smart port information platform. Consequently, the value and quality of information being disseminated to several port stakeholders plays an essential role in a successful system adoption.

14.7 Conclusion

The promising concept of a smart port offers new opportunities to improve the efficiency, flexibility, collaboration, and safety in port processes through an economic and efficient utilisation of information. This requires inter-organisational port-centric information management laying the foundation for an efficient coordination of port processes and actors as well as for supporting decision making and business intelligence in the age of digitalisation.

In this chapter, we have presented a framework for port-centric information management to categorise and explain the current and future use of IT and information systems in ports. The framework differentiates between the local and overall use of systems for supporting maritime, terminal, hinterland, and administrative port operations. We provide an overview on relevant systems from an institutional and functional information management perspective. The framework is further applied to discuss the implementation of a smart port application, predominantly from a functional information management perspective. We present an illustrative example, namely an implemented mobile cloud platform for interterminal truck routing, referred to as *port-IO*, to discuss important aspects of a port-centric information management. We are aware of the limitations of this work, in particular with respect to existing power structures among port actors and required governance structures and mechanisms. In this regard, we intend to investigate means to better secure information sharing in smart port environments and explore adoption challenges by conducting expert interviews with major port stakeholders. Finally, the development of standardised data formats and interfaces for harmonising the communication among involved systems is still an important field of work.

References and further reading

Chen, M., Mao, S., and Liu, Y. (2014) Big data: A survey. *Mobile Networks and Applications*, 19(2), 171–209.

Giuliano, G., and O'Brien, T. (2007) Reducing port-related truck emissions: The terminal gate appointment system at the ports of Los Angeles and Long Beach. *Transportation Research Part D: Transport and Environment*, 12(7), 460–473.

Gordon, J. R., Lee, P. M., and Lucas, H. C. (2005) A resource-based view of competitive advantage at the port of Singapore. *The Journal of Strategic Information Systems*, 14(1), 69–86.

Hamburg Port Authority (HPA) (2014b) *smartPORT logistics*. Retrieved from www.hamburg-port-authority.de/en/smartport/logistics/Seiten/Unterbereich.aspx

Hamburger Hafen und Logistik AG (HHLA) (2014) *PRISE Optimises Sequencing and Arrival of Mega-Ships on the River Elbe and at the Port of Hamburg*. Retrieved from http://hhla.de/en/press-releases/overview/2014/03/it-platform-optimises-harbour-processes.html

Heilig, L., Schwarze, S., and Voß, S. (2017) *An Analysis of Digital Transformation in the History and Future of Modern Ports*. Proceedings of the 50th Hawaii International Conference on System Sciences (HICSS), Bis Island, Hawaii, USA.

Heilig, L., and Voß, S. (2014) Decision analytics for cloud computing: A classification and literature review. In: Newman, A. and Leung, J. (eds.) *Tutorials in Operations Research*. Catonsville, INFORMS, 1–26.

Heilig, L., and Voß, S. (2017) Inter-terminal transportation: An annotated bibliography and research agenda. *Flexible Services and Manufacturing Journal*. 29(1), 35–63.

Hu, Q., Yong, J., Shi, C., and Chen, G. (2010) Evaluation of main traffic congestion degree for restricted waters with AIS reports. *International Journal on Maritime Navigation and Safety of Sea Transportation*, 4(1), 55–58.

International Port Community Systems Association (IPCSA) (2014) *Port Community Systems*. Retrieved from http://ipcsa.international/pcs

Keceli, Y., Choi, H. R., Cha, Y. S., and Aydogdu, Y. V. (2008) *A Study on Adoption of Port Community Systems According to Organisation Size*. Proceedings of the 3rd International Conference on Convergence and Hybrid Information Technology (ICCIT), 493–501.

Lai, K. H., Wong, C. W., and Cheng, T. (2008) A coordination-theoretic investigation of the impact of electronic integration on logistics performance. *Information & Management*, 45(1), 10–20.

Luftman, J. (2004) Assessing business-IT alignment maturity. In: Van Grembergen, W. (ed.) *Strategies for Information Technology Governance*. Hershey, Idea Group Publishing, 99–128.

Ma, X., and Liu, T. (2011) *The Application of Wi-Fi RTLS in Automatic Warehouse Management System*. Proceedings of the IEEE International Conference on Automation and Logistics (ICAL), 64–69.

Navis (2014) *SPARCS N4 the Industry Standard Terminal Operating System*. Retrieved from http://navis.com/solutions/container/sparcs-n4

Notteboom, T. E. (2004) Container shipping and ports: An overview. *Review of Network Economics*, 3(2), 86–106.

Notteboom, T. E. (2006) The time factor in liner shipping services. *Maritime Economics & Logistics*, 8(1), 19–39.

Pernia, O., and De los Santos, F. (2016) Digital ports: The evolving role of port authorities. *Port Technology International*, 69, 30–32.

Port of London Authority (PLA) (2014) *Thames AIS – What Is Thames AIS*. Retrieved from www.pla.co.uk/Safety/Thames-AIS

Posti, A., Häkkinen, J., and Tapaninen, U. (2011) Promoting information exchange with a port community system – Case Finland. *International Supply Chain Management and Collaboration Practices*, 4, 455–473.

Robinson, R. (2002) Ports as elements in value-driven chain systems: The new paradigm. *Maritime Policy & Management*, 29(3), 241–255.

Schäfer, R. P., Thiessenhusen, K. U., and Wagner, P. (2002) *A Traffic Information System by Means of Real-Time Floating-Car Data*. ITS World Congress, 1–8.

Spiekermann, S. (2004) General aspects of location-based services. In: Schiller, J. and Voisard, A. (eds.) *Location-Based Services*. San Francisco, Morgan Kaufmann, 9–26.

Stahlbock, R., and Voß, S. (2008) Operations research at container terminals: A literature update. *OR Spectrum*, 30(1), 1–52.

Steenken, D., Stahlbock, R., and Voß, S. (2004) Container terminal operation and operations research – A classification and literature review. *OR Spectrum*, 26(1), 3–49.

Tsen, J.K.T. (2011) *Ten Years of Single Window Implementation: Lessons Learned for the Future*. Global Trade Facilitation Conference, 1–30.

UNECE (2005) *Recommendation and Guidelines on Establishing a Single Window to Enhance the Efficient Exchange of Information Between Trade and Government*. Recommendation No. 33. United Nations, New York.

Van Baalen, P., Zuidwijk, R., and Van Nunen, J. (2009) Port inter-organizational information systems: Capabilities to service global supply chains. *Foundations and Trends in Technology, Information and Operations Management*, 2(2–3), 81–241.

Venkatraman, N. (1994) IT-enabled business transformation: From automation to business scope redefinition. *Sloan Management Review*, 35(2), 73–87.

Voß, S., and Gutenschwager, K. (2011) *Informations Management*. Heidelberg, Springer.

Wang, W., Yuan, Y., Wang, X., and Archer, N. (2006) RFID implementation issues in China: Shanghai port case study. *Journal of Internet Commerce*, 5(4), 89–103.

Zhao, W., and Goodchild, A. V. (2010) The impact of truck arrival information on container terminal rehandling. *Transportation Research Part E: Logistics and Transportation Review*, 46(3), 327–343.

15

INTERMODALITY AND SYNCHROMODALITY

Lóránt Tavasszy, Behzad Behdani and Rob Konings

15.1 Introduction

Ports usually offer a wide array of hinterland transport options. Next to road transport, usually rail or inland waterway networks are available, reaching deep into the hinterland. The availability of different modes of transport provides ports with important opportunities for service differentiation: besides allowing to easily accommodate shipments of different nature and scale, multiple modes allow for a differentiation in transport speed at different levels of cost. As rail and waterways generally exert less pressure on the environment than road transport, there is also a societal benefit in having alternatives available. Use of road transport is sometimes inevitable, however, because shippers (sending or receiving firms) may not have direct access to the rail or waterways networks. In these cases, road transport is used for the first and final leg of the journey, with transshipment occurring at an inland terminal (also called a dryport) near the location of the shippers. While this has been the traditional role of intermodal terminals, some have evolved into larger network hubs, providing trimodal transport options, also between rail and waterways services. More and more, the multimodal transport network has become connected into one intermodal network.

In recent years, we have seen a shift of emphasis in network development from physical connectivity towards service connectivity. Synchronisation of service schedules and operations among modes of transport and also with the inland terminals' services (including transshipment and storage of containers) aims to provide seamless operations, leading to reduced waiting times and intermediate storage, thus reducing overall transportation costs. Furthermore, this synchronisation provides the possibility to jointly optimise services and consequently present more customisation and responsiveness (leading to improved flexibility and resilience) in hinterland transport. This is, in fact, an important benefit of strategic value for the whole freight transport system. Synchromodality, or synchronised intermodality, can be briefly summarised as the vision of a network of well-synchronised and interconnected transport modes, which together cater to the aggregate transport demand and can dynamically adapt to the individual and instantaneous needs of network users.

Our position in this chapter is that synchromodality marks the next stage in port/ hinterland network development and deserves exploration and experimentation. The objective of the chapter is to introduce the idea of synchromodality, closely linked with intermodality. Its organisation is as follows. Section 15.2 discusses the current position and evolution of intermodal hinterland transport systems. Section 15.3 describes the main elements of a synchronised intermodal transport system. Section 15.4 treats the innovations that are necessary to arrive at synchromodal transport systems and the barriers for future development including technological, economical and institutional aspects. We conclude our chapter in Section 15.5.

15.2 Intermodality: basic facts and evolution

In Europe and in many other regions in the world, the truck is by far the most dominant mode in freight transport. In the European Union the truck accounts for 49% of all freight (in tonne-kilometres) carried in this region, and when the sea mode is excluded its share rises to more than 70% (Eurostat, 2015). Moreover, over the last decades the share of truck transport has continuously increased.

As freight flows are forecasted to grow significantly in the future and as more sustainable transport is being advocated worldwide, alternative modes like rail and barge are expected to increase their role in freight transport. In general, intermodal transport is considered as one of the most promising techniques for train and barge to regain market share, as it combines advantages of barge or train with those of truck transport. Volumes transported by intermodal rail and barge services in Europe have increased spectacularly over the last two decades. Intermodal rail volume (containers, swap bodies and truck trailers) handled in the European Union in 1996 was estimated at 8 million TEU[1] (Eurostat, 2002), while the volume amounted to 16.5 million TEU in the European Union (excluding Sweden and Finland, and including Switzerland, Turkey, Croatia, Serbia and Macedonia) in 2009 (International Union of Railways, 2010). Intermodal barge transport, which has a much shorter history and smaller land surface that is being served (i.e., The Netherlands, Belgium, France and Germany), recorded a growth of approximately 0.5 million TEU in 1985 to 2 million TEU in 1996 and more than 5 million in 2009 (Konings, 2009). Although these volumes are impressive, intermodal (rail and barge) transport still account for less than 5% of the total surface traffic (in tonne-km) of goods in Europe as a whole.

Intermodal transport allows access to large-scale modalities like inland waterways and rail, from areas that have only access to the road network. This can be important for unit load transport, to allow the bundling of freight flows from several origins and destinations, thus reaching lower costs. In practice, transshipment from one mode to another can be done at seaports and at inland ports, but in the latter case introduces extra transshipment costs and additional handling time (see, e.g., EIA, 2013). Therefore in the past, intermodal transport has mostly been successful in situations where transport costs could be kept extremely low (e.g., over very long distances or with double-stack container movements in the United States), where natural or regulatory barriers were present against road transport (e.g., the regulations limiting permits for road transport across the Alps) or where transshipment costs had to be incurred anyway (e.g., at major seaports).

On these segments, the share of intermodal transport in the total freight market has been significant, often exceeding the share of road transport. Nevertheless, the share of intermodal transport across the entire freight transport market has remained low.

In the future, however, this technology might become more important, as road transport is facing increasing congestion and risks higher tariffs due to rising wages, fuel prices and environmental costs. Rail and inland waterways could become a strategic alternative to road transport, more than they are at present. In addition, rail and waterways as well may become complements, rather than competitors. Especially in case of major traffic incidents that blocked an entire corridor,[2] these larger-scale modes may act as alternatives for each other and create a more robust freight network. In order to understand the potential future role of intermodal transport it is important to discuss its evolution during the past decades.

Since the start of intermodal barge transport on the Rhine River in the late 1960s, different evolutionary phases in the transport services can be distinguished (Notteboom and Konings, 2004). The first years of operations were characterised by irregular services, and due to the uncertainty about departure times the barge services were used for transport of empty containers only.

The implementation of scheduled line services during the mid-1970s offered customers reliability and punctuality through fixed departure times. This was a major step forward in the development of this new hinterland transport mode. The growth in transport volumes enabled to increase the number of terminals that were equipped to handle containers along the Rhine River and also led to the establishment of terminals as facilities to exclusively transship containers. Barge operators also became involved in establishing terminals: setting up single-user terminals that supported their line services. In addition, independent terminal operators emerged by setting up common-user terminals. The fact that these barge services were gradually also offered as integrated services, that is, including barge transport, transshipment and pre- or post-truck haulage, increased the interest of shippers in intermodal barge transport.

Notable during the next phase (mid-1980s until mid-1990s) was the willingness among barge operators to co-operate in raising the level of services and prevent ruinous competition, which resulted in joint operated services. As a result the frequency of services increased and the destinations to serve along the Rhine River extended (to more than 25 in 1995). In this period also container barge transport between the ports of Rotterdam and Antwerp came to prosperity, as the barge was discovered as an interesting mode for feeder traffic of the deep-sea lines.

From the early 1990s the type of barge services became more diversified and also the service area expanded to outside of the Rhine River corridor and the Rotterdam-Antwerp trade route. Domestic hinterland services were started in France (1994), The Netherlands (1995) and in Belgium (1996), and later on also in Germany (in the hinterland of Hamburg). At that time the vertical integration of the container transport chain had great significance. Barge operations, terminal operations, and pre- and post-truck haulage became increasingly vested in one pair of hands, while some of these activities were previously outsourced. In line with this integration strategy, it was noticeable that barge operators were also seeking possibilities to exploit the complementarity with rail transport in order to offer more flexibility and transport options to their customers.

From the mid-1990s barge operators also have been increasingly involved in offering complete door-to-door logistical solutions (value-added logistics). So actually they have extended their container-related activities to cargo-related activities (e.g., warehousing). As the merchant haulage market became of increasing importance for container hinterland transport, barge operators have obtained much stronger ties with the merchants, large forwarders and logistic service providers in particular.

Nowadays container barge transport has reached a development phase of full maturity in hinterland transport. In the major Western European container seaports that have accessibility by waterways (i.e., Rotterdam and Antwerp), the share of barge transport is about 33%, but there is a challenge to further increase its share. Such a challenge also exists in other Western European container seaports (like Hamburg, Zeebruges and Le Havre), but there the existing role of barge transport is limited (less than 10%) and the future expectations are modest.

A tendency of rationalisation in the container barge sector can now be noticed. The number of barge operators is reducing, but they are increasing in size and are even more intertwined with inland terminal operators. These circumstances have created conditions to improve the service network. The traditional line service network, however, remains a dominant type of service. In this network a few terminals in the hinterland are visited to limit the transit time of a service, while on the other hand these multiple calls enable to achieve a sufficiently high utilisation rate of barges as well as acceptable service frequencies. Only a few Rhine terminals generate sufficiently large volumes to enable a 'one stop' in the hinterland. The best example is the Duisburg Express, which is a service that only calls at ECT Maasvlakte and Waalhaven in Rotterdam and the DeCeTe-terminal in Duisburg (Kreutzberger and Konings, 2013). Outside the Rhine corridor the number of inland terminal visits is related to the distance to the seaport. If the distance is short, the transit time performance is more critical, because of heavier competition with road transport, and then calling at one inland terminal is the rule. In order to serve tributaries of the Rhine River more efficiently some more complex bundling principles are applied. Containers are shipped in regular services to crossroad (hub) terminals where these containers are transshipped to a feeder service to arrive at their destination terminal. These kinds of services have been developed for instance along the Mosel and the Main.

Recently a pilot has started with a hub-and-spoke service configuration in which the terminal of Nijmegen (The Netherlands) acts as the hub terminal. On the one hand this enables an improvement of the performance of the hinterland barge services (e.g., higher frequencies), but on the other hand it opens the way to develop continental barge services.

Developments in intermodal rail networks differ from barge networks as the roots of rail transport were in continental transport (i.e., transport of swap bodies and truck trailers). Nowadays about half of the intermodal rail volume is hinterland transport, while the other half consists of continental transport.

Starting from a situation where the so-called wagon load system was common practice, transporting containers was initially a cumbersome process requiring complex bundling (see Figure 15.1). When rail volumes increased, it became possible to introduce dedicated intermodal trains for containers and swap bodies, but the use of intermediate stops at shunting yards remained a requisite to achieve a sufficient loading degree of trains and hence to enable services to multiple destinations. Intercontainer, a joined subsidiary of the principal national railway companies in Europe, was established to organise and to boost the development of intermodal rail services; it has worked for a long time according to this bundling principle (i.e., it used the shunting yard to exchange wagon groups instead of individual wagons). However, the performance of the intermodal rail services appeared in many situations not competitive to road transport in terms of costs and transit time (mainly caused by the process at the shunting yards), and moreover services were not punctual at all.

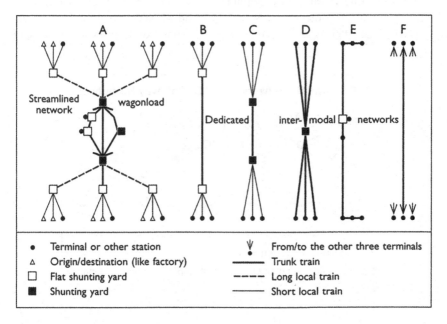

FIGURE 15.1 Streamlined and new rail networks

Source: Kreutzberger (2008).

A further streamlining of the rail networks was needed (see Figure 15.1). Direct services, also known as shuttle services, were introduced and became preponderant (since the early 1990s). Because trains run in a fixed formation of wagons from begin-to-end terminal, this train service could improve the cost performance as well as the transit time and reliability of the services. Hence much more and closer hinterland destinations could be served by rail as a competitive alternative to truck transport.

Although the shuttle concept is nowadays still dominant, the ideas and applications of more complex forms of bundling have been revived over the last decade. The shuttle concept works well if the cargo flows between the seaport and the hinterland region are substantial. However, it restrains the possibility to develop new rail services for which the flows are still too small to offer a service at an acceptable frequency. Line services have been introduced serving for instance two terminals in the hinterland.

In view of the challenge to have cost-efficient train operations and to offer many services and destinations in the hinterland, so-called gateway networks have been developed (see Figure 15.2). Hupac is the inventor of this concept and interconnects Italian service networks with North European ones at its gateway terminals near Milan. Other rail operators (like CEMAT and SOGEMAR) have also adopted the concept. The gateway concept essentially pushes two direct networks together at begin-to-end terminals without generating scale effects (see also Kreutzberger and Konings, 2013).

The increasing container throughput in seaports, and consequently the increasing number of rail and barge terminals in the port, adds a new dimension to the pressure to have an efficient organisation of hinterland transport services. The increasing

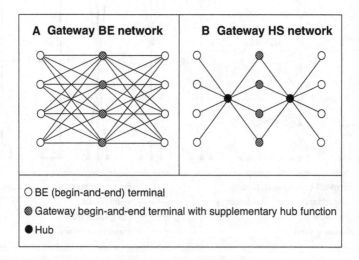

FIGURE 15.2   Gateway networks: connecting bundling networks

Source: Kreutzberger (2008).

number of terminals leads to fragmented container flows and hence more trains and barges that have to visit multiple terminals to load, which is a time-consuming and hence expensive process. In anticipating these problems, new service networks are and have been developed where a hub is used nearby the seaport (like the container transferium for barges outside the port of Rotterdam) or further away in the hinterland (like the rail company Distrirail did in 2014 in Duisburg to restructure its intermodal rail services to the hinterland in Germany).

15.3 From intermodal to a synchromodal freight transport system

Despite the fact that intermodality has been discussed for decades, unimodal road transport is still the preferred hinterland transport means for most shippers. Studies on quality perception of shippers show that although intermodal transport has advantages in terms of security or price (for long-distance transportation), single road transport is perceived as a superior alternative when considering several other key factors including lead time, reliability and flexibility of service (Cardebring et al., 2000). Meanwhile, global supply chain management has experienced numerous trends, like just-in-time, agility and Efficient Consumer Response, in the last two decades, which call for faster, more reliable and more flexible transportation services. To survive in the intensely competitive transportation market, intermodal transport has to fulfil the growing customers' needs and adapt to the changing business environment by improving the flexibility and presenting a more customised service. This requires new concepts that aim at new modes of operation and new arrangements to improve service quality and achieve the greatest cost reduction in hinterland transport.

A promising concept in this regard is "synchromodal freight transport." The cornerstone of this concept is an integrated view in the planning and management of different modalities to provide flexibility in handling transport demand. Because multiple modalities are involved in a door-to-door journey chain, the integration of service has always been an important issue for intermodal freight transport. However, the main focus has been on *vertical integration* of logistic services within one intermodal transport chain, which includes the transport and transshipment services.

One such example is the integration of hinterland transport operations and inland terminal processes in dry ports. The distinctive feature of a synchromodal transport system is through *horizontal integration* within a whole transport system (Figure 15.3). In other words, synchromodality seeks to integrate the transport service on different modalities to propose a "single transport service." This allows for optimisation of trade-offs between the quality and cost aspects of multiple modalities. For instance, although barge transport is cheaper, it is less flexible than trucking. Rail transport is even less flexible because there are specific constraints like shared infrastructure with passenger trains, which influences the timing of service.

By looking at the complementary nature of available transport modes, a synchromodal freight transport system provides a service that is no longer dependent

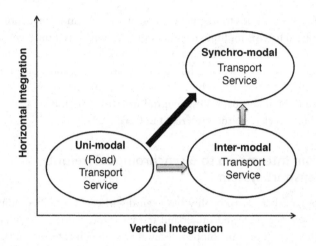

FIGURE 15.3   Dual integration in a synchromodal freight transport system

Source: Adapted from Behdani et al. (2016).

on the type of modality that is used for the main haulage. Instead, a range of customised services can be designed for shippers with different types of products and
with different sets of logistics requirements. Subsequently, depending on the specific
delivery time requirements for each container batch and the availability of each
modality in real time, the most appropriate mode can be selected. Such a service
can provide a satisfactory level of main factors that are important for each specific
shipper (Figure 15.4). A recent first estimation of benefits showed promising results
(Zhang and Pel, 2016). Depending on the specific synchromodal service definition,
a particular pricing scheme can then be set. This "service-based" pricing approach
is essentially different from the traditional "mode-based" fare design, but can be a
challenging task for the service provider (see, e.g., Ypsilantis and Zuidwijk, 2013).

The added value created by synchromodal services can be illustrated by a simple
two-dimensional service trade-off between transport time and price, as shown in
Figure 15.5. The graphs show how different transport services represent the service
differentiation within a network.

In this case, conventional networks with unimodal transport (on the left side of
Figure 15.5) would allow for three separate modes of transport, providing services
with different transport times and costs. The choice of preference among the three
options will depend on the relative importance of transport time and transport costs,
or values of time (VOT); road transport will be preferred by users with high VOT,
waterways by those who have a low VOT, and rail by those with an intermediate
VOT (see, e.g., Tavasszy and de Jong [2014] for a further explanation of these concepts). Some modes (like the zeppelin, in this figure) will not be a preferred mode
for any user. Here, road, rail and waterways form the so-called efficient frontier, or
range of non-dominated solutions. In intermodal networks, combinations of these

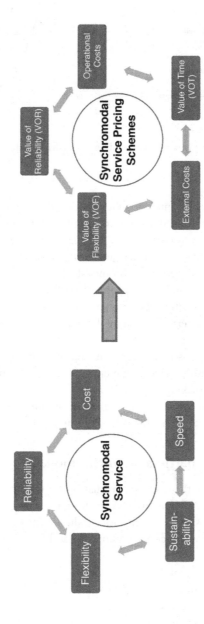

FIGURE 15.4 Synchromodal service definition and pricing

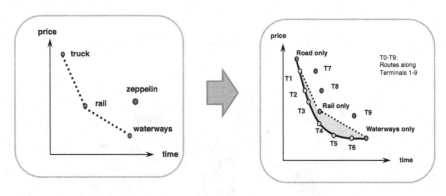

FIGURE 15.5   Added value of synchromodal services (shade surface)

modes become possible, by connecting the networks through intermodal terminals (here, noted by T1–T9). Intermodal networks will create additional points on this frontier. Synchromodal transport, due to flexible arrangements, will increase the number of combined routes dramatically, and create a continuum of non-dominated routing solutions. This range of synchromodal options (right) creates additional value added, compared to conventional networks (left), indicated by the shaded area between the old and new efficiency frontier. Understanding the demand volumes for each efficient choice alternative requires knowledge of shippers' preferences, and particularly of variations therein. Preliminary studies show that there are significant variations in the willingness to pay for time and reliability, for example, Wanders (2014), which would justify market differentiation through synchromodal service. This approach towards making the added value of new services explicit and measurable, can be extended towards other performance indicators, including external impacts of freight transportation and quality of information services.

15.4   Enablers and barriers of synchromodality

In this section we explore the main system changes for synchromodality. We distinguish system changes at different levels. Apart from the changes within the logistics system, there are changes in the framework conditions that allow the logistics layer to function. These framework conditions operate at different time scales:

- Transactions (days to months)
- Governance arrangements (months to years)
- Institutions (years to decades)
- Culture (decades to centuries).

At each of these levels, challenges have to be met to create a better synchronised intermodal transport system. We discuss these next.

15.4.1 Logistics layer

To provide this synchromodal transport service, the resources in a whole transport system must be continuously aligned with customer demands and needs. The resources include both "stationary resources" like transport infrastructure (e.g., roads, rails, and navigable waters) or transshipment nodes (e.g., inland terminals) and "moving resources" (e.g., trucks, trains, and barges), which provide the transport services between specific origins and destinations.

This alignment of resources with transport demand must be carefully maintained with an integrated view at three levels (Figure 15.6):

Integrated network design: At the strategic level, the synchromodal systems view, instead of a single modality standpoint, should be the reference for which transport infrastructure development/funding programs are based. In other words, the expansion of infrastructure networks for different modalities needs to be planned from an integrated view, based on the projected trading patterns, to build the missing links of a holistic synchromodal infrastructure network and support the parallel availability of modalities.

Integrated service design: Considering the expected shippers' needs, the packages of different synchromodal services (with different characteristics, as shown in Figure 15.6) are defined at the tactical level. Moreover, the service network must be designed, which includes the determination of routes within a network of multiple modalities as well as the timing and schedule of modality for each corridor, together with the capacity of each mode on the corridors.

Integrated operation and control: As the service network is designed, there is the need for an orchestrator (which can be one actor or an ICT platform) to match the actual demand and supply of service at the operational level. The decision may depend on the details of transport orders (e.g., the due date or size) and the state of transport resources (e.g., traffic information or the availability of a specific train/barge service).

Besides the point that integration should happen within these levels, typically, synchromodality addresses the challenge of between-level integration. Here,

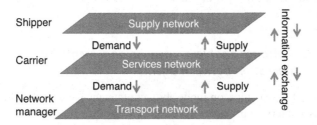

FIGURE 15.6 Levels of integration for synchromodal services

information exchanges between shippers, carriers and network managers is an even bigger challenge. The growing availability of information has enabled more integration and coordination of processes in the hinterland transport network. The challenge is, however, the fragmentation of data; the data that are needed to operationalise a synchromodal freight transport system are controlled by different actors in the chain. The synchronisation of multiple modalities requires integrated information solutions to accommodate the time-critical transfer of data and assign the best available service based on actual circumstances (e.g., traffic information and real-time availability of modalities). Moreover, unexpected events (e.g., the late arrival of barges, order cancellations, or the late releasing of containers) can frequently take place during the operation of a transport system. Consequently, real-time adjustments to transport arrangement need flexible, shared, and comprehensive ICT platforms.

15.4.2 Transaction enablers – contracts that allow synchronised transport

Transactions take place in markets, where demand and supply are aligned. The business value of synchromodality lies in the increased added value of transport services through a better alignment of supply to demand, based on dynamic information about preferences and expected transport service performance. The most elementary change in terms of transactions is that shippers, when they book transport services, do not yet fix the mode of transport. The demand for service is expressed by the required service quality, or service preference (this is also referred to in the literature as amodal booking, based on mode-abstract preferences). Once the service demand is known, the supply of services can be composed based on these preferences and the attributes of the available modes of transport. What follows is a cascade effect on all other contracts needed for realising the services, including the second- and third-order services, administrative handling of the shipping, the conditions of shipping, the division of responsibilities, the contracting of insurance services and so forth. Barriers for these changes in transactions primarily lie around the change of business models and the adoption of practices and instruments for data access, processing and sharing. Information systems are creating a new transaction layer for supply information (e.g., road conditions, ship delays), where the value of information is determined by the additional added value that can be obtained in services.

15.4.3 Governance enablers – new arrangements for co-operation between agents

The reliance of information on mode abstract preferences introduces the possibility for shippers and service providers to collectively optimise their systems. The service provider will provide several options, varying along the service/price continuum and propose choices based on the shippers' preference. This requires stronger collaboration between service providers and shippers, compared to the (still dominant) situation when shippers do not ask for options at all, but seek only low prices. Once such a collaboration is shown to be possible, shippers might look into alternative supply chain configurations, for example, through hybrid channels. Eventually, collaboration between service providers may emerge, to consolidate shipments from different shippers that have the same service requirements.

For these arrangements to become reality, collaboration structures need to develop that are not yet customary. Competing service providers are not used to work together, and often are not allowed to, given antitrust policies and regulations. Shippers and service providers often have a relationship that is not based on a shared awareness of all possible services that are available. Data availability is hampered by the absence of proper data standards, incompatibility of company information systems and a lack of sharing agreements. Possibly, governments could intervene to support the development of information markets by supporting standardisation of data and message formats, and a harmonisation of information systems.

The integration of modalities requires new organisational and legal arrangements. This is primarily because synchromodality may give rise to new business models and the change of roles in hinterland transport. This is already happening for other relevant concepts like extended gate service in which, instead of carriers or shippers, the terminal operator is responsible for the operation of inland transport: "terminal operator haulage" vs. "merchant haulage" or "carrier haulage" (Rodrigue and Notteboom, 2009). With these changes, of course, the legal responsibilities and liabilities would be essentially different. Additionally, a synchromodal transport system aims to integrate the resources in a network of multimodal chains (i.e., inland waterways, rails and roads). Each of these chains consists of several firms with different inter-organisational relations, different market positions, and different business models, consequently leading to differences in incentives, resources and capabilities. Coordinating multiple actors in a multimodal transport chain is a main challenge and has been the subject of many studies. This coordination can be even more challenging for a synchromodal system in which the operation of different chains must be simultaneously synchronised. Failure to coordinate may cause logistical problems and hinder the value of synchromodality. To achieve the optimal performance of a transport chain, new contractual settings need to be developed. These contracts provide the basis for partnership formation and the exchange of information between parties.

15.4.4 *Institutional enablers – from a centralised to a decentralised organisation*

For synchromodality, there are a number of relevant institutional issues. Shortcomings in current and governance arrangements concerning liability include the lack of incentives to disclose and share information about contents and the absence of an agreed approach to transparency of data in global trade lanes (Klievink et al., 2012). These shortcomings are deeply rooted in current institutions. Lately, distributed information brokerage with peer-to-peer information systems have emerged (see, e.g., the issues surrounding the advent of Uber and other peer-to-peer services). These may provide an alternative for current centrally and publicly regulated markets and will force both public and private institutions to rethink their roles and functions. Over the longer term, an interesting paradox is appearing, which may result in a clash of cultures in the logistics sector. One of the main barriers, as seen at current, is the fragmentation of the service market into a large number of small-scale companies, typically owning few assets and serving local markets. The SME culture with respect to innovation is very different from the culture of leading large forwarders and service providers. Due to their low capability to adopt technology and set standards, many SME firms are lagging behind in their use of information. This may change rapidly, however, as the landscape of information systems moves more and more towards decentralised, distributed, peer-to-peer-based control.

15.4.5   Cultural enabler – mind shift in transport planning and control

Synchromodality in hinterland transport requires a new mind set and calls for multiple paradigm shifts. The first essential aspect is "mode-free booking." Shippers must leave mode selection to the service providers and network managers. They only book the freight transportation service with predetermined price and quality requirements, and the synchromodal transport service provider has the freedom to decide which transport mode to use, according to the specifications of the customer and the availability of each mode. This "amodal booking" is a key requirement for a synchromodal freight transport system. If shippers book containers on specified transport modes (or routes), there is no possibility to integrate the modalities and provide flexibility by switching between different modes. The second aspect is the shift from a "mode-based" to a "service-based" hinterland transport. This paradigm shift is not only necessary for the shippers, as the demand representatives, but also for the transport service providers on different modes. Without this mental shift, the modalities can be simply seen as competitors; the complementary nature of them cannot be explored and the integration cannot happen. Finally, there is a need to go from a (dominated) "predict and prepare" hinterland operation towards a (complementary) "sense and respond" mind-set. In fact, the performance of hinterland transport is less defined by our "prediction capabilities" than by our "control-room capabilities" and how to react to sudden changes.

15.5   Concluding remarks

Transport networks have evolved from multimodal networks towards integrated networks allowing for intermodal transport, the carriage of a single load unit by consecutive modes in a transport chain. Interconnectivity between networks has been realised with intermodal terminals. Intermodal transport movements are becoming increasingly synchronised due to the availability of advanced information systems, allowing higher performance of the total freight network. This development of networks, along the line multimodality-intermodality-synchromodality, has been the main thread throughout this chapter.

We have discussed the salient features of intermodality. Despite its low share in all freight transport, it plays an important role for the movement of maritime containers in the hinterland of ports. Organisational issues arise in intermodal transport due to the need to (1) obtain sufficient containers to allow good utilisation rates of rail and waterways, (2) organise access to the networks by road transport and (3) create efficient transshipment of containers between modes.

After interconnectivity between modes, improved interoperability is the next step. This can be realised by synchronisation of operations in different networks, in a way that freight can shift from one mode to another in a flexible way, without pre-arranged and pre-booked modalities. This shifting can be useful to optimise capacity utilisation or to tailor services to the needs of shippers. Over the long term, such synchronised transport systems may even evolve into self-organised services, where flexible arrangements door-to-door will become analogous to the current internet (see, e.g., Montreuil [2011] for this vision on the "physical Internet").

Synchromodality will take time to develop. Major reorganisations are needed at several levels of the logistics system. At the operational level, transport contracts, planning and booking needs to be rearranged. At the governance level, operators of different modes have to learn to work together to adjust their operations for a better alignment of services. At the institutional level, operations will become more flexible when co-operative schemes are installed between operators based on improved standards, and control is not centralised. At the cultural level, a so-called mind shift may be needed to move from a paradigm where planning and design dominate transport operations, to one where only flexibility and responsiveness can guarantee that goods are delivered according to shippers' service requirements.

Notes

1 Twenty-foot equivalent unit.
2 The famous case of the inland waterway ship blocking the Rhine River for 33 days in 2011 cost the transport sector over 50 million euros, partly due to a lack of alternatives.

References

Behdani, B., Fan, Y., Wiegmans, B., and Zuidwijk, R. A. (2016) Multimodal schedule design for synchromodal freight transport systems. *European Journal of Transport and Infrastructure Research*, 16(3), 424–444.

Cardebring, P. W., Fiedler, R., Reynauld, Ch., and Weaver, P. (2000) *Analysing Intermodal Quality; A Key Step Towards Enhancing Intermodal Performance and Market Share in Europe*. Hamburg/Paris, TFK/INRETS.

EIA (2013) *European Intermodal Yearbook 2011–2012*. Brussels, European Intermodal Association.

Eurostat (2002) *EU Intermodal Freight Transport: Key Statistical Data 1992–1999*. Office for Official Publications of the European Communities, Luxembourg.

Eurostat (2015) *EU Transport in Figures, Statistical Pocketbook 2015*. Luxembourg, Publications Office of the European Union.

International Union of Railways (2010) *2010 Report on Combined Transport in Europe*. Paris: UIC.

Klievink, A. J., Van Stijn, E., Hesketh, D., Aldewereld, H., Overbeek, S., Heijmann, F., and Tan, Y. H. (2012) Enhancing visibility in international supply chains: The data pipeline concept. *International Journal of Electronic Government Research*, 8(4), 14–33.

Konings, R. (2009) *Intermodal Barge Transport: Network Design, Nodes and Competitiveness*. TRAIL Thesis Series No. T2009/11, TRAIL Research School, Delft, The Netherlands.

Kreutzberger, E. (2008) *The Innovation of Intermodal Rail Freight Bundling Networks in Europe: Concepts, Developments and Performances*. TRAIL Thesis series, Nr. T2008/16, Delft, TRAIL Research School.

Kreutzberger, E., and Konings, R. (2013) The role of inland terminals in intermodal transport development. In: Rodrigue, J. P., Notteboom, T. and Shaw, J. (eds.) *The SAGE Handbook of Transport Studies*, London, Sage.

Montreuil, B. (2011) Toward a physical internet: Meeting the global logistics sustainability grand challenge. *Logistics Research*, 3(2–3), 71–87.

Notteboom, T. and Konings, R. (2004) Network dynamics in container transport by barge. *Belgéo-revue Belge de géographie*, 4, 461–477.

Rodrigue, J. P., and Notteboom, T. (2009) The terminalization of supply chains: Reassessing the role of terminals in port/hinterland logistical relationships. *Maritime Policy & Management*, 36(2), 165–183.

Tavasszy, L. A., and de Jong, G. (2014) *Modelling Freight Transport*. London, Elsevier.

Wanders, G. (2014) *Determining Shippers' Attribute Preference for Container Transport on the Rotterdam – Venlo Corridor*. MSc thesis, Delft University of Technology.

Ypsilantis, P., and Zuidwijk, R. A. (2013) *Joint Design and Pricing of Intermodal Port-Hinterland Network Services: Considering Economies of Scale and Service Time Constraints (No. ERS-2013-011-LIS)*. ERIM Report Series Research in Management.

Zhang, M., and Pel, A. (2016) *Synchromodal versus Intermodal Freight Transport: Case of Rotterdam Hinterland Container Transport*. Paper 16-2237, Procs. Transportation Research Board Conference, Washington, TRB.

Suggestions for further reading

Besides the papers provided as references in the text, we recommend the following text as a suggestion for further reading.

Riessen, B., Van Negenborn, R. R., and Dekker, R. (2015) Synchromodal container transportation: An overview of current topics and research opportunities. In: Corman, F., Voß, S., and Negenborn, R. R. (eds.) *Procs. 6th Int. Conf. Computational Logistics*. Springer International Publishing, 386–397.

16

THE INDUSTRIAL SEAPORT

Bart Kuipers

16.1 Introduction

The skylines of most large seaports in the world are characterised by impressive container cranes, and chimneys and refining columns belonging to industrial plants. The industrial infrastructure in seaports is usually clustered in the vicinity of maritime basins and is associated with an impressive logistics infrastructure of tank-parks and bulk terminals. Spatial proximity is an important characteristic of the seaport-related manufacturing industries, in contrast to the location pattern of seaport-related distribution centres that show a spread towards the hinterland of most seaports. Although many modern seaports are associated with seaport-related industries, the structure and impact is different for each port (Table 16.1).

The port of Antwerp is an example of a seaport where 50% of direct port-related added value has been realised by manufacturing industries, with the chemical industry as the main sector. The chemical and oil refining industry is very important for the port of Rotterdam. The port-related industry as a whole in Rotterdam realised a total added value of €4.7 billion in 2013, some 38% of total added value of the port.

16.2 Mechanisms behind the seaport as a manufacturing location

The direct port-related industry in the two seaports illustrated in Table 16.1 is dominated by the process industry and by the metalworking and transport equipment industry, especially shipbuilding. Shipbuilding is an industrial activity that can only take place in seaports, because large ships and offshore platforms cannot be transported over land, with the exception of specific shipbuilding activities like the construction of luxury yachts or specialised maritime suppliers. Most of the other industrial activities are linked to seaports because of certain advantages, but facilities

TABLE 16.1 Direct port-related added value in the ports of Antwerp and Rotterdam, 2013, in billions of euros

	Port of Antwerp		Port of Rotterdam	
	billion	*%*	*billion*	*%*
Maritime node functions	**3.3**	**33**	**6.4**	**51**
Transport	0.8	8	2.5	20
Transport-related services	0.6	6	1.9	15
Handling/storage	1.4	14	1.9	15
Location-based activities	**6.7**	**67**	**6.1**	**49**
Manufacturing industry	*5.0*	*50*	*4.7*	*38*
Food	0.5	5	0.3	2
Oil refining	1.0	10	1.4	11
Chemical	2.9	29	1.8	14
Metalworking	0.2	2	0.3	2
Transport equipment	0.2	2	0.1	1
Utilities	0.3	3	0.5	4
Other	0.3	3	0.3	2
Wholesale	0.9	9	0.7	6
Services/port authority	0.7	7	0.6	5
Total	**10.0**	**100**	**12.5**	**100**

Source: NBB (2016); EUR (2016).

Note: Definitions between ports differ.

like oil refineries may also be located in the middle of Europe or the United States, such as in central Germany or in Kansas, especially because of the available pipeline infrastructures connecting refineries to oil terminals located in seaports and in the US because of local shale gas availability. Four important mechanisms are responsible for seaports as a location for performing manufacturing operations.

16.2.1 Seaports offer transport cost advantages

The first and classical mechanism for locating production facilities in seaports is related to transport cost advantages. When goods must be transshipped from a deep-sea mode into an inland mode, terminal handling costs and other seaport-related costs are incurred that destroy long-haul transport cost advantages. These extra costs may be decreased or avoided by locating the production facility at the 'break-of-bulk point' (Dicken and Lloyd, 1990). Economies of scale resulting from the use of very large bulk carriers further increased these transport cost advantages.

16.2.2 Seaports offer agglomeration economies

Next to transport-cost advantages, Winkelmans (1973) stressed the impact of agglomeration-economies as a driving force for the location of certain industrial activities in seaports, which is a second important mechanism. Economies of agglomeration are localised external economies of scale – savings that a firm gains from its connection with other plants or firms. By clustering in proximity to other firms, firms will benefit from low transport costs, lower levels of inventories, labour skills or by better inter-firm communication (Dicken and Lloyd, 1990), and in addition from low transaction costs as can be seen by vertically integrated systems of production, such as the chemical industry (Chapman, 2005).

16.2.3 Policy drive supporting seaport industrial development

The industrial linkages between firms in seaports and between the firms in the seaport towards the national economy were seen as 'engines of growth' for the national economy because of the multiplier effect of direct employment and added value. This mechanism was called 'growth poles' by the French economist Perroux (Dicken and Lloyd, 1990). A growth pole is an industry or a number of firms that (a) is large and therefore generates also large indirect effects, (b) is relatively fast growing, (c) has a high intensity of linkages by which the effect of growth could be transferred to the national economy and (d) is innovative. In the 1950s and '60s these characteristics were especially applicable to the chemical industry.

16.2.4 Availability of large sites for industrial development

Another mechanism for locating certain industries in seaports is the availability of large sites next to transport infrastructure having the ability to handle very big ships. Seaports underwent a spatial development in which the seaport infrastructure was moving out of the traditional historical city districts towards the sea, which is described by Bird (1971) in his 'Anyport model'. Specialised quayage is constructed for large-scale industry and the handling of large-scale bulk cargoes. Moving away from the port-city towards downstream locations means both the availability of large industrial sites and lower rent compared to locations near the urban core area. Especially the oil and chemical industry demanded those large sites because of growing demand for petrol and asphalt related to the postwar growth of auto-mobility and because of the commercial production of a large number of chemical products invented in the 1930s and '40s by certain German and US-based firms (Stobaugh, 1988).

16.3 Phases of port industrial development

The development of industrial complexes in most of the seaports in the world has been driven by technological and institutional developments. Five main phases of the industrial seaport are discerned that show a distinctive spatial development related to the port-city and reflect a development starting at the centre of the historical port-city and ending also in districts near the city centre.

16.3.1 Classical port industries

In the middle ages, port-cities imposed 'staple rights' or 'storage rights'; these rights required ships to unload their cargo at a port and to offer this cargo for sale during a certain period to the traders in the city. These rights offered advantages for the location of pre-industrial manufacturing in port-cities. Farmers also were forced to sell their surpluses to the port-cities. In addition to staple rights, port-cities also had mechanisms for closed shop systems or truck systems (forced sourcing) preventing the development of manufacturing in the rural hinterland, such as the cloth industry in Flanders or porcelain in China (Blussé, 2011). In addition, transport cost advantages related to large amounts of low-value raw materials resulted in the location of food industries in ports and, as stated before, traditionally shipbuilding is an important industrial activity in seaports. A characteristic is that mechanical handling equipment in seaports was yet absent: 'Products such as coal and grain were often shipped in bags, oil was packed in barrels, and ore was hardly shipped through the port at all' (Van Driel and De Goey, 2000: 7). Quay length was limited and ports often did not have direct access to the sea but depended on complicated routes with shallow barriers.

16.3.2 Modern industrial seaports

Four technological innovations in the first half of the twentieth century were underlying the rise of industrial seaports. The first innovation was the increasing scale of ships, resulting in economies of scale and strengthening the cost advantages of seaports for the location of large industrial facilities. These larger ships could only be used because of two other innovations: first the mechanisation of cargo handling technology, (floating) elevators, grabs, oil pumps and so forth (Van Driel and De Goey, 2000); and second, innovations in dredging and water-management infrastructure such as quays, locks or canals improving the connections of the port both to the foreland and hinterland. These three innovations are interrelated. But the fourth and perhaps the most important innovation is also an interrelated complex of innovations based on the discovery of oil, the invention of the internal combustion engine and the introduction of the telephone. These three interrelated elements 'gave rise to a new communication energy complex that would dominate the twentieth century' (Rifkin, 2014: 47). Rifkin uses the term 'Second Industrial Revolution' for this interrelated set of innovations. In addition to the internal combustion engine, important chemicals were invented in the 1930s together with chemicals production technologies like continuous refining and cracking technology (Stobaugh, 1988). The industrial seaport is an important location where these production technologies emerged, first in ports in the United States like Freeport/ Houston; then in European ports like Teesside, Rotterdam, Antwerp, Tarragona, Terneuzen and Marseilles; still later in Asian ports in Japan, South Korea, Taiwan and Singapore; and finally in ports in the Middle East, India and China. These petrochemical production complexes are usually highly integrated industrial sites with an oil refinery linked to producers of base and intermediate chemicals. But also steel production complexes (blast furnaces) and the production of automobiles were increasingly located in seaports; a prime example is Ford's River Rouge plant in Detroit, which was at the time the largest industrial site in the world.

The sites where these industrial seaport developments took place were referred to as maritime industrial development areas (MIDAs). Although the spectacular growth of the MIDAs in the 1950s and '60s stopped and economic development stagnated, most of these port-related industrial complexes were able to remain important, measured by added value or production volumes realised. The industrial added value of the port of Antwerp decreased from 57% to 50% of total direct added value of the port as a whole produced in the period 1995–2012; as a share of the total economy of Belgium, this is a decrease of 1.6% to 1.3%. The national economy of Belgium in this period was characterised by processes of de-industrialisation, growth of finance, IT services, and re-exports and other important structural changes, and the national economy realised a real economic growth of 36% in these years, but the position of the port-industrial complex remained strong despite of the continuous improvement of the efficiency of petrochemical operations. This continued strong position is also true for the industrial and seaport-related activities in the port of Rotterdam. Also in Rotterdam the share of industrial added value in total direct port-related added value decreased during the period 1995–2012 from 43% to 41%. However, as a share of total added value of the Dutch economy, the share of industrial added value in the port of Rotterdam remained stable at 0.8% throughout nearly three decades (Kuipers and Vanelslander, 2015). Three factors were responsible for the continued strong position of industrial complexes in seaports. First, these complexes still profit from the mechanisms responsible for their rise, such as an excellent location resulting in logistics cost advantages, large and integrated clusters producing agglomeration economies, although some of the smaller port-industrial clusters showed decline in the last decades, as the case study on Teesside at the end of this chapter makes clear. Second, continuous investment in petrochemical facilities: ExxonMobil and Shell announced important investment projects in the ports of Rotterdam and Antwerp in 2015–2016. Third, new industrial practises like industrial ecology and biobased operations, which are presented in more detail later, and the development of specialised and advanced market niches. In the port of Rotterdam, traditional shipbuilding moved to Asia, but the construction of highly specialised vessels for offshore and dredging and related advanced suppliers remained in the port region.

In the last two decades, however, the development of new large industrial seaports continued outside of the OECD countries, and especially China has shown impressive investment in a number of chemical industrial parks in the port-cities of Shanghai, Tianjin and Guangzhou. In addition, in Saudi Arabia Dow Chemical and Saudi Aramco at the moment are building a 20 billion USD chemical complex in Jubail, which was already a large industrial seaport before this project. The scale of these new facilities and the new technologies which are available, together with very low costs for feedstock and energy, further undermines the competitive advantage of the MIDA-type petrochemical complexes of the 1960s in the ports of Western Europe and Japan. In addition to these chemical complexes, there are also other impressive industrial activities that have emerged in the seaports of China, such as shipbuilding. However, some of these complexes have only thrived for a short period, such as the shipbuilding complex of Rongsheng Heavy Industries in Rugao, a port-city located in the Delta near Nantong in the Yangtze delta, once the largest shipyard in China and the location where a series of Valemax ore carriers – the world's largest – were built. The enormous complex started operations in 2005, but since 2011 a decline started and in 2015 the company was virtually bankrupt.[1] This is an illustration of stable practises in the ports of Western Europe and much more fluctuating developments in port complexes in other parts of the world.

16.3.3 Logistics-driven port industries: postponed manufacturing

The rise of the container in global trade, together with advances in production technology, IT and logistics innovations, resulting in the fragmentation and modularisation of supply chains, gave rise to a new form of industrial activity closely related to seaports: postponed manufacturing, also referred to as 'value-added logistics'. The basic production logistics concept in these logistics-driven port industries is the import of products from manufacturing locations overseas. Asia is the dominant origin for both Europe and the United States, the customs clearance and storage of these products in warehouses, the adaption of the product to the needs of certain markets or customer requirements in the warehouse, and finally the 're-export' of the product to other markets. The result of added value produced between the re-import stage and the re-export stage of the same cargo is in large part realised by postponed manufacturing. The main differences with the 'modern port industries' are the nature of the products produced and applied production processes. Most of the re-exported products are discrete industrial or consumer products such as office equipment, appliances, textiles, pharmaceuticals or machines instead of the bulk-oriented products in the modern port industries. The associated manufacturing may be characterised as 'light manufacturing' instead of heavy continuous process technology used in the MIDAs and modern port-industry. Postponed manufacturing is a form of final assembling, adding small parts to a product according to customer specification. An example is the final assembly of imported cars from overseas locations, often executed in seaports – such as in the ports of Zeebruges or Amsterdam. In seaport car terminals, value-added customisation activities and pre-delivery inspection is executed, making the car terminal a 'conjunctive pole' in inbound and outbound flows (Dias et al., 2010).

The added value related to this light-manufacturing character of postponed manufacturing is substantial lower than the added value related to other stages of the manufacturing process. The Dutch Statistics Office calculated that the production of 1 euro of re-export cargo realised on average only 7–9 euro cents of added value. In contrast, 'real' export of manufacturing products produced an added value of 58 euro cents on average. The very large share of re-export in countries like The Netherlands and Belgium results in a high impact of re-export as a share of the national economy, up to 4% for Belgium (Kuipers and Vanelslander, 2015).

Postponed manufacturing and the related re-export activities are closely associated with countries with large container ports, such as The Netherlands (Rotterdam), Belgium (Antwerp), Hong Kong or Singapore. An important characteristic of postponed manufacturing is that the warehouses or freight distribution centres that facilitate these operations are located near inland terminals in the hinterland of the port. Notteboom and Rodrigue (2005) call this the 'port regionalisation phase' in port-system development. Port-related activities are pushed out of the seaport region to regional freight clusters, usually linked to the ports by intermodal infrastructure.

16.3.4   Biobased industries and the circular economy

About 10 years ago, most industrial seaports started to develop a strategy towards a transition of the petrochemical industry towards the refinery of biobased products and the development of a chemical industry based on biofuels and on circularity principles. The biobased economy is a production system based on biomass as an input, with the exception of food used for human consumption and feed. An important concept underlying the biobased economy is the 'value pyramid'. At the bottom of the value pyramid, products produced in large volumes but with a low product value are distinguished, such as biofuels. Products produced in small quantities but with a very high product value are at the top of the value pyramid; think of biocomposites and fine chemicals produced by using biobased building blocks. In between are biochemicals, both fine and bulk chemicals. Activities at the top of the value chain are most attractive for a seaport region in terms of added value produced; the bottom of the value chain, however, might be more in accordance with the characteristics of most industrial seaports.

Seaports are an attractive location for the production of large volumes of biofuels for three reasons. First, most of the inputs for biofuels are imported from overseas locations and because of their bulky character, the mechanisms identified before for seaports as attractive locations are applicable. Second, most of the logistics assets for handling biofuels are already available in industrial seaports, such as storage locations and specialised handling equipment for bulk cargoes. In addition, the required skills for the production of biofuels is already available on most of the port-based petrochemical clusters; there is a high degree of 'skill-relatedness' and also of other cluster advantages for the production of biofuels available in seaports. Third, in most countries biofuels are added to traditional oil-based fuels. The location of large oil refineries in seaports means that this is also a location advantage for producers of biofuels. Seaports also offer location advantages for the production of biochemicals. The costs and availability of feedstock, access to markets and costs of labour, together with logistical factors, are very important for the location of commercial biochemical investment (Suurs and Roelofs, 2014). Seaports offer these location factors in abundance. The location of a new facility for the production of biobased FDCA by Dutch biochemicals producer Avantium on the site of BASF in the port of Antwerp, announced in the spring of 2016, is an illustration of this location tendency. FDCA is an important input for the production of the bioplastic PEF (polyethylene furanoate), an alternative for PET, used for bottles of soft drinks.

The biobased industry is still in an early stage of development, a take-off phase, and heavy investment in biobased facilities has not yet occurred, which has been mainly due to low oil prices and unstable government policies. The transition from oil to biofuels is an important strategy for the greening of industrial ports. Next to a number of important strategies, increasing energy efficiency of industrial operations, carbon capture and storage initiatives, and adaption of new technologies like hydrogen, a set of business practices associated with the circular economy is a second important 'grand strategy' for the greening of port-industrial complexes (Kuipers et al., 2015).

The 'circular economy' is an economic system that focuses on maximising the reusability of products and raw materials and minimising value destruction. Several reasons are underlying the emergence of a circular economy. First, the desire to reduce dependence on scarce primary resources. Future availability of these resources is increasingly uncertain, among others because of geopolitical developments. There is also a strong emerging social need to deal with sustainable materials. In a circular economy, raw materials are used as long as possible in cycles. By-products, which are very relevant for the chemical industry, may be given a new use in cascading developments. Products and processes thereby not only have a smaller ecological footprint, but are also less subject to availability and the price fluctuations common to scarce resources. Although the term 'circular economy' at the moment is a buzzword in policy making, in many sectors cycles are already nearly closed: metals such as iron, copper and gold, and materials such as glass, paper, plastic or textiles are already being reused largely or recycled.

When analysing the opportunities offered by circular economic practices for seaports, four important themes emerge (Kuipers et al., 2015):

> *Maintenance and repair.* Excellent maintenance activities and repair prevents products from becoming waste; the product-life of goods is increasing, preventing the demand for new goods. This certainly has an effect on global goods flows.
>
> *Reuse of goods.* The reuse of goods is used for a broad category of applications. Think of the reuse of a sea-container: a large number of alternative uses are documented, from the use for student housing to the use of a restaurant

inside a number of containers, but in general reuse gives a second life to an existing asset instead of disassembling or recycling.

Refurbishing or remanufacture of products. The life of a product may be increased by refurbishing or remanufacturing. A good example is the closed-loop supply chain practice of Fuji Xerox, a producer of copying machines. The first step is of course well-executed maintenance of the copier. But if the copier finally comes to the end of its life, the copier is returned to a production facility to be disassembled and cleaned. Fuji Xerox claims that a significant part (more than 45% and up to 70%) of the original parts of the copier are reused in a completely new copier and that only a small part of the machine is recycled (and that the production of waste is prevented altogether). The disassembly activities of Fuji Xerox are executed in two 'eco-manufacturing locations', one in Chonburi, Thailand, near the Laem Chabang container port, and one in Suzhou, China, near the container port of Shanghai.

Recycling and recovery of raw materials and energy, especially related to residual and by-products. This practice is very relevant for petrochemical clusters, related to the strong linkages of product flows inside these clusters. An important strategy to overcome the disadvantages of old facilities and high prices for feedstock and fuel is the tight integration of the chemical cluster by means of industrial ecosystem development (Hollen et al., 2015). This integration is realised not only in the industrial cluster, but also on the site level where different firms participate in co-operating in co-siting developments. This means that an industrial site is not managed by a single company performing all the activities but by a federation of different companies, each company having a certain industrial activity as its core function and often being a global market leader in the specific activity (see for further illustrations, the Teesside case presented later). These industrial activities and utilities are linked by flows of feedstock, energy and other utility flows and information. Linking these flows means that the amount of traditional rest – and by-products realised in most chemical processes is minimised. In addition to these industrial practices, linking industrial heat, cold and CO_2 from port users to non-port applications like greenhouses or urban areas is visible in an increasing number of ports.

16.3.5 New manufacturing: crossovers port and city

The last phase of the industrial seaport is related to a family of manufacturing processes referred to as Manufacturing 3.0 or 4.0. These are manufacturing practices strongly related to IT, such as additive manufacturing or 3D-printing and creative forms of new production practices. Old warehouses in historic port districts are often used for these creative new manufacturing industries. An important characteristic is the crossover of the new manufacturing between uses for both the port and the city: high-precision manufacturing for advanced maritime uses as well as the advanced construction of new materials like glass-fiber and carbon.

These five phases of industrial development may not be visible in all the industrial seaports on the globe. Quite a lot of the seaports are still struggling with the inheritance of 'modern' seaport development and are often showing strong locked-in effects, preventing further economic development. Teesside in the UK is an example of such a situation, presented to illustrate a number of the issues presented in this chapter.

16.4   Case study: port-industrial cluster in Teesside (UK)

The Teesside port region in North East England is a good example of the dynamics in industrial seaports since the last decades and is documented in detail by Chapman (2005), on which this case study is based. Teesside is an illustration of a port-based industrial cluster that has shown impressive economic development since the 1950s and was home to the chemicals producer ICI, the former British 'industrial champion' in the chemical industry. In the late 1960s employment in the metals and chemicals sectors in the region was 30,600 and 33,100 jobs, respectively, and in addition, the shipbuilding industry employed 27,000 workers. Teesside was seen as a national growth centre and was part of the national strategy to promote the modernisation of the UK economy (Chapman, 2005: 601). However, since its peak in the late 1960s, industrial employment in Teesside declined. In 2014 total regional industrial employment was 24,900 jobs[2] and employment in the chemical industry declined to 6,000 jobs in 2015. Declining industrial employment in the process industry is seen in most of the petrochemical clusters in the world because of increasing productivity and the outsourcing of non-core functions. The declining direct employment, however, results in a less declining volume of indirect employment because of the outsourcing processes. But in Teesside processes were strengthened by market developments in the 1970s, such as the effects of two oil-price shocks reinforcing the saturation of markets and of the global shift of investment to Asia. The result was a further declining employment, underinvestment resulting in low productivity of assets and in low new business formation in the Teesside region.

16.4.1   Commercial fragmentation and domino effect

The response by ICI, the 'leader firm' in the cluster, on the changing market developments had important repercussions for the Teesside chemical cluster as a whole. The cut in investment and labour costs reduced the significance of Teesside within ICI's corporate geography: Teesside became peripheral to the new strategic priorities of ICI (Chapman, 2005). ICI increasingly invested in high-value speciality chemicals capacity and in pharmaceuticals in overseas locations instead of in base chemicals produced in Teesside. In addition, ICI gave priority to investment overseas. This (geographical) diversification strategy was typical for nearly all producers of petrochemicals in the 1980s and '90s; firms in the chemical industry show a strong uniformity in their business strategies (Kuipers, 2000). The result was the obsolescence and shutdown of facilities and a process called 'commercial fragmentation' (Chapman, 2005). ICI sold its assets to different companies and the fully integrated site that was coordinated by ICI became fragmented into individual assets, related to the industrial networks of different firms. ICI is not visible in the Teesside complex anymore; instead a large number of firms form the cluster.

Instead of giving priority to the viability of the individual plant to the whole Teesside cluster, individual plants were seen in the perspective of the corporate geography of individual firms. Commercial fragmentation is seen as a risky direction for regional development because it may result in a 'domino effect', in which the closure of one plant may threaten the position of other plants in the complex. If production of a certain product in the complex stops, the product may be imported from overseas locations resulting in increased transport costs. Also by-products of different plants in the tightly knit input-output systems in chemical complexes may not be sold locally because of the closure of the former plant; these products may even become waste. The closure of firms also undermines the base for utilities and R&D activity in the cluster. This domino process becomes very serious for the cluster when critical plants in the cluster are affected, such as crackers or important aromates units. These facilities supply the whole chemical complex of a range of chemical building blocks. US chemical producer Huntsman acquired the cracker plant in Teesside in 1999 from ICI and sold it in 2004 to Sabic, a Saudi-based chemical firm. In 2014 Sabic announced a major upgrade of the cracker of £400 million and the dismantling of complete production units that belonged to ICI production units that stood idle before 2006 when Sabic bought the cracker. The investment by Sabic in cracker capacity in Teesside is seen as a very important development in the region and is even considered of strategic importance for the total UK chemical industry.

16.4.2   Co-siting: the former ICI site in the port of Rotterdam

In the Rotterdam petrochemical complex, the process of commercial fragmentation is seen from a different perspective. The fragmentation of the former ICI site in the port of Rotterdam is a process that mirrors the Teesside development of ICI, and also in Rotterdam the main ICI activities were acquired by Huntsman. However, commercial fragmentation in Rotterdam is translated into a process called 'co-siting', as indicated earlier. In this co-siting development, the large integrated site traditionally coordinated as a whole by the incumbent producer becomes fragmented into a confederation of business. In the former ICI site in Rotterdam, now Huntsman is the dominant firm, producing methylene diphenyl diisocyanate–based polyurethanes. The business on the site may be linked to the plants on the former site, but also to chemical plants in other sites, resulting in increased opportunities for 'industrial ecology' – the integration of different facilities by input-output relations. Co-siting is seen a process increasing the commercial opportunities of the different plants and of utilities, especially by enabling economies of scale. In addition, the plants that were traditionally coordinated by one diversified firm in the industrial complex are now coordinated by global leaders in the operations concerned and are linked to world-class industrial practises on a global scale, resulting in the transfer of manufacturing excellence practises 'outside in'.

16.4.3 Strategies for revitalising Teesside

Despite declining employment, underinvestment and a weakened research base, cluster policy remained the policy focus in Teesside. However, since the 1970s a number of different regional economic revitalisation strategies were proposed to bring back growth to Teesside. These strategies have also been used in other declining industrial regions and have a relation with the circular practices that were previously presented.

The first strategy is to strengthen the cluster by investing in chemical businesses that are 'missing gaps' in the industrial networks of chemical input-output relations. By filling these gaps, the complex as a whole becomes stronger and possibilities for increased use of by-products increase the efficiency of the cluster as a whole. This policy tries to strengthen the cluster by stimulating a reverse commercial fragmentation process called 'industrial ecology': the design of cascading industrial structures using inputs and by-products in the cluster, before this process was referred to as 'industrial ecosystem development' (Hollen et al., 2015). The diversification into downstream specialised chemical products and pharmaceuticals is also an important policy to strengthen the cluster, because of higher growth rates and because of being less sensitive to the effects of the business cycle. This strategy to strengthen the cluster by filling the gaps and diversification into specialties and pharmaceuticals was not successful in Teesside because these activities have other needs with respect to location choice; instead of a nearness to a seaport, short distance to important users (specialty chemicals) or R&D facilities (pharmaceuticals) are important location characteristics. In addition, the image of Teesside as 'Rustbelt Britain' or 'Britain's Detroit' are fatal for spatial decision making, a decision making process influenced by subjectivity and perception among other factors. Finally, other regions like the Middle and Far East and competing locations in Europe such as Antwerp, Rotterdam and Tarragona, are seen as more attractive for foreign direct investment in the chemical industry.

The second strategy aims to increase the number of start-ups, promote small and medium-sized enterprises (SMEs) in the region and aims to strengthen links between the region's university research base and the chemical industries. Chapman (2005) concludes that the opportunities for SMEs within the large-scale capital-intensive sectors of the industries found in Teesside have been limited and that impacts of connecting industries to the existing research base are incremental and long term. However, a decade after the analysis performed by Chapman and despite the rustbelt-image, Teesside developed into a prominent location in the innovation geography of the UK according to research performed by the Enterprise Research Centre (Roper et al., 2015), even outperforming London and Manchester.

The third revitalisation strategy is aimed at the greening of the business in the Teesside chemical complex by attracting investment aimed at green businesses and by a carbon capture storage initiative. The provider of industrial gases Air Products has invested in a £300 million renewable energy plant using gasification technology to produce energy from waste; a second investment of the same magnitude was announced in 2012. Based on serious engineering problems demanding a redesign of the facilities, Air Products however stopped the construction of this second plant and left the energy from waste business altogether because of high costs, and continued to concentrate on its core business, the production of gases.[3] This means a serious loss for the Teesside cluster.

The Teesside case illustrates the problems of the chemical industry to survive as an important growth industry in a seaport. The Teesside petrochemical cluster is different from other port industrial clusters, like Singapore, Rotterdam or Antwerp. It is an older and smaller chemical complex, it is located in a smaller urban region, it does not have a large and captive hinterland or is located on a major hinterland infrastructure, it has the historical influence of the dominance of one big firm, namely ICI, and the Teesside-region is British, with its own political and regional economic dynamism during the last decades (Chapman, 2005). Therefore, the continued strong position of the chemical industry observed in the ports of Antwerp and Rotterdam has not been seen in Teesside. However, there is economic growth in the region at the moment, coming from the digital economy and from innovation.

16.5 Conclusion

In this chapter, four forces underlying the location of industrial complexes in seaports were introduced. Also, industrialisation in five different types of seaports have been presented, with most attention given to the 'modern' port industry. Three dominant perspectives have emerged: first, the declining impact of small industrial clusters based on mature technology in small ports, like the case of the Teesside port industrial complex; second, the continued strong position of large integrated complexes like Rotterdam or Antwerp due to the continued agglomeration economics and superior logistics performance; and third, the rise of impressive new industrial port complexes in the Middle East and Asia. Industrial seaports will stay important for some time in the economic landscape, but these ports enter a period of transition towards a more sustainable economic model. The greening of the ports business and the potential the biobased and circular economies are offering, together with the realisation of the crossovers between port and city, will dominate the future agenda of industrial seaports around the globe.

Notes

1 'China's biggest shipyard is now a ghost ship', *CaixinOnline*, November 3, 2015.
2 www.nomisweb.co.uk/.
3 'Hopes of an energy-from-waste scheme on Teesside being revived are fading', *Chronicle Live*, April 27, 2016.

References

Bird, J. (1971) *Seaport and Seaport Terminals*. London, Hutchinson University Library.

Chapman, K. (2005). From 'growth centre' to 'cluster': Restructuring, regional development, and the Teesside chemical industry. *Environment and Planning A*, *37*(4), 597–615.

Dias, J.Q., Calado, J.M.F., and Mendonça, M.C. (2010) The role of European "ro-ro" port terminals in the automotive supply chain management. *Journal of Transport Geography*, 18(1), 116–124.

Dicken, P., and Lloyd, P. E. (1990) *Location in Space: Theoretical Perspectives in Economic Geography* (3rd ed.). New York, Harper & Row.

EUR (2016) *Havenmonitor. De economische betekenis van de Nederlandse zeehavens 2002–2015*. Rotterdam, Erasmus Universiteit Rotterdam.

Hollen, R.M.A., Van den Bosch, F.A.J., and Volberda, H. W. (2015) Strategic levers of port authorities for industrial ecosystem development. *Maritime Economics and Logistics*, 17(1), 79–96.

Kuipers, B. (2000) Flexible restructuring in the port of Rotterdam region. Case studies in the port-related petrochemical cluster. In: Johansson, I. and Dahlberg, R. (eds.) *Entrepreneurship, Firm Growth and Regional Development in the New Economic Geography*. Uddevalla, University of Trolhättan, 433–458.

Kuipers, B., de Jong, O., Van Raak, R., Sanders, F., Meesters, K., and Van Dam, J. (2015) *De Amsterdamse haven draait (groen) door. Op weg naar een duurzaam concurrentievoordeel door inzet op biobased & circulaire economie*. Rotterdam, Wageningen, Erasmus University (RHV/Drift)/Wageningen University (FBR).

Kuipers, B., and Vanelslander, T. (2015) De toegevoegde waarde van zeehavens. Antwerpen en Rotterdam: centra van hoogwaardige logistieke toegevoegdewaardecreatie? *Tijdschrift Vervoerswetenschap, Jaargang*, 51(3), 83–97.

NBB (National Bank of Belgium) (2016) *Economic Importance of the Belgian Ports: Flemish Maritime Ports, Liège Port Complex and the Port of Brussels – Report 2013*. Brussels, National Bank of Belgium.

Notteboom, T. E., and Rodrigue, J.-P. (2005) Port regionalization: Towards a new phase in port development. *Maritime Policy and Management*, 32(3), 297–313.

Rifkin, J. (2014) *The Zero Marginal Cost Society: The Internet of Things, the Collaborative Commons, and the Eclipse of Capitalism*. New York, Palgrave Macmillan.

Roper, S., Love, J., and Bonner, K. (2015) *Benchmarking Local Innovation. The Innovation Geography of the UK*. Coventry, Enterprise Research Centre.

Stobaugh, R. (1988) *Innovation and Competition: The Global Management of Petrochemical Products*. Boston, Harvard Business School Press.

Suurs, R., and Roelofs, E. (2014) *Quickscan investeringsklimaat voor biobased bedrijven*. Delft, TNO Earth, Life & Social Sciences.

Van Driel, H., and De Goey, F. (2000) *Rotterdam. Cargo Handling Technology 1870–2000*. Eindhoven, Stichting Historie der Techniek/Walburg Pers.

Van Oud Alblas Blussé, J. L. (2011) *Aan de oevers van de grote rivieren: de Rijn en Yangzi delta's 1350–1850. (On the banks of the great rivers: The Rhine and Yangtze deltas 1350–1850)*. Address Delivered by the Occasion of His Retirement as Professor in the History of Eurasian Affairs at the University of Leiden, June 6, 2011.

Winkelmans, W. (1973) *De modern havenindustrialisatie*. Rijswijk, Nederlands Vervoerswetenschappelijk Instituut.

Suggestions for further reading

Besides the papers provided as references in the text, we recommend the following texts as suggestions for further reading.

Chapman, K. (1991) *The International Petrochemical Industry: Evolution and Location*. Oxford, Blackwell.

Hollen, R.M.A. (2015) *Exploratory Studies into Strategies to Enhance Innovation-Driven International Competitiveness in a Port Context. Toward Ambidextrous Ports*. Rotterdam, Erasmus Research Institute of Management/ERIM.

PART 3

Ports and networks

Perspectives

17

PORT COMPETITION IN HISTORICAL PERSPECTIVE, 1648–2000

The ports in the Hamburg–Le Havre range

Hein A. M. Klemann

17.1 Introduction

Port competition is hinterland competition. When a region can be supplied from more than one port, the adequacy of the services in the port and their costs as well as the connections between these ports and their hinterland will be decisive for the question which port will be chosen. Notteboom (2008) emphasises that it is difficult to say exactly what is the hinterland of a port, that the hinterland can differ in time and for different kinds of cargo, and that 'market dynamics makes it dangerous to have a static concept of ports hinterlands as being God-given and everlasting.' Nonetheless, when looking to continental Western Europe, transport markets are characterised by the competition between a limited numbers of ports at least from the seventeenth century.[1] As an economist who just analysed the last few years, Notteboom is right when he comes to the conclusion that fluctuations are characteristic, but it may be useful to abstract from all these nervous day-to-day instabilities and draw a long line through the centuries. This chapter will try to do this for the ports of continental Western Europe. Thus it can be seen which long-term developments were decisive for the position of the different ports within the Hamburg–Le Havre range.

From late medieval, early modern times, Dutch ports – during the heydays of Dutch trade in the seventeenth century, especially Amsterdam with Dordrecht a far-off second – had an enormous advantage in their competition: they were connected with the hinterland by the major inland transport route of Europe, the Rhine. Of course, compared with all modern means of transportation, even on this major river the scale of transport was extremely small, but in the lower Rhine, from Cologne to the coast, barges could transport up to 150 tons of cargo. A horse-drawn cart on a well-paved road – and such roads were rare exceptions in early modern Europe – could only transport 1.5 tons per horse. Only a limited number of special roads allowed heavy transport with Hessen wagons, which were covered wagons pulled by more than two horses and using a larger track width.[2] Consequently, Rhine barging had a huge scale compared to all alternatives. It made transport possible that for practical reasons or prohibitive costs otherwise was impossible. Nonetheless, given the transport costs even in Rhine barging, cheap products were not worth transporting. They had to be made locally or were simply not available. As there hardly was any bulk transport – bulk goods were transported in barrels or boxes, and handled as if it was general cargo – transport of cheap mass products simply did not exist. The fact that as a consequence almost everything was produced almost everywhere in small amounts, is one of the causes of pre-industrial Europe's low productivity. As Rhine barging seemed in many cases the only way of transport, any competition for the Dutch ports at the Rhine estuary from other ports in the Hamburg–Le Havre range seemed unlikely. Nevertheless, there was such competition. In the seventeenth and eighteenth centuries, long before railways made large-scale inland transport possible when there was no navigable waterway, Bremen was often used as the port of Cologne or Frankfurt, while Le Havre in France was an important port for Switzerland and Southern Germany, places and regions that seemed the natural hinterland of the Dutch Rhine ports.

From the mid-nineteenth century, when the Prussian Rhineland and Westphalia started to industrialise, it seems logical that Rhine transport became more important. However, in the German industrialisation rail transport was dominant. When one can build a completely new railway network, the traditional geography of connections at once is not relevant any longer. It can go in all directions. For many in Belgium and Germany, it was clear that this was an opportunity to get rid of the Dutch, who according to especially Cologne traders had exploited and undermined German trade by tolls, taxes and regulations, destroying all new opportunities for the Rhineland economy for decades if not for centuries. Therefore, the best rail connections were created to Antwerp and the German ports. In the 1860s and 1870s the Dutch ports, the most competitive of the pre-industrial period, seemed to lose their position. The Dutch government did everything to recover the dominant position of its ports, but notwithstanding enormous investments in infrastructural projects, it hardly was successful. Only from the late 1880s the situation improved. Then the completion of the canalisation – mostly called normalisation – of the Rhine created a new situation. The enormous scale in Rhine barging that now became possible resulted in a sharp decline of freight rates. Consequently, just at the moment that in the iron and steel production new large-scale techniques were implemented demanding the cheap transport of enormous quantities of iron ore, pit wood and coal, the costs of Rhine transport dropped dramatically. Thus from the early twentieth century the Dutch ports, now in the first place Rotterdam, became dominant for this type of cargo. For general cargo a railway connection was much more efficient, and so for this Antwerp and the German ports became leaders.

Nowadays dry bulk is no longer the main cargo. From the 1950s wet bulk, oil and oil products prevailed. As long as such products went by rail or barge tankers, the port of Rotterdam was the principal one just as with all bulk, but when the scale of transport increased, pipelines became a cheap alternative for inland transport of such products. As new pipelines in their design stage are footloose and still were able to go anywhere just as railways in the mid-nineteenth century, it was not immediately clear whether these should go from Rotterdam to the German industrial centres. At first, the port of Marseilles near the oil production areas in the Middle East seemed a logical alternative. When it was clear that oil should go by sea tanker to Western Europe, the northern German ports protected by the German federal government had a good chance to become leaders in the transshipment and hinterland transport of these products. In the end however, it was Rotterdam again that took the main part of this type of bulk cargo. Finally, it even built up a very strong position in modern container transport. In the last few decades the German ports and Antwerp seemed to win some of their position back, however. The question is, of course, why the Dutch ports lost their traditional dominance in the mid-nineteenth century; how they – especially Rotterdam – could come back in the late nineteenth century, and how it could keep this position for almost a century, notwithstanding fierce competition supported by protectionism of the national governments of much larger countries.

17.2 Rhine shipping and port competition 1648–1806

After the Peace of Westphalia in 1648, the Emperor of the Holy Roman Empire of the German Nation – the old German empire – lost much of his authority, while the autonomy of the German princes, of which there were about 350, as well as of the Free and Imperial Cities (40), substantially increased. Because the princes were always in need of money and the cities were every time trying to increase their privileges, the fragmentation of the Reich resulted in more guild regulations, artificial monopolies, staple markets, increased taxation and more and higher tolls. This can be seen along the Rhine but also, for example, along the Elbe River. Taxation and regulation were all but fatal for Rhine shipping, and even for the competitiveness of the entire German Rhine region.

In Cologne and Mainz, the skipper of every passing barge was obliged to enter the port and offer his cargo for sale. After it was sold, the new owner could transport it further along the Rhine, but only if he used the services of the local skipper guilds. The rates of those skippers were set by these cities, just as those of the horses needed for upstream towing. As the stream was often stronger than the wind, up to six horses were needed to pull the barges. Around 1850, when steamers became more common in Rhine transport, there were still about 2,000 horses used in Rhine shipping. Consequently, the transport market was monopolistic, as local governments and guilds suppressed competition. Thus freight rates were kept high. The staple markets of Cologne and Mainz were founded in the medieval period, but various other forms of regulations and taxation undermining Rhine shipping were introduced only after the 1648 peace treaty. Their legal basis was controversial, but the emperor was far away in Vienna and his power was limited. There were some attempts by the most important regional princes to coordinate regulation and give the skippers some air. Every few years representatives of the region's electors – princes with a seat in the electoral college that decided who would be the next emperor – met at Bingen, a town near the Lorelei, to discuss such issues in the so-called Shipping Chapter. However, these electors, the bishops of Trier, Mainz and Cologne and the Count of the Palantine (Pfaltz), could not solve the problems at hand, if only because they defended their interests just as jealously as most lower-ranking German princes.

Because Rhine shipping fluctuated with Dutch Rhine commerce, some eighteenth-century observers had the impression that non-Dutch ports like Bremen and Hamburg were to blame for the decline of navigation on the river. However, these observers mixed up results and causes of the problem. The fact that traders using Dutch ports and Rhine connection for their trade with the German hinterland felt more and more competition from the ports of Le Havre, Bremen, Hamburg and Antwerp should be attributed to the many barriers to Rhine shipping. As a consequence of these obstacles, transport by horse-drawn cart was relatively cheap. From Le Havre the route went for instance by barge to Paris along the Seine and from there by horse and wagon to southern Germany or Switzerland. Another often-used track went via Bremen and the Weser, and from that river by horse-drawn cart to Cologne. As the scale of transport by horse-drawn wagons was extremely small compared to that of barge shipping, Rhine transport should have been much cheaper on a free market. In pre-industrial Europe markets were not free, however. It was normal that local princes defended old privileges, raised tolls where they could, chartered taxes and tolls to private tax collectors or gave this privilege to favourites.

The success of land routes competing with Rhine shipping and so of alternatives for the Dutch ports at the Rhine estuary can only be explained by taxation and regulation obstructing river navigation. Of course this was known at the time, but short-sightedness and the self-interest of the princes and cities undermined the main inland transport route of Europe. Every now and again there were people among the princes and their representatives who publicly declared that giving up the narrow-minded policy of local interests would result in an improved situation for everyone. Free navigation was already discussed at the Peace of Westphalia (1648), Baden (1714) and Rijswijk (1797), but for small princes the local Rhine toll often provided a substantial part of their income, while their more powerful colleagues needed the money for warfare or court extravagances and the cities along the river defended their vested interests.

Notwithstanding the fact that in theoretical discussions many understood the importance of liberalisation not only in the general interests of the Rhine states, but even in the interest of economic development, such vested interests were too strong. This became clear when in 1699 in Cologne a conference was held of all Rhine states, including France and the Dutch Republic. There these problems were discussed, but even the initiative to strengthen the competitiveness of Rhine barging, and thus of the Dutch ports by also taxing road transport, failed completely. There were so many roads and paths that it was always possible to find one without a toll office and land tolls were ineffective. However, they raised transport costs even higher. As a result, late eighteenth-century transport between Amsterdam and Frankfurt am Main was 15%–35% cheaper by horse and wagon than by Rhine barge. This not only resulted in more transport by land and thus via non-Rhine ports like Bremen, Antwerp or Le Havre, but above all in a substantial decline in transport in general. Products from the German Rhineland lost their competitiveness on the international markets. For example, in the Amsterdam trade French wine supplanted Rhine wine, simply because French wine was cheaper. The undermining of Rhine commerce thus affected the market for the products of the riparian states.

Partly because it was always possible to pay bribes or otherwise corrupt toll officials, Rhine transport never entirely collapsed. Nevertheless, according to Dutch sources, by the end of the eighteenth century, only 60 barges a year passed the Dutch-Prussian border in upstream direction. In The Netherlands this extremely low level of transport was mainly attributed to Prussian tolls just across the border. The Dutch data suggest that only 8,600 tons of cargo went upstream across the border. In other words, as far as Dutch ports were considered, the German hinterland was all but lost. German sources were much more optimistic and as substantial quantities of Dutch colonial products like sugar, tobacco and coffee were traded in the German Rhine region, it seems that there was something in their optimism. It suggests that trade and transport went on by clandestine methods. Probably the figure of 60 ships per year is inaccurate, although on the basis of the revenues of the tolls at Emmerich it can be calculated that in 1806, just before the introduction of Napoleon's Continental System, the total volume of the ships – upstream and downstream together – was no more than 66,000 hundredweight, or 3,331 tons. Again clandestine trade is not included, and there are indications that this was substantial. Nonetheless in the lower Rhine region only 88 barges were active, most of them only in local transport. Probably some of these small ships – not necessarily an indication of the size of the ship but of trade from a village to the nearest market place – were also illegally active, however, on large-scale transport across the Dutch border. Contemporaries correctly concluded that very little was known of the actual importance of Rhine shipping and thus of the relevance for the German hinterland of transport over Dutch ports. Taxes were everywhere, but it was almost always possible to corrupt the underpaid officials or to bribe their higher-ranking supervisors; the same is true for all kinds of other transport. Leaving the road or the river just before a toll, and returning at the other side by making a detour was common practice if one can believe the scarce information. Anyway, statistics for this period are not just rare, but also very unreliable. It is clear however that although Rhine shipping was confronted with many obstacles, it never was completely lost. These obstacles not just shifted trade from Dutch ports and the Rhine to alternative ports in Germany, France and the southern Netherlands, but in the first place destroyed trade. Only the French Revolution and the resulting territorial shifts in the Rhine area would change that situation.

17.3 The long nineteenth century

In 1792 the Conseil Exécutif, the French revolutionary government, stated that any obstruction of trade was a breach of natural law. Hence, in 1794, when France conquered the entire left bank of the Rhine, diplomats of the new republic pleaded for liberalising Rhine barging. Yet it was not until 1802 – after the downfall of the Second Coalition against France – that Paris used its power to implement the principle of free trade on the river. In 1804 it forced the collapsing Holy Roman Empire to accept that the authority over the Rhine from Switzerland to the Dutch border was transferred to a director-general. This official would collect a tax that replaced all tolls and would guarantee that these were considerably reduced. Eliminating the staples in the by then French cities of Mainz and Cologne was impossible, however, if only because Paris tried to undermine the position of the Dutch ports in favour of its by then French-occupied Antwerp. Notwithstanding this and the Continental System, the French Revolution and the Napoleonic Era were a period of liberalisation. After the Napoleonic time, this was picked up at the Vienna Congress where it was decided that trade on international rivers should be free and the Rhine should be liberalised. Again, however, political conflicts between the Rhine states – apart from Switzerland, France and The Netherlands, by then only five German states – made it impossible to implement such ideas. Only in the mid-nineteenth century when a new transport situation appeared, as railways made large-scale inland transport possible in regions without navigable waterways, did this change as well.

In the early nineteenth century, there were severe conflicts between the kingdom of The Netherlands (including Belgium) and Prussia, then two states of more or less the same size. Both were trying to improve their positions. The resulting conflicts concentrated on the Rhine, as according to the Agreement of Vienna of 1815, the river should be liberalised. The Dutch wanted to use their position at the estuary of the river however, to control trade with the hinterland. For them transit was not enough; they hoped to revive the position of the seventeenth-century Dutch staple market, when Dutch traders bought all overseas imports, and resold these to foreign traders wherever needed. In fact, the German hinterland did not need intermediate trade any longer and cut out Dutch middlemen. The Dutch king taxed direct transit without Dutch intermediators, however. For Prussia this transit tax was a reason not to terminate the remains of the Cologne staple as was agreed in Vienna. All ships had to transship there on barges of the local guild. For Hessen this was a good argument to do the same in Mainz. Consequently, the modern liberal ideas on free trade and economic development, which by then were dominant in the Prussian Rhineland and Westphalia, were confronted with conservative ideas on vested interests in The Netherlands. Especially in Cologne, the most important economic centre of this part of Germany, important businessmen and liberal politicians tried to find a way to circumvent the Dutch. Wild ideas on digging a canal from the Rhine to the German coast were replaced by more realistic possibilities when railways appeared. It was not just Friedrich List who thought that the railway was a solution for Germany. Already in 1825, a year in which he came up with plans for a railway from Cologne to Minden, one of the first Ruhr industrialists, Friedrich Harkort (1793–1880) wrote: 'The railways will cause many a revolution in the world of commerce. It will connect Elberfeld, Cologne and Duisburg with Embden or Bremen and the Dutch tolls will not exist anymore!'

The first rail link between the German Rhineland and the sea was only opened in the 1840s. Until then, Dutch transit levies dominated trade. As a consequence in 1831 Rhine freights were again on the same level as before the French Revolution. Transport by horse wagons from German and French ports was again a realistic alternative for river navigation via the Dutch ports. Only after the opening of the railway Antwerp-Cologne in 1844 did the Dutch have to adapt their policy. As Antwerp was in Belgium, it became possible to circumvent the Dutch with the most modern way of inland transport, the train. The seventeenth-century staple trade was lost forever. Transit was all there was left, but from the moment Antwerp had its railway, even in transit, Dutch ports could hardly compete any longer.

Railways were fast, modern, accurate, and their scale was much larger than even that of the largest Rhine barge. Barges, especially old-fashioned sailing barges, were extremely slow, especially when towed, small scale, and in wintertime ice obstructed navigation completely. Finally, although Rhine shipping was cheaper than rail transport per ton/kilometre, in the end the price difference seldom was favourable, as most of the time the cargo of a barge needed to be transshipped a number of times, while railways often could reach the final destination. Until the 1840s Prussia was trying to break the Dutch monopoly on the exit to the sea, but at the same time Rhine transport was extremely important for this massively industrialising German state. After the power of the guilds was broken – with the Rhine Charter of Mainz in 1831– barging created a competitive transport market, while nineteenth-century railways owned by private companies all had a monopoly on their tracks. As railways almost everywhere destroyed inland navigation – in England rail companies even bought canals to fill them – this was most unfavourable for a country whose major industrial centres were relatively far from the sea. For its supply with raw materials and exports of coal, the Ruhr area needed good port connections. The most efficient ports were the non-German Antwerp or the Dutch Rotterdam. In the 1860s and '70s, when railways became dominant, Antwerp seemed to become the most important port for the most important industrial centre of Europe, the Ruhr area. Consequently, in The Netherlands there was some panic. Enormous investments were needed to improve the position of the Dutch ports. Both Amsterdam and Rotterdam needed a better connection with the sea. Even more important seemed the development of a port with a good railway connection with the hinterland. Everybody knew that railways had the future, so getting a good railway port was absolutely necessary to remain important in transit. For that reason huge investments were done to modernise the old navy port of Vlissingen – the only absolute ice-free North seaport – and connect it by rail to Germany and Paris. This was extremely expensive as Vlissingen was on an island, and this island – Walcheren – could only be connected with the mainland via another island – Zuid Beverland. As it cut off two waterways needed by Antwerp, connecting these islands resulted in the obligation to dig two new expensive canals. It seemed, however, that it did not matter. Vlissingen – a town with fewer than 10,000 inhabitants – should become the most important port of Western Europe, so everyone in The Netherlands agreed in the 1860s.

Actually Rotterdam became the most important port of Europe, although a better connection with its hinterland was vital, for quite a different reason. Rhine transport never disappeared completely, as did inland water transport almost everywhere else, and even made a spectacular comeback in the late nineteenth century. Prussia feared the monopolies of the railways and high transport costs for its landlocked industrial centres. In 1847, on the difficult track Strasbourg-Basel a new railway destroyed Rhine navigation within a year. Therefore, in 1847, the Prussian commissioner in the Central Commission for the Navigation of the Rhine (CCNR), the 1815 founded organisation of all Rhine states with the duty to liberalise navigation and control that all Rhine states kept the channel and tow paths in a good condition, made clear that it was necessary to make the Rhine navigable for modern large-scale steam barging. A technical commission elaborated that the Rhine was hardly navigable for modern barges. Thereupon an enormous canalisation process started that would last 50 years and was hindered by technical, financial and organisational problems.

To make sure that in the end the Rhine would be navigable for large-scale barging, Prussia used its growing power to put pressure upon smaller Rhine states. The tiny Nassau, that controlled an important but difficult and thus expensive to regulate part of the river, was simply absorbed by Prussia after the 1866 internal German war. Notwithstanding all these problems, at the end of the nineteenth century, the Rhine was navigable from Rotterdam to Mannheim without meeting any locks for steam-pulled trains of ships of up to 400 metres and with a capacity of 6,000 tons. As barges were now of iron and the steamer a modern ship with a compound engine and a screw – no longer the traditional Watt engine and radar – the use of fuel (coal) had not increased since the mid-century when the trains were only 40 metres and the capacity 400 tons. In the second half of the nineteenth century, the traditional way of loading barges also disappeared, and it became common practice to fill the entire ship with one type of bulk good. Rhine shipping became much cheaper.

In 1880 the freight rates of Rhine shipping as well as the railways were 79 (1870 = 100). In 1913, those of the German railways were 64, but those of Rhine shipping only 19. Of course the overall price level fluctuated, but it was in 1913 more or less the same as in 1870. As a consequence of the enormous decrease of the costs of Rhine shipping at the moment that the demand of the German industry for raw materials grew massively, the port of Rotterdam became dominant in continental Western Europe. In 1913 almost 25% of all German exports and 22% of its imports crossed the Dutch-German border on a Rhine barge. In the last decades of the nineteenth century and the first of the twentieth century, there was a statistically significant relationship between the freight rates of Rhine shipping in percent of that of the railways, and the share of German trade crossing the border on a Rhine barge. Furthermore, a significant relationship between Rhine shipping and the incoming cargo in the port of Rotterdam was demonstrated. Until the early twentieth century the relationship between the outgoing cargo from Rotterdam and German exports correlated less, due to the policy of the Rhenian Westphalia Coal Syndicate. The competition between the ports was significantly influenced by the cost of transport to the hinterland. As a consequence of the recovering of Rhine shipping, in this part of the world transport markets remained competitive. As a result Rhine shipping became dominant for bulk, and Rotterdam as the port at the Rhine estuary the bulk port. For general cargo, railways were a better way of inland transport, and as Antwerp had better rail connections with the hinterland, this port became dominant in that type of transport.

17.4 The twentieth century

In the period until the Second World War, when coal was a major export product of the Ruhr area, Rotterdam as a bulk port had more or less the same amount of incoming as outgoing cargo. Coal went downstream, while iron ore, pit wood and cereals went upstream. This balance was destroyed when from the 1950s coal disappeared as the main source of energy and oil became dominant. In the same period new modes of transport (trucks and pipelines) and new types of cargo (oil, chemicals and containers) became relevant. This completely changed the situation. Of course iron ore and cereals still went upstream along the Rhine, but when the demand for the new commodity, oil, reached such a scale that pipelines were needed, the main advantage of Rotterdam, its cheap waterway connection, was worthless in a stroke. Other ports also tried to become the oil hub for Germany. The oil companies decided that there were two possibilities: the German Wilhelmshaven or Rotterdam. As especially Esso wanted to use the biggest tankers available for transport to this part of Europe, the question which port would have the capacity to facilitate these became vital. In 1957 in Rotterdam the Port Authority reacted with the largest postwar plan for expansion: Europoort. Wilhelmshaven received financial support from the German government to develop a pipeline.

The only oil company that did not want to go to Wilhelmshaven was Royal Dutch Shell, as it had already an enormous refinery in Rotterdam. That was, however, not the only problem the Shell group had with the developments. Shell did not believe in national solutions. It thought the European market too small to fall apart in a number of countries. Therefore it initiated already in 1956 a European pipeline plan. The economies of scale won, but the pipeline would be destroyed by economic nationalism resulting in a number of national pipelines. Therefore it came with a plan to pump oil through a pipeline from Marseilles all over Europe. Even Rotterdam would get the oil it needed through that pipeline. The increasing scale of oil tankers and the limited capacity of the Suez Canal were already a problem before the 1956 Suez crisis, when such ideas were finally wiped out. Oil would come from the Middle East by large tankers that went via the Cape of Good Hope and so would come ashore somewhere in north-western Europe. Thereupon Royal Dutch Shell decided in 1957 to build a pipeline from Rotterdam to Cologne. At the same time the Rotterdam Port Authority – which had close connections within the Dutch Shell group – assured that even the largest tankers could enter the port. By guaranteeing this in a very short period of time, Rotterdam took the lead in this market. With this pipeline and infrastructure it had the strongest connection with the hinterland again, also for this new commodity. It would be the start of the development of Rotterdam as the main oil port in Europe and for some decades even the largest port in the world.

17.5   Conclusions

Competition between ports is in substantial part competition between the connections with the hinterland. These connections are dependent on technical and political developments. For centuries the Rhine was the only more or less efficient transport route of continental Europe. Therefore the ports at its estuary were most likely to become predominant. Exactly for that reason it also was most interesting to raise tolls and organise monopolies along the river. These obstructed Rhine shipping, although there are strong indications that river transport never was destroyed completely and that clandestine transport compensated partly for these developments. Nonetheless, ports like Bremen, Antwerp and Le Havre received chances they would never have had as their connections with the hinterland were completely based on a very small scale and thus inefficient transport by horse-drawn carts. Only the obstructions of Rhine transport by tolls and staples created the opportunity for these ports to become of more than local importance.

New modes of transport created new opportunities, and from the mid-nineteenth century the railways created a completely new situation. At the start of the development of this new mode of transport, it was possible to create railways wherever you wanted them. Traditional transport routes were not relevant anymore. Belgium and the German Rhineland had good reasons to circumvent The Netherlands, and so it seemed that the Dutch ports would lose their strong competitiveness. The new railways went to Antwerp and the German ports, leaving the Dutch ports with their old-fashioned, small-scale waterway connection with the German hinterland. That transport via the Rhine came back was hardly a credit one could give to the Dutch. Prussia wanted to keep the transport markets competitive in this part of Europe, as it needed a cheap means of transport for its huge developing industries. For that it needed the Rhine, as river transport by then was liberalised and skippers competed with each other, while on railways every track was a monopoly. As Prussia became the dominant power in the region, it could press the other Rhine states to participate in its enormous canalisation project, which was only completed in the 1890s. From then on, in bulk, Rhine transport became dominant and as the German industrial centres needed ever more bulk transport, the Rhine became the most important transport route of Europe. With that Rotterdam became the dominant port. However, as Notteboom wrote already, the hinterland of a port is different for diverse types of cargo. As for general cargo, railways were a better way of hinterland transport, the German ports and especially Antwerp became dominant in that.

What happened with railways happened again with the new modes of transport that developed after the Second World War. Trucks and containers have not been discussed here, as the space is too limited for that, but for oil all possibilities seemed open when the scale became so large that pipelines could be used. The reasons why in the end it was Rotterdam again that developed into the most important port for this type of cargo, and thus in the main port of Europe, are diffuse. When the discussion started what should be done, there were a number of other possibilities. When the pipeline was there, and Rotterdam made sure in a short period of time that it could accommodate even the largest tankers, it won and created again a stable connection with the hinterland.

Notes

1 Theo Notteboom, *The Relationship Between Seaports and their Intermodal Hinterland in Light of Global Supply Chains*. Discussion Paper OECD 2008–10 Research Round Table, Seaport Competition and Hinterland Connections, Paris, 10–11 April 2008, 4.
2 Probably the name had nothing to do with the German region Hessen, but rather with an old Germanic word for horse.

Reference

Notteboom, T. (2008) *The Relationship Between Seaports and their Intermodal Hinterland in Light of Global Supply Chains*. Discussion Paper OECD 2008–10 Research Round Table, Seaport Competition and Hinterland Connections, Paris, 10–11 April 2008, 4.

Suggestions for further reading

Besides the papers provided as references in the text, we recommend the following texts as suggestions for further reading.

Boon, M. (2014) *Oil Pipelines Politics and International Business. The Rotterdam Oil Port, Royal Dutch Shell and the German Hinterland, 1945–1975*. Unpublished thesis, Erasmus University Rotterdam.
Klemann, H., and Schenk, J. (2013) Competition in the Rhine delta. Waterways, railways and ports, 1870–1913. *Economic History Review*, 66(3), 826–847.
Paardenkooper, K. (2014) *The Port of Rotterdam and the Maritime Container. The Rise and Fall of Rotterdam's Hinterland (1966–2010)*. Unpublished thesis, Erasmus University Rotterdam 2014. Retrieved from http://repub.eur.nl/pub/51657?_ga=1.152349243.300584682.14 17590831

18

SUSTAINABILITY

Harry Geerlings and Tiedo Vellinga

18.1 Introduction

As shown in many chapters of this book, transport plays a crucial role in modern societies. A well-functioning maritime transportation system facilitates the process of globalisation and ongoing economic growth; it is a motor for economic processes. Also the port-cities benefit directly, not only in terms of employment and well-developed infrastructures, but also with respect to related service industries and so forth. In the EU-28, these activities represent about 10% of the employment, but for Singapore and Shanghai this is even above 20%. This explains why the transport sector, for more than a century now, has experienced unprecedented growth. At the same time, transport has undesired side effects. We see that the maritime transport sector (including ports activities and hinterland connectivity) is increasingly associated with negative feelings: there are serious concerns related to emissions (at the regional, national and global levels), safety, public health issues and so forth. To illustrate: more than half of the SO_2-emissions in Hong Kong are ship related, about 85% of all truck traffic on certain highway sections in Los Angeles is port related, the port area in Antwerp encompasses one-third of the city's land mass and so forth. All these concerns are encompassed in the concepts of sustainability. To improve the environmental performance of shipping, port activities and hinterland transportation, and to fulfil environmental standards are nowadays preconditions to operate.

In this chapter we first look to what is meant by sustainable transportation, the leading concept embraced to come to a (more) environmental friendly transport system. This is the central notion of this chapter. In Section 18.2, the definition of sustainable development (SD) is discussed. In this book (Parts 1 and 2) we consider a port as a nodal point in a global network that starts with shipping, next there are the port-related activities, and from the ports goods are transported to the hinterland (or vice versa). For all three links in this chain there is a need for sustainability.

Section 18.3 deals with the operationalisation of the concept of SD for shipping; Section 18.4 puts emphasis on the port-related activities; and Section 18.5 deals with the impacts of production activities. In Section 18.6, as an illustration, the case of standard settings in the United States, Europe and China is discussed and finally conclusions are drawn.

18.2   The concept of sustainable transportation

Over the last century, the transport sector has been characterised by an unprecedented growth. This growth can also be observed in maritime transport as well, and it is a trend that is occurring all over the world. In general, the growth of freight transport has been faster than the economic growth. Projections up to 2020 indicate a further growth in transport, particularly in freight transport and concentrated in Asia: freight and passenger transport is predicted to increase by 52% and 35%, respectively, between 2000 and 2020.

The success of the transport sector has also a negative impact in several respects. Initially, the attention was on the impacts of emissions (such as SO_2, CH_4 and NO_2) and on health and safety, but at present a new concern has been added to the political agenda: climate change (CO_2). The concern for the environmental burden caused by human action has been studied for many years and is reflected in numerous reports and policy documents; nowadays it is covered under the concept of "sustainable development." This term was first mentioned in 1980 in the report *World Conservation Strategy* of the International Union for Conservation of Nature and National Resources (IUCN). It has since then become central to thinking on environment and development for a small group of scientists and policy makers. The discussion has gained new momentum by the introduction of the concept of sustainable development in the 1980s.

In 1987, the World Commission for Environment and Development (WCED) described the concept of sustainable development more extensively in their report "Our Common Future" (the "Brundtland Report" (WCED, 1987)). In this report, sustainable development is defined as "a development that meets the needs of the present without compromising the ability of future generations to meet their own needs." It is interpreted as "a process of change in which the exploitation of resources, the direction of investment, the orientation of technical development, and institutional change are all in harmony and enhance both current and future potential to meet human needs and aspirations."

There is, however, as yet no universally accepted definition of sustainability, sustainable development or sustainable transport. This is definitely also the case for sustainable maritime transport or sustainable ports. Sustainable development distinguishes several, sometimes seemingly opposing goals, which makes it a very difficult task to find synergy between them. Three interrelated systems within sustainable development include the ecological system ('planet'), the economic system ('profit'), and the socio-cultural system ('people'). This is called the Triple Bottom Line (TBL; Figure 18.1).

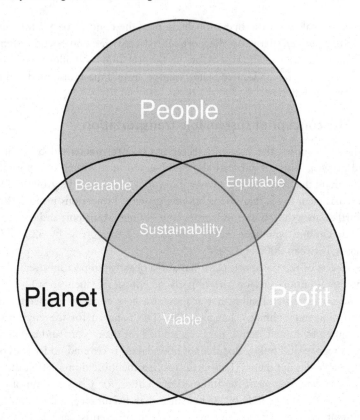

FIGURE 18.1 Triple Bottom Line

Especially in matters where finding the right balance between ecological and economic aspects is concerned, development in harmony with sustainability will be more difficult to put into operation.

From a "profit" perspective we can distinguish several economic impacts of ports, for instance, port-related value-added growth, port-related employment growth, port-related labour productivity, moderate economic impact with relatively large spillover effects and so forth.

From a societal perspective (the "people"), attention is paid to issues such as equity, human health, quality of life, education, public participation and so on. It is expected that the focus on the impacts of the port (and its related) activities on quality of life (QoL) will increase, and this will have implications for the transport sector.

Concern for the environmental impact of ports (the "planet") has been studied for many years. This is not an easy topic to address as the characteristics of the external effects are quite substantial and differ significantly. In terms of time scales, some are instantaneous (e.g., noise), some are pervasive (e.g., hydrocarbons), some are permanent (e.g., visual intrusion) and others are cumulative (e.g., CO_2). Spatially, there is a clear distinction between adverse effects that have a direct impact at the place where they are generated (e.g., emissions of lead) and those that are carried through the air on a continental or even a global scale (CO_2). Some of the impacts are associated with direct physical effects and can easily be measured (SO_2); others, particularly those affecting public health and the quality of life, are more subtle and less susceptible to objective measurement. Sustainability is sometimes defined narrowly; other studies of sustainability focus on long-term resource depletion and climate change, on the grounds that they represent the greatest risk and are prone to being neglected by conventional planning. It is also important to stress that the concept of sustainable mobility is not a static situation. The concept of sustainability has to be achieved over time, and is part of a process (temporal aspects) which also manifests on a spatial scale (spatial aspects; see Geerlings, 1998).

Therefore, sustainability is an essentially contested notion; it is intrinsically complex, normative, subjective, ambiguous and inherently context-specific. It is one of those concepts which "inevitably involve endless disputes about their proper uses on the part of their users," and "to engage in such disputes is itself to engage in politics." In summary, the concept of sustainable development implies that it is a subjective, dynamic concept with different degrees of freedom. It provides a framework for directions of change and as such it can be used as a constructive tool. Still it is up to the decision makers to operationalise sustainability and a change of attitude is therefore a crucial precondition (Geerlings, 1998).

The concept of sustainability is translated and applied in many different sectors, such as agriculture, production processes and energy conservation. It is also being applied in the transport sector. Sustainable transport, according to the US Transportation Research Board, is a transport system which:

- Allows the basic access and development needs of the people to be met safely and is consistent with human and ecosystem health, and promotes equity within and between successive generations;
- Is affordable, operates fairly and efficiently, offers a choice of transport mode and supports a competitive economy, as well as balanced regional development;
- Limits emissions and waste, and uses resources at the level of generation or development, while minimising the impact on the use of land and the generation of noise.

But in the development of the transportation sector and all its external effects, the optimum situation rarely occurs. There is, for instance, a tension between improving accessibility and the decreasing emissions, or between economic growth and the trade-off with congestion.

Still it is expected that the environmental interest will increase in importance, and will have implications for the maritime transport sector. We observe, partly because of the need to address the very complicated policy challenges such as global warming, a new sense of urgency. This requires a structural change and a redefinition of policy institutions, and will affect the balance between transport policy, environmental policy and spatial policy. The EU was leading when it introduced the so-called Cardiff Agreement, initiated at the EU Council meeting in Cardiff in 1998, which was developed to integrate environmental concerns into transport policy. Since then the idea of sustainability has been implemented in the EU Treaty, and the European Union's Sustainable Development Strategy was published at the Stockholm Summit in 2001. The White Paper on European Transport states that "a modern transport system must be sustainable from an economic and social, as well as an environmental viewpoint" (European Commission, 2011: 3).

In the field of transport, there is a general awareness among governments and other stakeholders that new approaches and policy measures are needed. It is an observation that the environmental performance of liner shipping is far behind road and air transport. Urgent action is needed.

18.3 Sustainable shipping initiatives

Environmental performance of shipping is an area of increasing intensive attention. Shipping has the potential to contribute to sustainable transport because it can be exploited as a cost- and energy-effective transport mode. But in practice shipping has a substantial negative impact on the environment. Due to economic growth, the transport sector has become the largest energy consuming sector in the EU, accounting for 31% of the total energy consumption in 2013.[1] The contribution of the maritime transport sector to these emission gases in the last years has grown and vessels are becoming the biggest source of air pollution in the EU. The operation of deep-sea vessels creates substantial negative effects on air pollution and the marine ecosystem. Ship emissions have the potential to contribute disproportionately to air quality degradation in coastal areas; in addition they contribute to global air pollution. In this context it is common to make a distinction between two types of emissions: gases such as CO_2 with a global impact; and NO_x, SO_x and particulate matter (PM) with a regional/local impact. In European coastal areas, shipping emissions contribute to 1%–7% of ambient air PM_{10} levels, 1%–14% of $PM_{2.5}$, and at least 11% of PM_1. Contributions from shipping to ambient NO_2 levels range between 7% and 24%, with the highest values being recorded in The Netherlands and Denmark. Impacts from shipping emissions on SO_2 concentrations were reported for Sweden and Spain. Shipping emissions impact not only the levels and composition of particulate and gaseous pollutants, but may also enhance new particle formation processes in urban areas (Viana et al., 2014).

From the shipping perspective, the International Maritime Organisation (IMO; a United Nations specialised agency) is responsible for shipping and deals with the prevention of pollution from ships. IMO operates through a number of committees, including the Marine Environment Protection Committee (MEPC). In the 1970s MEPC developed the International Convention for the Prevention of Pollution from Ships (the MARPOL Convention), with the objective of providing a comprehensive range of measurements to limit the discharge of oil, chemicals, sewage and garbage into the world's oceans and seas. In shipping, sustainability can

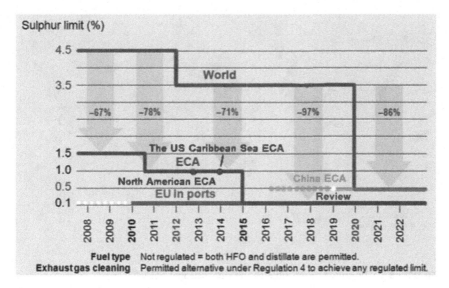

FIGURE 18.2 Sulphur limits for marine fuels

Source: Xu and Geerlings (2016).

go together with green and lean, as was illustrated when oil topped record-breaking prices in 2008. Even without the focus on reducing their emissions, businesses have spent time working on reducing fuel consumption through more efficient routing, better placement of distribution centres, new technologies that help trucks run more aerodynamically. All of these solutions reduce costs while also reducing carbon emissions. In the maritime industry this is all embraced under the heading of "green shipping,"[2] for example, transportation and logistics activities that strive to limit the negative external effects. And of course this must go beyond the low-hanging fruit of fuel efficiency.

The first legislation for the reduction of ship emissions dates back to 1993, when the European Union with the adoption of council directive 93/12/EEC decided to set limits for the amount of sulphur in marine fuels. This council directive has been amended several times, with the last amendment made in 2005. In 1997 the IMO adopted Annex VI of MARPOL, which was created to minimise several emissions from ships, such as SO_x and NO_x emissions and particulate matter. Annex VI entered into force in 2005. Furthermore, lower fuel oil sulphur limits were introduced for special areas, termed emission control areas (ECAs). In these areas a higher level of protection was demanded due to the proximity of shipping routes close to populated areas, the national susceptibility of acid depletion and so forth (see Figure 18.2). At the moment (2017) there are four established ECAs: the Baltic, North Sea and (from 1 August 2012) the North American. The US Caribbean Sea ECA is in effect from January 1, 2015 (see Figure 18.3).

A complication in the introduction of ECA zones is that a level playing field is missing and that the concept is only successful for areas where higher levels of

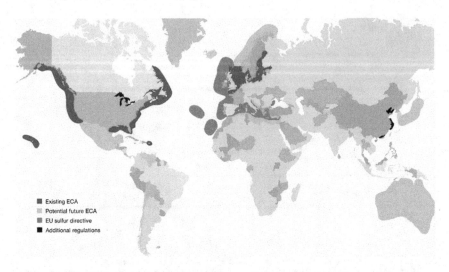

FIGURE 18.3   Existing and possible future international ECA

Source: Xu and Geerlings (2016).

protection are affordable. The approach taken within MARPOL Annex VI to SO_x is a staged reduction scheme. This should provide the industry and bunker suppliers time to plan for and adjust to the changes.

The solution for ship owners in meeting new sulphur limits varies across fleets, vessel classes and routes. For example, a ship owner operating vessels almost exclusively in an ECA could find an option with high capital expenditure viable and cost-effective if it offered reduced running costs while inside the ECA, whereas an owner with vessels trading in and out of the ECA zones might prefer to accept higher running costs on occasional visits to the ECA in exchange for lower or no up-front investment costs. Some owners are making changes to their entire fleet, some to certain vessels. While considering which solution is best, operators need to consider any limitation the solution may have on the short term and on the long term. The Baltic and International Maritime Council (BIMCO) is the largest international shipping association representing shipowners. Its members from over 120 countries control around 65% of the world's tonnage. BIMCO agrees with the ECA policies but stresses the need for a clear level playing field in regulation and enforcement. Currently, the most commonly selected ECA compliance option is fuel-switching – changing over to a lower-sulphur fuel, a process which must be complete before entry to an emissions-regulated zone and changing back to the regular fuel after the vessel has left the zone.

With respect to the global pollution, IMO has adopted the Ship Energy Management Efficiency Plan (SEEMP) and the Energy Efficiency Design Index (EEDI). The SEEMP establishes a mechanism for operators to improve the energy efficiency of currently operational ships. The newly adopted EEDI sets technical standards for improving the energy efficiency of certain categories of new ships, which in turn lead to fewer CO_2 emissions with estimated reductions of between 25% to 30% by 2030. The EEDI became mandatory from 2015, and requires a minimum energy efficiency level for different ship types and sizes. The EEDI requires, in its first phase (2015–2019) an efficiency improvement of 10% and will be tightened every five years to keep pace with technological developments and reduction measures. The EEDI will most likely be used through the Efficiency Incentive Scheme (EIS) that has recently been presented by the World Shipping Council. This instrument will force shipbuilders to construct ships with lower CO_2 emissions by imposing penalty fees on ships that do not meet the emission requirements for their class. These fees are collected by fund administrators that will possibly be controlled by national or local governments in the future. The funds are then invested back again in company assets, creating a return on investment.

From a governance perspective these efforts do not express high ambitions. The best results are reached at the moment by the efforts taken by the shipping liners which introduced "slow steaming" due to the high energy prices. In the 1990s it had already become evident that further issues, including air pollution, needed to be addressed, and that if the IMO did not take action, individual states would start to enact national measures which could result in a patchwork of differing, and potentially contradictory, requirements. In recent years we see that international bodies such as the EU, national governments and port authorities have a growing amount of influence in the policy making and legislative processes. Non-governmental organisations (NGOs) such as Greenpeace and the WWF have an influential role in these processes by putting environmental issues on the EU agenda and pressuring for stricter legislation and regulations for emissions. Also at the national and local levels various NGOs are in constant interaction with each other and with national and local governments to influence the decision-making processes, for instance on the discharges of oily bilge and ballast water; spills of oil, toxic or other cargo or fuel at sea or in ports; the noise from ships; the impact on marine animals and so forth. Finally attention is needed for what happens to a vessel at the end of its life cycle. In the coming decades a large number of ships are expected to be scrapped since they will reach their end-of-life period due to age, overcapacity in tonnage and the fast change of regulations with regard to construction and exploitation. Even today most vessels are scrapped by the beaching method: simply sailing a ship on the beach by high tide. The ship is stripped under very poor working conditions and the ecological, health and social hazards are obvious.

18.4   Sustainable port initiatives

Ports are nodal points or hubs in the transport network. Its industrial and business areas add value to the transported goods, and ports are increasingly becoming centres of energy production (and consumption). Increasing environmental awareness creates new challenges for the development of ports. Besides the economic activities, environmental issues, and the impact of port activities on public health, also climate change adaptation and mitigation measures are nowadays important in port policies. All these different elements are merged into the concept of the "green port" as an answer to the new challenges. But like the concept of sustainability, here too there is no clear and comprehensive description of what a green port actually is. However, there is a general understanding that a sustainable or green port strategy should be a strategy that accommodates the future development of the port in harmony with the region and the natural system. Although, how this should be reached, and which stakeholders should be involved, are still undefined.

The impacts of ports are quite significant. Sources of port-related emissions include oceangoing vessels, heavy-duty trucks, harbour craft, cargo-handling equipment and railroad locomotives that emit diesel particulate matter (DPM), nitrogen oxides (NO_x) and sulphur oxides (SO_x), all of which have been known to affect human health and contribute to the formation of smog. In 2006, the ports of Long Beach and Los Angeles together adopted the Clean Air Action Plan (CAAP). The CAAP focuses on strategies to reduce health risk to communities surrounding the ports and targets significant reductions in DPM, NO_x and SO_x. Since 2005, the port of Long Beach has cut diesel particulates by 81%. In addition, nitrogen oxides were down 54% and sulphur oxides were down 88% over the same period. These results, from data collected through 2012, represent six straight years of improving air quality in the harbour area.

Table 18.1 illustrates that ports are subject to a wide variety of policy initiatives. The relation between ports and the regulatory framework is worldwide far from uniform: in some regions, ports tend to regulate themselves as much as possible (e.g., the Hanseatic ports in Europe), while in other regions, ports rely on a strong national legal framework. In any case, the relations between the port and the local authorities have their main component in common sustainable issues. Public authorities include those at the national, regional or local level, which include the city. They are ruled by their own laws and regulations covering most activities of citizens and companies operating within the boundaries of their jurisdictions; in some cases they are shareholders of the port. Hereby the government is responsible for making public policies at all levels: local, regional, national and supranational. It is important to realise that national authorities are members of international organisations and have subscribed to many conventions that became part of their national legislation. In some cases, the national authorities are the local enforcement authority of the international conventions, for instance the IMO conventions. This is illustrated by a wide variety of policy instruments to mitigate air emissions in ports.

So we see that sustainable port management will benefit from governance at all levels and many international, national and local stakeholders are already involved when it comes to decision making and implementation. Governments are crucial, but more actors can be involved varying from multinational corporations to media, lobbyists, NGOs, finance institutions, public communities and so forth.

In the latter case, three of the latest ways forward are presented that contribute to more sustainable port management.

TABLE 18.1 Policy instruments

Policy types	Instruments	Intervention level	Examples
Regulation	Global emission cap	Global	IMO Marpol Annex VI
	Emission control areas	Cross-boundary	North America, Europe
	Technical standards	Global	IMO: EEDI, SEEMP
Information	Emissions inventory and monitoring	Local	Los Angeles
	Port-state control co-operation	National	Antwerp
	Compliance monitoring		US
Incentives	Bunker tax/emission trading	Global	
	Fuel switch	Local	Hong Kong
	Slow steaming	Global, local	Long Beach
	Cleaner ships	Local	ESI, Sweden
	Modal shifts	Local	Alameda corridor
	Truck retirement	Local	Los Angeles
Technology upgrade	Clean bunkering	Local	Rotterdam
	Shore power	Local	Gothenburg
	Electrification equipment	Local	Busan
	Renewable energy	Local	Zeebruges

18.4.1 Reporting

Some ports are reporting, within the framework of corporate social responsibility (CSR) policy, about a number of indicators that cover economic, social and environmental issues. The cornerstone of CSR reporting is the interaction with different stakeholders. Through identification of and reporting about relevant (environmental) issues, a basis is provided for new initiatives contributing to the license to operate, the basis for development and operations at each and every port. Some countries, like Spain, are promoting the obligation to report regularly on their sustainable performance.

The Global Reporting Initiative (GRI) is a network-based organisation that produces a comprehensive sustainability reporting framework that is widely used around the world. This reporting framework is based on the principles and performance indicators (PI) that organisations can use to measure and report their economic, environmental and social performance. Its cornerstone is the Sustainability Reporting

Guidelines. The third version of the Guidelines – known as the G3 Guidelines – was published in 2006, and is a free public good. GRI have developed a thoroughly revised version, the G4 Guidelines, which are now available. Next to these general guidelines on sustainability reporting, a number of sectoral guidelines are also available (e.g., for airports). There are no sectoral guidelines yet for ports. An international working group, however, in collaboration between the International Association of Ports and Harbours (IAPH) and the World Association for Waterborne Transport Infrastructure (PIANC), have presented a guidance for ports on sustainability reporting.

One of the lessons learned is that the guidance for sustainability reporting for ports should aim at a tailor-made approach, keeping in mind the possibilities of an individual port as well as the stakeholder interaction, and it can help to create more transparency regarding sustainable performance. It is important that it includes a process in which the port authority and its stakeholders participate in and jointly draft the key performance indicators that are relevant in their specific situation with regard to sustainability. Ports that are producing sustainability reports include the port of Amsterdam, port of Los Angeles, port of Antwerp, port of Auckland, port of La Coruna, port of Rotterdam, port of Sines (Portugal), port of Sydney and Transnet (which includes Transnet National Ports Authority of South Africa), and port of Metro Vancouver, but this list is rapidly growing.

18.4.2 Environmental management systems and certification

Certification refers to the confirmation of certain characteristics of the organisation, in our case a port. This confirmation is usually provided by some form of external review, evaluation, assessment, or audit. With the advent of the International Organization for Standardization (ISO) and the promulgation of ISO9001 (Quality Management System), ISO14001 (Environmental Management System) and ISO26000 (Social Sustainability), international ports were early adopters of a systematic approach to port operations and development certifications.

There is significant implementation of environmental management systems (EMS) or a facsimile thereof at seaports worldwide. The majority of seaports that utilise EMS limit the systems to specific properties, operations or programmes. In the United States, the American Association of Port Authorities has sponsored an EMS programme for over 25 port authorities to help realise these benefits. In Europe, with the full support of the European Sea Ports Organisation (ESPO), the concept of port environmental management has developed markedly during the last 15 years. The progress was driven by mutual collaboration between the port sector, research institutions and specialist organisations. The framework for this mutual collaboration was developed through joint activities instigated and funded by primary port partners and partly funded by EC Research and Development Programmes. The cooperation between port professionals, academic researchers and specialist organisations has proved to be a potent mix in terms of delivering a functional framework of cost-effective solutions developed to implement policies and produce continuous improvement of the port environment.

Over the last 10 years, the Port Environmental Review System (PERS) has been established as the only port-sector specific environmental management standard. The scheme effectively builds upon the policy recommendations of ESPO and gives ports clear objectives at which to aim. The EU funded project PORTOPIA created an integrated knowledge base and management system of port performance to serve the industry's stakeholders in improving the sustainability and competitiveness of the European Port System. Several seaports were also certified by the European Union's Eco-Management and Audit Scheme (EMAS). This is a voluntary instrument which acknowledges organisations that improve their environmental performances. Port Compliance is a US port-sector tool developed in a partnership between the National Center for Manufacturing Sciences (NCMS) and Environmental Protection Agency (EPA; see www.portcompliance.org/index.cfm for more information).

18.4.3   Environmental permits

As stated before, a port is also subject to existing regulations which include compliance with environmental permits. Around the world permitting procedures are different in their appearance, but ports could, together with the permitting authorities, proactively promote that the permitting instrument is transparent and includes stakeholder involvement including contractors, and that the instrument is used to ensure:

- Integrated assessment of port activities
- Integrated monitoring and evaluation of port activities.

It should be realised that in many situations transparent agreements with operators or the listing of requirements up-front in contracts can be very effective instead of the permitting procedures. When used well, this approach can be a welcome instrument for the landlord port manager (see Chapter 4) to ensure long-term sustainability and improve the transparency of the footprint of all the industrial and terminal-related activities in the port area, including the footprint of its related transport processes, when supported by integrating monitoring and evaluation processes.[3]

18.5   Hinterland connectivity

Hinterland connectivity is also an important issue for ports. The impact of port activities also reaches the hinterland. The modal share (also called mode split, mode-share, or modal split) is the percentage of goods using a particular type of transportation or number of trips using said type. In freight transportation, this may be measured in mass or TEU. Modal share is an important component in developing sustainable transport within a city or region. In Rotterdam, approximately 30% of the containers that enter the port continue their voyage across the sea. This transshipment does not put any pressure on the connections with the hinterland. The remaining 70% of the cargo finds its way to the hinterland via inland shipping, road and rail. Moreover, for the chemical industry, pipeline transport is of crucial importance. This division according to the type

TABLE 18.2 Modal shift objectives of Rotterdam between 2005 and 2033

Road	from	47%	to	35%
Water	from	40%	to	45%
Rail	from	13%	to	20%

of hinterland transport is called the modal split. Agreements have been made with all parties involved to reduce road haulage in favour of inland shipping and rail. Cutting back road transport is an important environmental objective. The modal shift objectives of Rotterdam between 2005 and 2033 are seen in Table 18.2.

The ever-increasing volume of traffic filling Europe's roads and towns is a huge challenge for citizens, for the economy and for the industry. In stark contrast, coastal shipping, inland waterway and rail options are not used anywhere near their full potential.

If we are to address this imbalance and the environmental impacts it creates, we must develop integrated strategies for the whole transport chain. As part of this, we need to optimise the interface and links between maritime transport and hinterland transport operators – including road, rail and inland waterways. It is also vital to make more efficient use of trucks. Making improvements in the logistics chain does not just mean investment; it requires the knowledge and expertise of the partners. Public authorities at various administrative levels have put in place a wide range of policy instruments to limit negative environmental impacts from ports in relation to the transport of the goods to the hinterland. Besides emission standards for vehicles used in the transport, and investments in better road and rail infrastructure, it can also be via financial measures (road pricing, taxation) or soft measures such as maximum transport speed limits, groupage, dedicated lanes for freight and so forth. This challenge is key to decision makers in the economic and management fields as well, whereby an integrated approach will contribute to a change in the operations and priorities of the logistics world.

18.6 An illustration: the role of standard setting

The earth's climate is predicted to change over time, in part because of human activities. Virtually all human activities have an impact on our environment, and transportation is no exception. While transportation is crucial to our economy and our personal lives, as a sector it is also a significant source of greenhouse gas (GHG) emissions – primarily carbon dioxide, methane, and nitrous oxide.

- Carbon dioxide (CO_2) is released to the atmosphere when fossil fuels (oil, natural gas, and coal) are burned.
- Methane (CH_4) is emitted during the production and transport of coal, natural gas, and oil.
- Nitrous oxide (NO_x) is emitted during combustion of fossil fuels as well as industrial activities.

Carbon dioxide (CO_2) is not a pollutant but a greenhouse gas which contributes mainly to global warming effects and which is associated with climate change. Greenhouse gases that are not naturally occurring include by-products of foam production, refrigeration, and air conditioning called chlorofluorocarbons (CFCs), as well as hydrofluorocarbons (HFCs) and perfluorocarbons (PFCs) generated by industrial processes. The heat-trapping property of these gases is undisputed. Although uncertainty exists about exactly how the earth's climate responds to these gases, global temperatures are rising.

On 12 December 2015, the heads of state of 195 countries agreed in Paris upon a framework of the United Nations Framework Convention on Climate Change (UNFCCC) named COP21 (Conference of Parties #21), dealing with greenhouse gas emissions mitigation, adaptation and finance starting in the year 2020. The aim of the convention is described in Article 2:

> Holding the increase in the global average temperature to well below 2°C above pre-industrial levels and to pursue efforts to limit the temperature increase to 1.5°C above pre-industrial levels, recognizing that this would significantly reduce the risks and impacts of climate change.

Transportation accounts for about 30% of global CO_2 emissions and is one of the few industrial sectors where emissions are still growing. Therefore, transport is one of the sectors that has been targeted where effective public interventions are being called for to reduce CO_2 emissions. There is widespread agreement to reduce CO_2 emissions from transport by a minimum of 50% at the latest by 2050. Still, the Paris Agreement is excluding the shipping and aviation industries. This is remarkable, as shipping accounts for about 3% of global warming and aviation accounts for 5%. And in recent years their emissions have grown twice as fast as those of the global economy; according to the European Parliament, they could make up 39% of world CO_2 emissions by 2050 if left unregulated.

To achieve the COP21 goals, an imminent peak in GHG emissions is required, followed by sustained emissions reductions. This means that all sectors of the global economy must reduce their GHG emissions by 40%–70% compared to 2010 levels. However, if no action is taken in the aviation and shipping sectors, these "bunker" emissions are expected to increase by between 50% and 270% by 2050. To achieve the 1.5- or 2-degree scenario, international shipping emissions must peak in 2020 and then start to decline sharply (see Figure 18.4).

Emission reduction targets for international aviation and shipping need to be urgently agreed upon so that these sectors can begin to contribute to the objective of avoiding a temperature increase of more than 1.5 or 2 degrees.

Transport ministers have addressed in a number of international conferences the need for CO_2 abatement and improved fuel efficiency in the transport sector, among other measures, mainly through:

- Innovative vehicle technologies, advanced engine management systems and efficient vehicle powertrains.

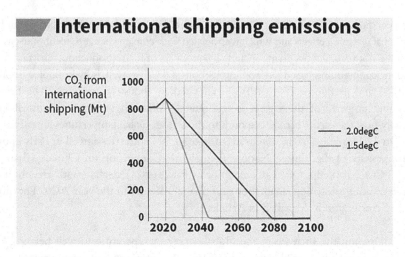

FIGURE 18.4   Required reduction international shipping emissions

Source: Transport & Environment (T&E) based on study by University College London and University of Manchester (accessed September 2016). https://www.transportenvironment.org/road-paris-climate-deal-must-include-aviation-and-shipping.

- The use of sustainable biofuels, not only of the first generation (vegetable oil, biodiesel, bio-alcohols and biogas from sugar plants, crops or animal fats, etc.), but also of the second generation (biofuels from biomass, non-food crops including wood) and third generation (biodegradable fuels from algae).
- An improved transport infrastructure together with intelligent transport systems (ITS) to avoid traffic congestion and to foster the use of intermodal transport (road, rail and waterways).
- Legal instruments (such as tax incentives for low-carbon products and processes, taxation of CO_2 intensive products and processes, etc.).

Emission standards for port-related transport are a promising legal instrument, as they can set specific limits on the amount of pollution emitted into the environment. Therefore standard-setting can be considered as an efficient way of direct regulation to realise objectives on the global, national, regional, and local levels, as it not only imposes limits (a constraint on behaviour), but also sets standards which can function as a positive impetus to stimulate technological development. This illustration focusses on the role of standards from a port and hinterland transport perspective, and describes the various policy initiatives undertaken in Europe, the United States and China (which represent the biggest ports in the world) with respect to standard-setting in transportation, with the aim to improve the air quality in their ports and hinterland connections.

Transport activities, excluding international shipping and international air transport, account for about 60% of the CO_2 emissions from total global fuel combustion (World Bank, 2016) and contribute significantly to the CO_2 emissions every day.

This also explains why so much attention is paid to inland waterway transport and railway transport. These two modes are considered to have relatively low carbon emissions. As discussed in this chapter, for those ports which deal with global trade flows, it is quite reasonable to give attention to the modal split and modal shift of hinterland transport. Governments and other stakeholders who work in the domain of port management are generally aware that policy measures are needed to find a balance between accessibility and sustainability. An important tool to achieve this goal, besides the modal shift policies, is to improve the performance of each separate modality. Emission standards are requirements that set specific limits to the amount of pollutants that can be released into the environment (Geerlings, 1998). Standard setting can therefore be considered as a way of direct regulation to realise targets on the global, national, regional and local levels, as it can be a good way to stimulate technological development. This is extensively discussed in Part 2 of this book.

Many emissions standards focus on regulating pollutants released by automobiles (motor cars) and other powered vehicles. The United States was ahead of the rest of the world in introducing emissions standards. The Air Pollution Control Act of 1955 introduced in the United States was the very first legislation which involved air pollution. This act provided funds for federal research into air pollution. And it was the Clean Air Act of 1963 that was the first federal legislation regarding air pollution control. It established a federal program within the US Public Health Service to control air pollution on the basis of emission standards on a national level. The 1970 amendments greatly expanded the federal mandate, requiring comprehensive federal and state regulations for both stationary (industrial) and mobile pollution sources. The State of California has special vehicle emissions standards, and other states may choose to follow either the national or California standards. California's emissions standards are set by the California Air Resources Board, known locally by its acronym "CARB." Given that California's automotive market is one of the largest in the world, CARB wields enormous influence over the emissions requirements that the major automakers must meet if they wish to sell into that market.

In Europe, the acceptable limits for the exhaust emissions of new vehicles sold in the EU member states are subject to increasingly stringent European emission standards, defined in a series of European Union Directives. Standards have been set for all road vehicles, trains, barges and "non-road mobile machinery" (such as tractors). In 2008, the EU planned to introduce the Euro 4 standard; for 2010 the Euro 5 and Euro 6 standards. These Euro 6 standards became effective on 31 December 2013. These dates to become effective were all three (for Euro 4, 5 and 6) moved forward for two years to enable the oil refineries to modernise their plants. For instance, for passenger cars, EU Regulation No. 443/2009 sets an average CO_2 emissions target of 130 g/km, which was gradually phased in between 2012 and 2015. From 2021, a stricter target of 95 g/km will apply.

In China, the number of cars on roads has rapidly grown as a result of increased wealth and prosperity, with a concomitant increase in pollution. Its first emissions controls on automobiles, equivalent to the Euro I standard, came into force in 2000. China's State Environmental Protection Administration (SEPA) upgraded these emission controls to the Euro 2 standard on 1 July 2004. An even more stringent emission standard, National Standard III, equivalent to the Euro 3 standard, came into effect on 1 July 2007. Euro 4 standards took effect in 2010. Beijing introduced the Euro 4 standard in advance on 1 January 2008 and was the first city in mainland China to adopt this standard. No standards exist for the CO_2 emissions of car use in China. Actually, in China, the relevant policies and measures for the control of carbon emissions were formulated and implemented by the Department of Climate Change, which is attached to the National Development and Reform Commission (NDRC) of the People's Republic of China; meanwhile, the local transport and port authority, as well as the local DRC, fully cooperate with the NDRC. At present, although China has no specific laws and regulations for CO_2 emissions, the Chinese government has taken many measures to control CO_2 emissions. According to the "Twelfth Five-Year Plan" for Energy Saving of Road and Waterway Transport, which was promulgated by the central government in June 2011, during the year 2015 the CO_2 emissions of road transport and inland waterway transport in China were decreased compared with 2005 (so in 10 years' time) by, respectively, 13% and 15%.

The most striking conclusion is that the increasing awareness of the sense of urgency in terms of public health, the widespread public concern, and the willingness of the government in China to address air pollution has recently led to policy acceleration with respect to standard setting. This seems in contrast to Europe, where clear standards have been formulated for vehicle emissions, even for the period after 2020, but where the European Commission seems to be slowing down its ambitions on account of economic interest. Overall, the standards applied in California are the most strict standards and function very well. This should be an inspiration for the global level playing field.

18.7 Conclusions

Increasing environmental awareness creates new challenges for the development of ports. In addition, climate change calls for adaptation measures that aim at not only minimising impacts of, for example, rising sea levels and increased floodwater heights but safeguarding accessibility of ports and waterways and sustainability for the social and natural environmental conditions. The definition of sustainability contains the concept of "future" as its time scale. In other words, the mentioning of sustainability should remind people of the "future" already. Green ports or sustainable ports are widely regarded as "the answer" to the aforementioned challenges (see for instance "Green Ship of the Future" at www.greenship.org/omos/). In the green port, based on a green growth strategy, sustainability is an economic driver in order to accommodate the future development of the port in harmony with the region and natural system. Sustainable thinking includes long-term thinking. In general we can see a shift of thinking away from a reactive "ports or nature/environment" approach towards a proactive "ports and nature/environment approach" that adds value through stakeholder participation. This green port concept does not only change the role of the port authority, but also the way in which operations are carried out. Under this concept, the port operates proactively and beyond legislation in a way based on a long-term vision.

The key issues that sustainable ports need to deal with are:

- Energy efficiency and energy transition (from fossil towards clean fossil and renewables)
- Sustainable materials and waste management (circular economy)
- Protection of habitat and integrity of ecosystems
- Climate change mitigation and adaptation
- Sustainable environmental quality (soil, water, air and noise)
- Stakeholder inclusive development and corporate social responsibility
- Co-operation with the private sector, public authorities, NGOs, academia and other ports.

To conclude, ports are in a unique and privileged position in the global logistics chain to capture and evolve their roles to initiate and consolidate the needed change, for their own benefit and the prosperity of the region that they serve. Frontrunners in port development and port management pick up this challenge and implement the growth paradigm, using innovative and state-of-the-art processes and technological developments. Also in shipping we see an increasing pressure, from the business community and consumers, to improve their environmental performance. This trend has to lead to new stakeholders' inclusive strategies that yield responsible innovation and create value for all parties that are involved. This is what sustainability is about and will determine the license to operate for all stakeholders involved.

Notes

1 For frequently updated information, see http://ec.europa.eu/eurostat/statistics-explained/index.php/Consumption_of_energy.
2 IMO proposed to introduce the concept of "Sustainable Maritime Transportation System" for shipping. See www.imo.org/en/MediaCentre/PressBriefings/Pages/41-WMD-symposium.aspx#.V1FV66P8LvU.
3 For an illustration, see www.portofantwerp.com/en/my-poa/services/port-one-stop-shop.

References

European Commission (2011) *White Paper Roadmap to a Single European Transport Area – Towards a Competitive and Resource Efficient Transport System.* COM/2011/0144 final. Office for Official Publications of the European Communities, Luxembourg.

Geerlings, H. (1998) *Meeting the Challenge of Sustainable Mobility: The Role of Technological Innovations.* Berlin/Heidelberg, Springer-Verlag.

Viana, M., Hammingh, P., Colette, A., Querol, X., Degraeuwe, B., De Vlieger, I., and Van Aardenne, J. (2014) Impact of maritime transport emissions on coastal air quality in Europe. *Atmospheric Environment*, 90, 96–105.

World Bank (2012) *Data Bank 2012.* http://data.world bank.org/data-catalog/.

World Commission on Environment and Development (WCED) (1987) *Our Common Future*. Oxford, Oxford University Press.

Xu, M., and Geerlings, H. (2016) *The Challenge of SECA Regulations in a Dynamic Environment; A Comparative Study of the EU, US and China*. Proceedings TRAIL-Congress 2016. PhD School TRAIL, Delft.

Suggestions for further reading

Besides the papers provided as references in the text, we recommend the following texts as suggestions for further reading.

Banister, D. (2008) The sustainable mobility paradigm. *Transport Policy*, 15, 73–80.

Elkington, J. (1997) *Cannibals With Forks: The Triple Bottom Line of Twenty First Century Business*. Mankato, MN, Capstone.

Meadows, D. L. (1972) *The Limits to Growth: A Report for the Club of Rome's Project on the Predicament of Mankind*. New York, Universe Books.

Suggestions for further reading on standards

Euro 1: See "Directive 1991/441/EEC Council Directive 1991/441/EEC of 26 June 1991 amending Directive 70/220/EEC on the approximation of the laws of the Member States relating to measures to be taken against air pollution by emissions from motor vehicles." Eur-lex.europa.eu.

Euro 2: See "Directive 2002/51/EC of the European Parliament and of the Council of 19 July 2002 on the reduction of the level of pollutant emissions from two- and three-wheel motor vehicles and amending Directive 97/24/EC." Eur-lex.europa.eu.

Euro 3 and Euro 4: See "Directive 98/69/EC of the European Parliament and of the Council of 13 October 1998 relating to measures to be taken against air pollution by emissions from motor vehicles and amending Council Directive 70/220/EEC." Eur-lex.europa.eu.

Euro 5 and Euro 6: See "Regulation (EC) No 715/2007 of the European Parliament and of the Council of 20 June 2007 on type approval of motor vehicles with respect to emissions from light passenger and commercial vehicles (Euro 5 and Euro 6) and on access to vehicle repair and maintenance information." Eur-lex.europa.eu. People's Republic of China (2011) *Twelfth Five-Year Plan for Energy Saving of Road and Waterway Transport*. Beijing, Ministry of Transport of PRC.

US Environmental Protection Agency (2016) *Emission Standards Reference Guide for On-road and Non-road Vehicles and Engines*. Washington, EPA.

19

PORT SECURITY

Changqian Guan and Shmuel Yahalom

19.1 Introduction

Before 9/11, the supply chain infrastructure and government regulations were mainly designed for economic efficiency. Security was not a major concern. The 9/11 attacks changed this paradigm. Supply chain vulnerability, especially in container shipments, was brought to the forefront of national attention in the United States and worldwide. A wide range of security legislations were enacted by Congress. Subsequently many security measures and regulations were implemented with the objective to enhance security across the supply chain. These have significant impacts on the supply chain community with respect to information sharing and reporting, cargo documentation, business practices, monitoring, tracking of cargo, and personnel training. Because most of the supply chain activities of container movements and cargo handling converge in ports, port security became a focus.

To address port security issues, the starting point is risk analysis or risk assessment. The analysis is focused on issues of the port. Treatment of risk takes into consideration the *probability* of an occurrence of an incident and the monetary *value* of the cost of the incident (direct and indirect). Because it is impossible to eliminate risk totally, there is a need to determine the *expected value* (probability X value) of an incident and what an acceptable risk is. The lower the risk tolerance, the larger the cost in terms of risk-reducing measures. The combination of the two determines the need of measures to be taken.

In general, since the cost of an incident in a port could be very large – estimated at about $1 billion a day in a major US port – the objective is to reduce its probability of happening. Addressing the cost of the incident is not practical because there is no knowledge of an incident's location. Therefore, there are different measures that are designed to reduce the probability of an incident in order to meet a social tolerance level. Thus, the chapter identifies various measures designed to reduce the probability of a terrorist incident in a port. These measures are designed to be forward-looking along the supply chain to the source of the goods' movement before loading. Once approved, the goods move onto the transit stage, which is followed by discharge and storage before reaching the customer. Security vulnerability exists at every link along the supply chain.

The modern-day freight system was designed with efficiency as its main focus. Given the added security concerns, how does a port keep the balance between maintaining freight movement efficiency and meeting security requirements? This chapter provides an overview of port security issues, concentrating on container security in order to mitigate vulnerability. After discussing port security issues and major US regulations/initiatives, the chapter addresses collaborations, risk-based approaches to port security, and emerging threats of cyber attack on ports.

19.2 Background

The global supply chain comprises a complex network of raw material supplies, manufacturing, warehousing/distribution, and end users linked by transportation services. This involves three main flows: information, materials, and money. Responding to global supply chain needs, a sophisticated and highly efficient transportation system was developed. The most prominent one is container shipping services. Since its inception in 1956, the container trade has been experiencing rapid growth in major trade lanes: North America–Asia, Asia–Europe, and North America–Europe. Container shipping services are highly efficient, low-cost, convenient, flexible, and multimodal. The container service network reaches every corner of the world with direct and feeder services provided by container shipping lines (Figure 19.1). There is a variety of container services provided to customers, or shippers. There are many partakers along the supply chain for a container moving from origin to destination. Each participant plays a different role depending on market needs. In general, the major supply chain entities that participate in container trade include:

- Importers and exporters
- Container carriers (container shipping lines)
- Non-vessel operating common carriers
- Freight forwarders
- Port authorities
- Terminal operators
- Trucking companies
- Railroads
- Warehouses/distribution.

The port-to-port segment of the container movement is provided by container shipping lines (including feeder services). The landside portion is provided either by trucking firms or a combination of trucking firms and railroads.

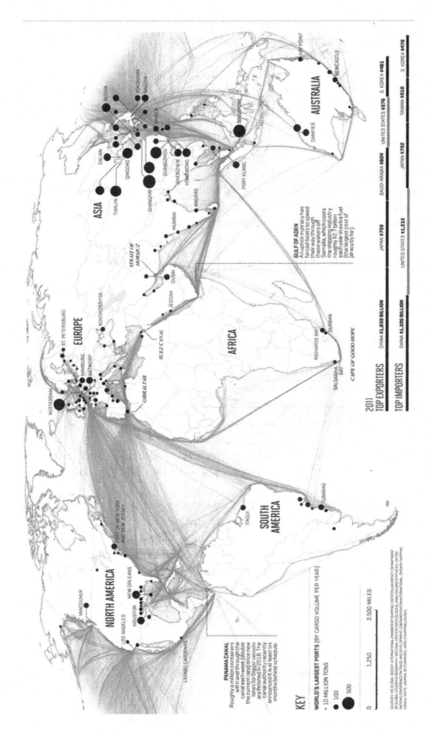

FIGURE 19.1 2014 global container shipping routes

Source: www.geographypods.com/2-changing-space---the-shrinking-world.html; credit: NASA Earth Observatory.

The container transportation system enables nations to send their goods across the globe at a low transport cost. The container itself, a standardised metal box, has become a symbol of globalisation. The container system is an open system that anyone can participate in. Thus, the system's openness and easy access makes it vulnerable to a terrorist attack; ports around the world are connected by global container shipping services (Figure 19.1) through direct or transshipment services.

Global connectivity includes regions that are politically unstable and hostile towards each other, and in particular towards Western countries, especially the United States. The 9/11 attacks raised serious issues about the security of the global supply chain due to multiple vulnerability aspects along the supply chain. Ports are the place where ship movements, container flows, and landside logistics activities interact. Terrorist attacks on port(s) have local, national, and international ramifications. As a result, port security issues are complex, multidimensional, and multilayered.

19.3 Dimensions of port security

Though no ports are exactly alike, they do share many common characteristics that make them vulnerable and a good target for a terrorist attack:

- Sprawling and easily accessible by water and land;
- Close to crowded metropolitan areas and interwoven with complex transportation networks designed to move cargo and commerce as quickly as possible;
- Container terminals include loading/discharging and storage facilities;
- Container terminals are frequently close to facilities critical to a nation's economy, such as refineries, factories, and power plants.

These complex multidimensional characteristics are challenges to port security in light of the complexity of the global supply chain (Bichou, 2009), including:

- *Physical assets*: Physical assets of the transportation and logistics system such as ships, trucks, aircraft, rail locomotives, railcars, containers, trailers, and freight facilities (warehouses, distribution centres, and buildings) are targets and can be the means to carry out terrorist attacks. For example, vehicles and vessels, goods and cargo all can serve as weapons for terrorist acts.
- *Information and monetary flow*: Information flow is the blood line for commercial transactions and supply chain activities. Monetary flow provides financial resources for both the commercial transactions and the supply chain movement of goods. These are also used by terrorists for plotting, organising, coordinating, directing, and controlling attacks. Financial resources provide the means for terrorist movements and the acquisition of weapons.
- *Global scale*: Terrorist organisations have a global cell network and global reach, using widely available Internet resources as do multinational corporations.
- *Economic impacts*: Terrorist attacks have global implications for international trade, production system, and supply chain. A terrorist attack could cause widespread disruptions impacting manufacturing, purchasing, outsourcing, transportation,

and service operations. The economic and financial costs could be extremely high if attacks are held against vital strategic interests such as oil or gas facilities, major freight transportation hubs and networks, such as large container ports, key intermodal transfer facilities, transshipment ports, and maritime choke points (e.g., Singapore Strait, Suez and Panama Canals, Strait of Gibraltar).

- *Public-private partnership*: Although security issues are usually the responsibility of the public sector, the private sector is also concerned with terrorist attacks. Ports are where public and private sectors' interests meet. Government agencies such as port authorities in the United States play key roles as policy maker, planner for infrastructure development, regulations executor and security provider. The private sector plays the role of freight operator, service provider, and business developer. Both sides have a common interest in keeping the supply chain secure. As a result, a partnership of both on port security is essential.

19.4   Objectives of port security measures

The 9/11 incident in New York brought to the forefront US initiatives to mitigate port security threats. Initially, these measures were unilateral because the United States was the target of terrorists. The United States took the lead to defend itself from terrorists. Realising the multidimensional port security vulnerabilities, a layered approach of protection was developed which ultimately became the global approach.

The US security measures implemented after 9/11 focus on three key areas: deterrence, prevention, and response.

- Deterrent measures are designed to discourage an attack or make an attack cost-prohibitive.
- Prevention measures are designed to stop or disrupt a planned attack.
- Response measures are designed to assess consequences and damages, and develop a response to mitigate those effects while returning to normal.

These measures require information sharing among government agencies and supply chain operators/entities. Along with these measures, there is also a need to update and standardise the means of recording and tracking cargo contents, movements, and chain of custody (O'Brien, 2010b). The improvement of supply chain security is based on cooperation among government agencies, international partnership, and private industry with the objectives to:

- Ensure authorised shippers know the exact cargo content inside the containers bound for the United States and report their contents accurately to the US authorities.
- Provide complete and accurate electronic shipment data and protect containers from infiltration with special seals and monitoring technologies.
- Reduce the risk of shipments being intercepted or compromised while in transit.

In short, the primary goals of efficiency of the global supply chain should be maintained and enhanced. According to Stephen Flynn, a US maritime security expert:

> Ultimately getting seaport security right must not be about fortifying our nation at the water's edge to fend off terrorists. Instead its aim must be to identify and take the necessary steps to preserve the flow of trade and travel that allows the US to remain an open, prosperous, free and globally engaged society.
> *(Gillis, 2002, p. 26)*

19.5 Overview of US port security regulations and initiatives

Port security measures have a silent face and a public face. The silent face is associated with intelligence and profiling. It is silent because very little of it is discussed in public. Intelligence gathering is responsible for collecting information through the global intelligence agencies in order to identify general and specific threats of individuals and organisations. This is a massive global undertaking with global cooperation of intelligence agencies. The silent approach also profiles the risk by region, country, individuals, organisations, and so forth. The public face is associated with regulations, inspections, reports, and so on.

The United States, taking the lead on this issue, implemented the Maritime Transportation Security Act of 2002 (MTSA) and subsequently the SAFE Port Act of 2006. These two legislations cover three key areas and 19 programs (Appendix 1: U.S. General Accounting Office, 2007). The centrepiece dealing with security of physical maritime assets is the International Ship and Port Facility Code (ISPS), which was passed by the International Maritime Organization (IMO) and adopted by the United States in MTSA in 2002. The following are several initiatives dealing with cargo security:

- The 24-Hour Rule
- Container Security Initiative
- Customs–Trade Partnership Against Terrorism (C-TPAT)
- Free and Secure Trade program
- A congressional mandate that requires that all containers inbound for the United States be scanned in foreign ports.

All these security programs are a layered approach to improve maritime security, which has significant impacts on the supply chain community.

19.5.1 ISPS Code

On 12 December 2002, contracting governments to the IMO adopted amendments to the International Convention for the Safety of Life at Sea (SOLAS), 1974, to enhance the security of ships and port facilities. Chapter XI-2, "Special Measures to Enhance Maritime Security," also approved a new International Code for the Security of Ships and of Port Facilities (ISPS Code). On 1 July 2004, the SOLAS amendments and ISPS requirements became effective (American Bureau of Shipping, 2005). The new requirements established an international framework through which governments, ships, and port facilities can cooperate to detect and deter acts which threaten security in the maritime transport sector. The ISPS Code provides a standardised, consistent framework for managing risk and permitting the meaningful exchange and evaluation of information between contracting governments, companies, port facilities, and ships (IMO, 2002). The government agencies have an active role in this effort, to assess and evaluate the threat and risk or potential unlawful act.

With the mandatory requirements and a structured risk management process in the ISPS Code, governments, ports, and shipping companies are required to implement measures to enhance the security of the world's maritime transportation system. Under the code, each signatory or contracting government is required to establish a three-tier security system and assess the risks their ships and ports face:

- Level 1 is a normal level of security threats.
- Level 2 is a medium threat level.
- Level 3 is a high threat level.

The Code also requires that ports, ship operators, and vessels establish a three-tier security plans that correspond with these three levels of security assessment. These security plans have both operational and physical measures that will be executed based on the threat assessment. These resulted in the establishment of security officers for a port, ship operator, and vessel, responsible for ensuring compliance with security requirements (O'Brien, 2010a).

19.5.2   The 24-Hour Rule

In order to identify early high-risk containers bound for the United States, the 24-Hour Rule was implemented. On 2 February 2003, the 24-Hour Rule went into effect. This is an important first-layer measure implemented by US Customs and Border Protection (CBP) to screen US-bound containers for cargo content before loading on vessels in foreign ports. It requires shippers/ocean carriers to electronically submit cargo manifests 24 hours before loading in foreign ports through CBP's Automated Manifest System (AMS). There are 14 elements to be submitted to CBP (Gillis, 2003):

1 Foreign port of departure
2 Standard Carrier Alpha Code (SCAC)
3 Voyage number of the vessel
4 Date of scheduled arrival in the first US port
5 Number and quantity and packages from the bill of lading

6 First port of receipt by the vessel operator
7 Precise description (or first six digits of the Harmonized Tariff Schedule number) of the commodity supplied by the shipper, and weight of the goods, or for a sealed container, the shipper's declared description and weight of cargo. Generic cargo descriptions, such as "freight all kinds" (F.A.K.) and "said to contain" (S.T.C.) are not acceptable
8 Shipper's names and addresses, or identification numbers assigned by CBP, from all bills of lading
9 Consignee names and addresses or identification numbers assigned by CBP, from all bills of lading
10 Name of the vessel, national flag, and vessel number
11 Foreign port where the cargo is taken on board
12 International hazardous goods code when such material is being shipped
13 Container number
14 Numbers of all seals affixed to the container.

The 24-Hour Rule has changed the way shippers, intermediaries, and ocean carriers engage in US-bound trade. However, some cargo manifest information provided in the 24-Hour Rule comes from a third party that passes along information, that is, origin and contents from the customer. Verifying the information is lacking. There is a need for more data in order to improve the risk assessment of cargo security and data availability from the point of origin, that is, manufacturer/factory and exporter, as well as supply chain intermediaries (warehousing and export packer). More data will help improve targeting the high-risk container shipments. Recognising gaps in risk assessment and in an effort to improve targeting, the CBP, the global leading security entity, sought more real-time cargo data in July 2004. The effort was designed to dig deeper into the supply chain for information on the true origin of cargo and its trails of activities through various parties that handle the cargo (Kulisch, 2004). As a result, CBP proposed the 10+2 Rule in 2008.

19.5.3 Import security filing (ISF or 10+2 Rule)

Under the SAFE Port Act of 2006, the US Congress requires regulations for advance electronic data to protect against terrorist smuggling of weapons or materials. Such data feeds input to the CBP's sophisticated Automated Targeting System that determines which inbound containers should be automatically screened (Kullisch, 2008). Importers and their agents are required to file 10 data items as follows (US Customs and Border Protection, n.d.):

1 Importer of record number
2 Consignee number
3 Seller (owner) name and address
4 Buyer (owner) name and address
5 Ship-to party

6   Manufacturer (supplier) name and address
7   Country of origin
8   Commodity six-digit code of Harmonized Tariff Schedule
    a   (The above data must be filed 24 hours before loading in foreign ports.)
9   Container stuffing location
10 Consolidator (stuffer) name and address
    a   (The above data must be filed as soon as possible, but no later than 24 hours
        prior to arrival.)

The two additional carrier data requirements are:

1   Vessel stow plan (filing no later than 48 hours after vessel departure)
2   Container Status Message (CSM) data (filing within 24 hours of creation or
    receipt.)

This 10 + 2 information is in addition to the data filed in the 24-Hour Rule,
improving the ability to identify containers that may pose a risk of terrorism for
additional scanning or physical inspection. This is based on an October 2009 report
by the GAO on Supply Chain Security.

The 10 + 2 Rule is part of the layered enforcement strategy in working collabor-
atively and collectively with other government agencies to achieve the highest level
of security and safety while facilitating legitimate trade. The 10 + 2 Rule became
effective on 26 January 2009 (Berman, 2015).

19.5.4   Container Security Initiative (CSI)

Even though CBP obtains advance cargo manifest information through the
24-Hour Rule, the information cannot be verified until the cargo reaches US ports.
This is another layer in the layered approach to provide CBP with the ability to
inspect and stop high-risk containers before they reach US ports. The Container
Security Initiative (CSI) was launched in early 2002. It is based on the general security governing concept of pushing out the US border to reduce the gap between the
time information about the cargo is obtained and the time it is verified/inspected.
Through bilateral agreements with other countries, CBP inspectors are stationed
at foreign ports to identify suspicious containers flagged by the CBP's Automated
Targeting System. Under the reciprocal agreement program, host countries' customs
authorities conduct the actual cargo inspection using high-tech X-ray equipment
to search for contraband and other materials that do not match the contents provided in the manifest (Kulisch, 2005).

There are three main elements in the CSI (US Customs and Border Protection, 2005):

*   *Identify high-risk containers.* CBP uses automated targeting tools to identify con-
    tainers that pose a potential risk for terrorism, based on advance information
    and strategic intelligence.

• *Prescreen and evaluate containers before they are shipped.* Containers are screened as early in the supply chain as possible, generally at the port of departure.
• *Use technology to prescreen high-risk containers to ensure that screening can be done rapidly without slowing down the movement of trade.* This technology includes large-scale X-ray and gamma ray machines and radiation detection devices.

The CSI serves as a deterrent to terrorist organisations that may seek to target any foreign port. This initiative provides a significant measure of security for the participating port as well as the United States. CSI also provides better security for the global trading system as a whole. If terrorists were to carry out an attack on a seaport using a cargo container, the maritime trading system would likely grind to a halt until seaport security is improved. Those seaports participating in the CSI handle containerised cargo far sooner than other ports that haven't taken steps to enhance security. To date, CSI is operational at ports in North America, Europe, Asia, Africa, the Middle East, and Latin and Central America. CBP's 58 operational CSI ports now prescreen over 80% of all maritime containerised cargo imported into the United States.

19.5.5 C-TPAT

Because the global supply chain involves mostly private sector business activities, engaging the private sector is essential in order to increase cargo security. This is another layer in the anti-terrorism approach by CBP. The objective of C-TPAT is to protect the world's trade industry from terrorists, maintaining and preserving the economic health of the United States and its trading partners. The partnership develops and adopts measures that improve security but do not have an adverse effect on trade – a challenging balancing act.

Pushing out the borders: Extending the United States' zone of security to the point of origin enables CBP to conduct better risk assessment and targeting, freeing CBP to allocate inspectional resources to more questionable shipments. The partnership establishes clear supply chain security criteria for members to meet and in return provide incentives and benefits like expedited processing. The outcome is to extend the partnership of anti-terrorism principles globally through cooperation and coordination with the international community. In 2005, the World Customs Organization (WCO) created the Framework of Standards to Secure and Facilitate Global Trade, which complements and globalises CBP's and the partnership's cargo security efforts.

How it works: When companies join the C-TPAT program, they sign an agreement to work with CBP to protect the supply chain, identify security gaps, and implement specific security measures and use best practices. In addition, partners provide CBP with a security profile outlining the specific security measures the company has in place. Applicants must address a broad range of security topics and present security profiles that list action plans to align security throughout their supply chain. C-TPAT members are considered low-risk and are therefore less likely

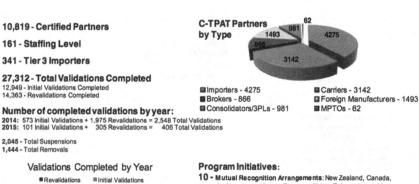

10,819 - Certified Partners

161 - Staffing Level

341 - Tier 3 Importers

27,312 - Total Validations Completed
12,949 - Initial Validations Completed
14,363 - Revalidations Completed

Number of completed validations by year:
2014: 573 Initial Validations + 1,975 Revalidations = 2,548 Total Validations
2015: 101 Initial Validations + 305 Revalidations = 406 Total Validations

2,045 - Total Suspensions
1,444 - Total Removals

Validations Completed by Year

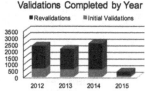

Program Initiatives:

10 - Mutual Recognition Arrangements: New Zealand, Canada, Jordan, Japan, Korea, European Union, Taiwan, Israel, Mexico, Singapore

2 - Mutual Recognition Projects: China, Switzerland

12 - Technical Assistance Projects: India, Turkey, Jamaica, Dominican Republic, Honduras, Panama, Colombia, Chile, Peru, Uruguay, Brazil, Costa Rica

4 - Partner Government Agencies: FDA, TSA, USDA, Coast Guard

FIGURE 19.2 C-TPAT achievements

Source: www.cbp.dhs.gov/ctpat.pdf.

to be examined. This designation is based on a company's past compliance history, security profile, and the validation of a sample international supply chain.

Growing partnership: As of 1 May 2015 (Figure 19.2), there are close to 11,000 certified partners spanning the gamut of the trade community that have been accepted into the program. These include US importers, US/Canada highway carriers, US/Mexico highway carriers, rail and sea carriers, licensed US Customs brokers, US marine port authority/terminal operators, US freight consolidators, ocean transportation intermediaries and non-operating common carriers, Mexican and Canadian manufacturers, and Mexican long-haul carriers. These nearly 11,000 companies account for over 50% (by value) of what is imported into the United States.

Mutual Recognition Arrangements: CBP has numerous Mutual Recognition Arrangements with other countries. The goal of these arrangements is to link the various international industry partnership programs so that together they create a unified and sustainable security posture that can assist in securing and facilitating global cargo trade.

The goal of aligning partnership programs is to create a system whereby all participants in an international trade transaction are approved by customs as observing specified standards in the secure handling of goods and relevant data.

19.6 Risk management in port security

In the aftermath of the 9/11 attack, several major port security programs and regulation initiatives were developed by the US government, intended to provide security layers to address supply chain vulnerabilities, in particular in container shipping.

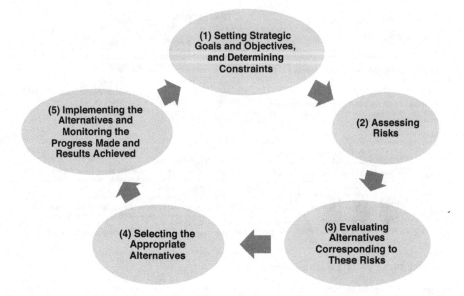

FIGURE 19.3 Risk management framework

Source: Created by the authors based on information from www.gao.gov/assets/130/120506.pdf, page 4.

All stakeholders have implemented security measures to meet US government requirements or collaborate with agencies on improving security. Simultaneously, the view of security risk professionals was that security should be a part of a system-wide function as one of the critical business tasks. This approach requires that port security management get thoroughly involved in the strategic management as a productive entity and integrate security measures.

Security has a significant impact on the business of the port. For example, an increase in the level of security checks on truckers arriving for pickup and delivery of containers can have significant impacts on operational efficiency of the port, that is, higher congestion level and waiting cost. Similarly, an increase in access control requirements such as additional permits or the escort of visitors in a port facility can have an economic impact on the bottom line of a port. Ultimately, a port's ability to apply such measures to security should lead to positive benefits and develop the physical security plan (Christopher, 2015).

It is widely recognised that no amount of money or effort can fully protect against every type of threat. As a result, an approach was developed that considers the risks of various threats posed and determines the best use of limited resources to mitigate them by responding effectively if they occur. Figure 19.3 is a combination of the findings of industry experts and best practices. The framework divides risk management into five major phases (GAO, 2007). This evolving and dynamic process changes as threats change from time to time.

There are several examples of risk management tools and applications used to assess and improve port security.

- US Coast Guard used the National Risk Assessment Tool (N-RAT) to evaluate benefits gains from increased security for vessels, port facilities, outer continental offshore facilities, and port areas. The Coast Guard used the N-RAT to determine risks associated with specific threat scenarios against targets within the US Marine Transportation System. This allowed the Coast Guard to systematically consider all segments of the commercial maritime community to evaluate their potential for being involved in a transportation security incident. The results provided assessments on vulnerability, risk reductions, and costs of implementation of security measures (US Coast Guard, Department of Homeland Security, 2003).
- Port stakeholders are taking preventive steps to reduce port vulnerabilities. For example, port stakeholders installed fences, hired security guards, and purchased cameras. To finance the costs of implementation, a port security grant program was provided. The US Department of Homeland Security used risk management tools to allocate port security grants based to risk scores. Using the DHS' risk analysis tools, it assesses the relative risk posed to ports throughout the nation and to help determine Port Security Grant Program (PSGP) eligibility and funding levels. The model consists of three variables: *threat* (the relative likelihood of an attack occurring), *vulnerability* (the relative exposure to an attack), and *consequence* (the relative expected impact of an attack). Data for each of these variables are collected from offices throughout DHS and from other data sources, and then, using the model, each port is ranked against one another and assigned a relative risk score. Based on risk, each port area[1] is placed into one of three funding groups – Group I, Group II, or Group III. Ports not identified in Group I, II, or III are eligible to apply for funding as part of the "All Other Port Areas" Group. The two highest risk groups receive the bulk of grant funding (U.S. General Accounting Office, 2011).
- The National Center for Risk and Economic Analysis of Terrorism Events (CREATE) undertook a research project, funded by the DHS, to develop a risk assessment and security resource allocation system (PortSec) for seaport operations. Although initially focused on the ports of Los Angeles and Long Beach, the project addressed a major challenge faced by operators of all seaports: determining the proper balance between increasing seaport safety, maintaining or maximising business throughput, and minimising the impact on the environment. Often these three objectives are at odds with one another (e.g., increasing safety by implementing additional scanning technologies can lead to container/truck traffic slowdown with a resultant increase in air pollutants). Figure 19.4 shows the risk management methodology used, applying both risk assessment and economic assessment. When fully implemented, PortSec will support both tactical day-to-day security allocation decision making and long-term strategic security planning (Orosz, 2012).

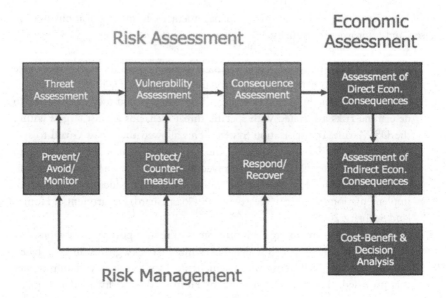

FIGURE 19.4 PortSec risk analysis methodology

Source: Orosz, M., CREATE, USC, www.orau.gov/dhssummit/2009/presentations/March17/plenary/Orosz_Mar17.pdf.

Risk management tools are widely used in port risk assessment, emergency response, incident recovery, and funding allocations. There are varieties of risk assessment models. Each model has its own strength and weakness. As threats evolve with the changing dynamics in regional conflict, political instability, and economic turmoil around the world, risk assessment tools need to keep up with the changing environment to develop proper security measures.

19.7   Cyber security

An emerging threat is the cyber security threat on port security. As global economy and trade expand, the maritime industry has become more and more dependent on information and communication technology (ICT) to manage operations, improve efficiency and e-commerce, and service clients. Ports, by virtue of their role, have a critical infrastructure whose well-being has significant impacts far beyond their vicinity, and carrying economic, political, and sometimes military significance. To deal with growing trade volume and increasing vessel size, ICT has become an indispensable tool in business transactions, inventory management, maritime operations, terminal operations, customs clearance, cargo tracking, and more. IT system malfunction could cause large-scale disruption, economic and financial losses, and even the potential loss of lives. For example, vessels could be directed to the wrong location. Container terminal operations could come to a halt, causing severe congestion and confusion, paralysing the port, leading customs to stop clearing cargo, and so forth. The stakes are high. There is increasing security vulnerability resulting from the extensive dependency on ICT.

The awareness of cyber security for ports was relatively low. The US ports are vulnerable to cyber attack as indicated by a Brookings Institute research report of 2013 (Homeland Security News Wire, 2013). The issue was the vulnerability of computer networks that help move energy, food supplies, and other goods. A cyber attack on US ports could cause significant damage to the economy. Similar conclusions were reached by European Network and Information System Administration (ENISA) and the U.S. General Accounting Office in their respective research reports. To deal with emerging cyber security threat, IBM recommends 10 essential practices (Kouwenhoven, 2014):

1 Build a risk-awareness culture.
2 Implement intelligence analytics and automated response.
3 Defend the workplace.
4 Design cyber security functionality as part of the system at the beginning.
5 Manage upgrades and updates properly.
6 Control network access.
7 Secure it in the cloud.
8 Implement the best practice.
9 Protect the company's critical assets.
10 Keep track of people's access and permission.

In short, this is a new dimension in port security. Its significance is becoming more and more pronounced. Any major cyber incident on ports will have severe impacts on the supply chain; such impacts could be global. However, the challenge is dealing with a cyber threat and improving security in conjunction with all other security measures in a diversified global business.

19.8 Conclusion

This chapter provides an overall description of the multidimensional security challenges facing ports. The approach used is a layered approach to enhance security in protecting physical assets and targeting high-risk containers in advance. The approach uses a variety of programs (ISPS, the 24-Hour Rule, the 10 + 2 Rule, CSI, C-TPAT, risk assessment, and cyber security) while maintaining the functionality and integrity of the global supply chain and efficient trade flow.

There are costs associated with increased security, but the stakes are high as global trade keeps expanding and the world economy is more connected. So far, there has been no major disruption to global container trade due to terrorist incidents. However, security regulations also forced supply chain entities to change the way they do business. Many take advantage of the early available information in order to streamline the supply chain. As the security threat environment evolves, there will be new challenges facing ports around the world. Lessons can be learned from the past to improve security while facilitating trade, which is the ultimate goal.

Note

1 At the recommendation of the Coast Guard, DHS considers some ports as a single cluster – known as a port area – due to geographic proximity, shared risk, and a common waterway.

References

American Bureau of Shipping (2005) *Guide for Ship Security (SEC) Notation.* Houston, American Bureau of Shipping.

Berman, J. (2015) *CBP's 10 + 2 is Now in Full Enforcement Mode.* Retrieved May 20, 2015, from Logistics Management: www.logisticsmgmt.com/article/cbps_102_is_now_in_full_enforcement_mode.

Bichou, K. (2009) Port security. In: Bichou, K. (ed.) *Port Operations, Planning, and Logistics.* London, Informa, 259–261.

Christopher, K. (2015) *Port Security Management* (2nd ed.). Boca Raton, CRC Press, Taylor & Francis Group.

Gillis, C. A. (2002) Flynn keeps focus on transport security. *American Shipper*, 26.

Gillis, C. A. (2003) NVOs: Manifest or perish. *American Shipper*, 28.

GlobalSecurity.org. (n.d.) *Container Security Initiative.* Retrieved May 10, 2015, from www.globalsecurity.org/security/ops/csi.htm.

Homeland Security News Wire (2013) *U.S. Ports Vulnerable to Cyber Attack.* Retrieved May 11, 2015, from http://www.homelandsecuritynewswire.com/dr20130707-u-s-ports-vulnerable-to-cyberattacks.

International Maritime Organization (IMO). (2002, December) *The International Ship and Port Security Code (ISPS Code).* Retrieved May 10, 2015, from IMO: www.imo.org/Our-Work/Security/Instruments/Pages/ISPSCode.aspx.

Kouwenhoven, N. (2014) *The Implications and Threats of Cyber Security for Ports.* Retrieved May 8, 2015, from https://www.porttechnology.org/technical_papers/2016_update_the_implications_and_threats_of_cyber_security_for_ports.

Kulisch, E. (2004) Homeland insecurity. *American Shipper*, 8–12.

Kulisch, E. (2005) Honeymoon over? *American Shipper*, 22–25.

O'Brien, T. (2010a) Ports. In *Intermodal Freight Transportation.* Washington, DC, ENO Foundation, 292.

O'Brien, T. (2010b) Understanding the supply chain. In L. A. Hoel (ed.) *Intermodal Transportation.* Washington, DC, Eno Transportation Foundation, Inc., 55–58.

Orosz, M. A. (2012, September). *National Center for Risk and Economic Analysis of Terrorism Events, USC.* Retrieved May 15, 2015, from http://research.create.usc.edu/project_summaries/110.

U.S. Coast Guard, Department of Homeland Security. (2003) *Implementation of National Maritime Security Initiatives. Federal Register.* Washington, DC, General Printing Office. Retrieved May 10, 2015, from General Printing Office: www.gpo.gov/fdsys/pkg/FR-2003-07-01/pdf/03-16186.pdf

U.S. Customs and Border Protection. (2005) Container security initiative: Just the facts. *Port Technology International*, 37, 109–111.

U.S. Customs and Border Protection. (n.d.). *U.S. Customs and Border Protection.* Retrieved May 20, 2015, from 10 + 2 Presentation: www.cbp.gov/sites/default/files/documents/10%2B2%20presentation.pdf

U.S. General Accounting Office. (2007) *The SAFE Port Act: Status and Implementation One Year Later.* Washington, U.S. General Accounting Office.

U.S. General Accounting Office. (2007) *Port Risk Management.* Washington, DC, U.S. General Accounting Office.

U.S. General Accounting Office. (2011) *Port Security Grant Program.* Washington, DC, General Accounting Office.

Suggestions for further reading

Besides the papers provided as references in the text, we recommend the following texts as suggestions for further reading.

Burns, M. (2015) *Logistics and Transportation Security.* Boca Raton, CRC Press, Taylor & Francis Group.

Christopher, K. (2014) *Port Security Management* (2nd ed.). Boca Raton, CRC Press, Taylor & Francis Group.

McNicholas, M. (Ed.). (2016) *Maritime Security: An Introduction.* Oxford, Elsevier.

20

TECHNOLOGICAL INNOVATIONS

Harry Geerlings and Bart Wiegmans

20.1 Introduction

Traditionally, the process of technology development has been considered an activity belonging to the domain of the natural sciences. Nowadays the process of innovation and diffusion of technological developments cover studies in economic and environmental science, business administration and other disciplines as well. Innovative capacity has even become a major indicator for the economic success of companies and sectors, and also for the sectors of shipping, ports and supply chain management. The meaning and potentials of innovation are central in this chapter.

Section 20.2 sketches the historical context of technology development and provides an overview of the various streams of analysis from the classical approach towards the development of the present innovation theories and concepts. In Section 20.3 it is stated that governments can play a crucial role in the development and implementation of innovations in the port and shipping sector. At the same time there is an increasing awareness within private firms that continuous innovation is a critical factor for success. These complexities and interrelationships, and the lack of a generic and strategic approach by government in partnership with the private sector when it comes to maximising the opportunities offered by innovations, are the leading items in this section. In Section 20.4 a classification for innovations in ports, shipping and supply chains is presented. Sections 20.5 and 20.6 describe the cases of the MARIN research agenda and synchromodal transport. In Section 20.7, challenges for the future are presented and finally, in Section 2.8, conclusions are drawn.

20.2 The history of innovation: from technology to system innovation

Before attempting to define technology, it is important to stress that a distinction has to be made between a technique and a technology. A *technique* is a concrete artefact or handling activity such as a new engine based on the principle of the fuel cell (shipping), sensors for safety control (ports) or a sophisticated IT system to optimise container operations at a terminal (supply chain). The working of a technique is closely related to principles of the natural sciences. Technology has a much broader meaning. Technology should not be equated with a machine or an artefact, but is rather some form(s) of knowledge; technology is directed by 'perceptions' of solving a problem. Second, technology should not be seen as a one-cycle process with separate rules, but rather as a network of intermeshing rules in an ongoing process; it is significant to judge technology as something that relates to and interacts with a (social) environment.

Although the development and meaning of technology will be handled more extensively later in this chapter, we define technology as 'a set of cognitive and social concepts and techniques employed by a community in its problem-solving and consists of a combination of current theories, goals, procedures, handling and using practice' (Geerlings, 1998, p. 41).

Technology has to be introduced to the market and has to be accepted by the potential users (this is named adoption). A classical approach to the adoption of technology is the invention-innovation-diffusion model based on the work of Schumpeter. This model is based upon a specific interpretation of technology developments, namely the identification of three different stages in technology development: invention, innovation and diffusion. In a very simple model a distinction is made in the first stage of the *invention* of a technology: a new idea or technique is developed. In this stage it is not a necessary condition that the technology will ever be applied or be a commercial success; the focus is on the science aspects that can have a fundamental scientific impact. The scope is mainly focused on the engineering part of the technique. The second stage is the *innovation* stage. In this stage the attention will be concentrated on the successful introduction of the technology to the market. The introduction could take place for commercial reasons, but it might also be because of other reasons (energy policy, environmental legislation, etc.). The last stage of the model is the *diffusion*[1] stage, where we see that the innovation will spread out over a large number of adopters. This conceptual distinction is hardly found in practice as a clear moment in time: the invention is in many cases introduced as an innovation, diffusion entails in many cases further innovation, and so on. Classical economists, like Smith (1776), Malthus (1798) and Ricardo (1819), had already implicitly linked economic growth to innovation. In the mainstream of these classical economics, technology has always been considered as an exogenous factor. Schumpeter (1935, 1939) expanded the idea of technological development. He developed the proposition that important innovations occur at the beginning of an economic recovery and maintained that, particularly in times of economic crisis, firms are willing to take risks and open new channels for trade. Scientific inventions and discoveries are turned into commercial innovations through the intuitiveness of entrepreneurs in a process he called *creative destruction*. A clustering of innovations and a diffusion of radical technologies cause the long economic cycles.[2]

In addition to the opinion that technology develops as a process of major technological changes and technological breakthroughs, there are other scientists who state that the process of innovation is a much more discontinuous process of small (incremental) improvements. The development of technologies is in their opinion more a process of 'adoption' and an 'innovation permanence'. Actually, the first approach does not exclude the second approach, and vice versa; there are more approaches towards innovation.

Geerlings (1998) states that the cumulative result of small technological innovations could be of more significance than the result of a sudden large technological break-through. He places great emphasis on the incremental and evolutionary character of technological development. Innovations are, according to Geerlings, not automatically superior to old technologies. New techniques have to prove they are superior after invention. In a process of learning, further improvements are introduced rendering the new technology suitable for the specific environment in which it must be used. To describe this process, he introduces the concepts 'learning by using' and 'learning by doing'. He also stresses the complementarity between innovations.

Technology is not freely transferable from one situation to another but has to be acquired and appropriated. Technological development is specific to time, sector and location. Technology development depends to a large extent on cultural as well as organisational factors. In addition, Geerlings puts great emphasis on the incremental and evolutionary characteristics of technological development. Abernathy and Utterback (1988) make a distinction between technological innovations (technique or hardware) versus process innovation (technology or orgware), which leads to the conclusion that technological development goes beyond the artefact, and that technology is the result of, and interrelated with, human action.

In summary, technology development may be viewed as the culmination of a number of processes in which technology develops in certain stages (Table 20.1).

Technological development in the transport sector can be explained as a process of small incremental changes: 'Evolution is the result of a sequence of replacements.' Often, new technologies create new 'niches' that lead to products and services hith-erto unavailable. More frequently, successful innovations can pre-empt an established niche by providing improved technical and economic performance or by promot-ing the social acceptability of existing services through new ways of fulfilling them.

TABLE 20.1 Stages in technology or process development

Process	Outcome
Discovery	Idea
Fundamental research	Invention
Proving technological feasibility	Basic innovation
Product or process oriented	Innovation
Research and development	Prototype
Product of process demonstration	Product/method
Product or process implementation	Diffusion

Source: Geerlings (1998).

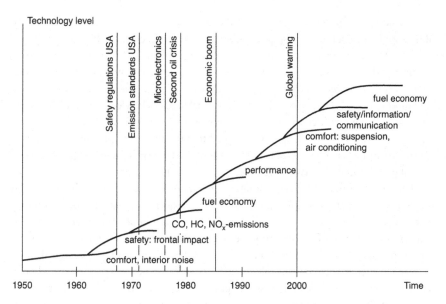

FIGURE 20.1 Dominant heuristics in relation to the technological development of the car 1950–2010

Circumstantial evidence shows that many pervasive systems have evolved through both of these evolutionary paths. They replaced older technologies, and then created new and additional market segments that did not exist before. This process is illustrated in Figure 20.1 that elaborates on the work by Seiffert and Walzer (1989), who show that the technological development of a basic innovation changes over time due to the dominant heuristic.

20.3 Transport innovation policy making in networks and alliances

The picture that the technological potential presents is complex. Technology contributed to an efficient transport system, which is a crucial precondition for economic development and an asset in local, regional and international mobility. Free transport of goods is considered an essential element for a modern society. With the integration of the world market, economic growth and higher levels of income, transport has become a major economic sector which is characterised by quantitative as well as qualitative growth. The benefits of transport and technological developments, however, come at a high price. No mode of motorised transport is environmentally friendly, for instance. Some modes of transport, notably deep-sea transport compared to air transport, have lower impacts than others. Some scholars see technology as the cause of many problems; at the same time technology is considered a potential answer to new needs and new problems, such as the reduction of environmental impacts and for energy supply, but also when it comes to the improvement of the efficiency of shipping.

The relationship between firms and the government is becoming an increasingly dominant phenomenon for transport technology development, and this is also reflected in the technology policy of governments; there is substantial literature on the extent, motives and process of technological collaboration. Linkages, which can range from formal joint ventures, strategic alliances, and joint research and development (R&D) and production sharing projects down to informal linkages between peers, known variously as 'collaboration', 'cooperation', and 'networking', are not a new phenomenon but in present times are very relevant. It is argued that the impact of present technological development is both a stimulus of, and focus for, increased collaborative behaviour. Successful participation in networks requires special skills and motives.

The growing number of collaborative agreements between private firms and the increasing rate of transnational alliancing in the field of technology imply a reorientation for governments. At the global level there are significant differences in the experience of forms of collaboration between government and business where steering technology into the desired direction is concerned. Generally speaking there is a clear separation between the role of business and that of the government when it comes to innovations. For business, the point of departure is that it would be more capable of designing a better technology policy without an intervening government. This has also to do with the fact that companies quite often have specialised knowledge and experience about technology in certain disciplines. This has also to do with the fact that companies act in very competitive environments and innovation can create a comparative advantage. In the Far East (Japan, Korea), government and business are traditionally more interwoven. The development and implementation of a technology policy is done in close cooperation between the government and private firms. The same counts even stronger for the Chinese R&D policy where the central governments is not only programming the R&D, but many important companies have strong ties with (or belong to) the central government. In the United States the picture is more business oriented. The image of the European technology policy is much more confusing. There is little coordination that exists between the technological policies of the individual European Member States; the European system is characterised by nationally fragmented production systems and standards, with a strong dependency on a national policy programme. Establishing a common and coherent EU technology policy is difficult partly because there is no clearly defined common interest. A clear EU technology policy was lacking, but with the implementation of programme Horizon 2020 cooperation it received a new impetus. Divided among several sectors there are a number of (independent) programmes. Ports and (maritime) logistics and sustainable transportation are two priorities in these programmes.

20.4 Innovative capacity in ports, ships and hinterland connection

There is not one single approach to classify innovations in the transport sector. A very common approach is to classify innovations based on the stratification of the transport sector (Geerlings, 1998). It is not too complicated to identify within the maritime transport or port system at least eight different *levels of aggregation*: within the whole traffic system (level 1) we can identify, for example, global networks. Scholars identify within the system the subsectors as road, water and air transport (level 2), which are based on traffic elements. The maritime sector is subdivided into the sectors of passenger transport (cruise ships, ferries, etc.) and freight transport (vessels and tankers) (level 3). Within the freight sector, based on the criteria 'demand', a distinction is made between international transport, short sea shipping and transportation to the hinterland (level 4). The next level is the distinction between private transport and commercial transport (level 5). Commercial transport is subdivided into regular (liner) and irregular (charter) transport services (level 6). Moreover, even within the clusters 'regular transport services', yet two more levels (levels 7 and 8) of the subsystems can be identified (e.g., couriers and van services, and fixed-route van-services and free carriers). All these levels have unique characteristics and therefore unique requirements and demands.

Another, and very common way, to classify technology is on the basis *of technological characteristics.*

20.4.1   *Vessel design and manufacture*

Vessel design determines the way in which shaft power is converted to a useful transport product. Thus vessels with lower mass and lower aerodynamic drag will convert shaft power more efficiently, hence reducing fuel consumption and the emissions of greenhouse gases. Vessel design includes choice of materials, construction methods and the actual body shape and size of the vehicle. The shape size and styling of the vessel body affects its water resistance performance as well as its mass. The method of construction also affects the vessel's mass, which in turn determines the available freight payload.

The process of manufacturing is closely related to the use of materials during the life cycle of the ship. It is noted that within a life cycle approach environmental implications are various and very complex. Special attention is given to the so-called sensitive materials and the upstream-downstream effects associated with the extra vessel operations and new transport technologies. Two more specific comments are possible, however. First, alternative design possibilities include the use of new lightweight material. In addition, computer-aided design with finite element stress analysis can improve the designer's ability to minimise the weight of material used for a given function.

20.4.2   *Engine technologies*

Engine technologies which improve the efficiency of fuel consumption reduce levels of carbon dioxide emission. Other gases emitted from fossil fuel combustion can also be suppressed. Both cars and trucks, vessels and aero engines may also be designed for noise reduction. Engine performance is not only affected by the patterns of load or speed, but also by a wide range of factors including the type of fuel used, the compression ratio, the method of introduction of fuel and of air, the ignition systems and the engine cooling systems. The three areas of most potential within this cluster appeared to be measures to reduce energy loss in internal combustion engines, suppression of emissions, noise reductions and the very special developments of aero engines.

20.4.3 Fuel technologies

It is obvious that the climate change debate will have an impact on the question of what kind of fuels are desirable for transport. Fossil fuels are under pressure and scrutiny. 'Clean' energy sources are electricity and hydrogen, however these sources need to be derived from other sources. The benefit derivable from hydrogen produced by electrolysis of water, or electric propulsion, depends crucially on the achievement of a breakthrough in the creation of electric energy from renewable sources. Hydrogen from renewable electricity sources is likely to be very expensive when produced from biomass. Indeed biomass might be better employed to produce heat and electricity in combustion, rather than in conversion into alcohol fuels. Any net benefit from biomass, however, clearly must depend on the energy required to produce the biomass, and the environmental effect of biomass production and consumption also needs to be taken into account. Alternative fuels can produce lower emissions of greenhouse gases per unit of useful energy. Some could also provide the opportunity to offset CO_2 emissions in use by carbon fixing plant growth. Moreover, they offer the opportunity to reduce a dependence on non-renewable petroleum-derived fossil fuels which at present form the vast majority of the fuel used for transport. In the short term, commercial applications are likely to be limited to those fuels that are based on the traditional fossil fuels, like gasoline, diesel, liquified petroleum gas (LPG) and combinations of these fuels. For the others, like liquified natural gas (LNG), ethanol, hydrogen, battery and fuel cell vehicles, and hybrid vehicles (diesel electric for barge is being financed), research is at a less complete stage of development. The four main areas of potential are improvement of hydrocarbon fuels, biofuels, hydrogen and electric propulsion.

20.4.4 Other areas of technological potential

Traffic control is a general term to comprehend the real-time process of maintaining a smooth flow of traffic, in response to information available to traffic controllers. Public traffic control systems may be implemented in order to improve traffic flows, but also to address specific environmental aspects. Individual companies may introduce their own internal systems to control fleet movements. In traffic management, fleet management, demand management and modal transfer technologies, the balance of the problem is needed.

20.4.5 Infrastructure/port development

Infrastructure and port development can help protect urban quality of life both by providing the capacity to enable traffic to be separated from other activities and by screening inhabitants against the adverse effects of shipping and port activities. At the same time, the infrastructure itself can detract from the quality of life by its visual and activity separating characteristics. There are some possibilities for a better reconciliation of these effects through new technology, in particular through the use of new materials for structures and surfaces.

20.4.6   The integration of technology and orgware

New technology (logistics concepts, information systems, fuel, etc.) requires new arrangements (coalitions, agreements, mandates, rules) and vice versa; new arrangements call for new technology. This requires vision, decisiveness and precision for the stakeholders involved in transport development. The biggest challenge for a transition to a sustainable transport system is not only to develop new technology (the 'techware'), but also to develop new governance arrange- ments (the 'orgware'). This is also mentioned by Corman and Negenborn in Chapter 8 of this book. The attention for orgware involves new contracts, coalition models and institutional arrangements in the relationship between shippers, freight forwarders, shipping companies, terminal operators and transport companies, market players and governments that are dealing with different levels of government, citizens and road users. In contrast to the hardware/technology oriented approach, orgware focusses on the process aspects of ports and transportation services.

20.5   The case of a strategic maritime research strategy and the R&D agenda; an illustration of technology pull

It is very important that R&D is embedded in a framework of activities with clear vision, objectives, targets, work plans with milestones and a budget. These conditions are very well reflected in the strategy plan of the Dutch Maritime Research Institute Netherlands (MARIN). MARIN has a dual mission: to provide industry with innovative design solutions and to carry out advanced research for the benefit of the maritime sector as a whole. In this way, the institute strengthens the link between academic research and market needs. It is a unique interaction that benefits all parties concerned. MARIN intends to be innovative, independent and a reliable partner in a wide diversity of networks. By taking this position MARIN created a powerful synergy with the maritime industry. Their network consists of commercial ship builders, fleet owners, navies, naval architects and offshore companies the world over. Nowadays, the maritime industry is confronted with shorter cycle times and increasing global competition in challenging environmental and economic conditions. By becoming involved in projects as early as possible, MARIN contributes to meet these challenges by feeding back the results of advanced research programs into commercial projects.

In 2011 the Dutch government launched the initiative to establish innovation contracts for leading industries, the so-called top sectors. The top sectors are considered very important for the Dutch society in terms of job creation and societal challenges. One of these top sectors is water, of which the maritime industry is a subsector. The maritime sector has been subdivided by the government in the following innovation themes:

- Winning at sea: (renewable) energy and raw materials;
- Clean ships: fuels, fuel saving and emission reductions;

- Smart and safe sailing: special ships with specialised operational systems;
- Effective infrastructure: effective use of ships, ports and waterways.

These themes demand excellent research in research areas such as hydrodynamics, maritime operations, structures and materials, systems and processes, design and building technology, and impact on the marine environment. In 2011, MARIN established a new strategy for the coming five-year period. In this strategy, hull-ice interaction and concept design are identified as new market opportunities. Furthermore, a shift towards the joint use of numerical simulation techniques, complex model tests and special measurement techniques is foreseen (for more information, see www.marin.nl).

20.6 The case of synchromodal transport; an illustration of technology push

In the inland waterway sector, many different innovations are being developed. Quite a number of innovations focus on the barge itself and the propulsion of the barge. This means that the outcomes of these innovations in general aim for the existing market and existing products of inland waterway transport and are therefore mainly incremental in nature. Several innovations aim for the increased usage of the possibilities of ICT in inland waterway transport. A final group of inland waterway innovations aims for smaller networks for the transportation of palletised goods or for city distribution by inland waterways and canals. However, the most promising innovation is synchromodal transport. Synchromodal transport is a concept where different transport modes and terminal handling (pre-haulage, terminal and main haulage by rail or inland waterway) are offered to the client whereby all service components are planned in an integrated way. The shipper offers its loads to the service provider and the service provider decides if in a certain transport solution rail, inland waterway or trucking is the preferred transport mode. Whereas in intermodal freight transport rail, inland waterways and trucking compete with each other, in synchromodal transport, these transport modes operate in a coordinated way.

Advantages of synchromodal transport are the use of possibilities offered by advanced IT packages, better service quality towards potential customers, increased utilisation of transport means and terminals, more optimal use of all transport modes leading to a stronger position of intermodal freight transport, good alignment with the policy goals of the European Commission, possible greater flexibility, improved reliability of intermodal freight transport and improved sustainability of the transport system.

Disadvantages of synchromodal transport are that IT systems of the respective actors must be aligned which can prove to be costly and therefore it might prevent synchromodal transport from being implemented. Furthermore, rail and inland waterways might be preferred over single-mode trucking transport, leading to opposition from the traditional trucking companies. Another disadvantage might be the governance of the different transport modes; the competition between the different transport modes should take place under equal conditions. The actual sharing of the realised benefits of synchromodal transport might be difficult to implement. The final

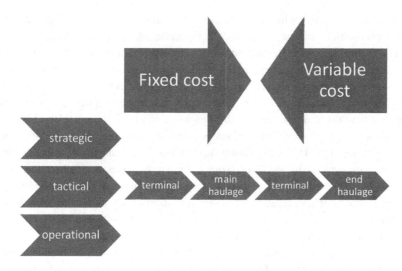

FIGURE 20.2 Coordination and decision making in synchromodal transport

Source: Based on Wiegmans and Konings (2015).

disadvantage refers to competition issues between the respective companies within the different transport modes. The respective synchromodal solutions might eliminate too much competition and therefore might not be allowed by the European authorities.

The core aspects of synchromodal transport are shown in Figure 20.2. At the core are the respective elements of the transport solution (terminal, main haulage, another terminal and end-haulage). All these elements of the transport solution possess fixed and variable cost aspects. Finally, the coordination and decision making in synchromodal transport carries three levels: strategic, tactical and operational.

20.7 Expected future developments: the Internet of Things

In this chapter, the role of techniques, system innovations and technology policies in relation to shipping and ports are discussed. One lesson that can be drawn from this is that fundamental future developments and changes are difficult to predict. But there are signals indicating that we are at the forefront of a major paradigm shift in global ports and transport development. Heilig and Voß (Chapter 14), Kuipers (Chapter 16), and Guan and Yahalom (Chapter 19) all refer to new potentials offered by the internet in the near future, and when all promises become reality it seems we might be at a crossroads in technology development.

This is also argued by Rifkin, who published a book entitled *The Third Industrial Revolution* (2012). According to Rifkin, there have been already two industrial revolutions. The adoption of the steam engine, steam locomotive and steam liner in the nineteenth century marked the First Industrial Revolution. New technologies enabled humans to conduct systematic management and operations for coal-fired transportation and factories. This First Industrial Revolution developed the handcraft industry into machine-based industry, which contributed to flourishing ports and growing trade, created opportunities for dense urban zones and stimulated large-scale production.

The Second Industrial Revolution is characterised by the wide application of the internal combustion engine in the twentieth century, which ushered in the era of the oil economy and the automobile, bringing prosperity to suburban real estate and industrial zones. In his book Rifkin comes to the conclusion that we are at the beginning of a new era of major changes: the Third Industrial Revolution (TIR). The TIR is characterised by the exchange of information and data offered by opportunities of the internet. This TIR will be dominated by digital manufacturing, turning away from mass manufacturing and towards much more individualised and distributed production, like 3D printing. Some scholars share the opinion that the transition might lead to a shift from 'globalisation to continentalisation' of the global economy. In his book, Rifkin refers to Peter Bakker, a former chief executive officer of TNT, as saying, 'Globalisation is dying.' According to Bakker, soaring oil prices in the world markets make it increasingly problematic to send freight across the oceans. Government efforts to tax CO_2 emissions only adds to logistics costs. The economic current, he said, is shifting from globalisation to continentalisation. He argued that growth in commerce and trade is going to become increasingly drawn to continental markets. The logistics business, he said, is already redirecting much of its focus to a continental world.

Furthermore, Rifkin states that communication and data flows will change fundamentally. He identifies three new appearances of the internet: the communication internet, the energy internet and the logistics internet. All are part of the concept of the 'Internet of Things' (IoT; Rifkin, 2012). In his opinion, the IoT is the key driving factor (and game-changer) for the transformation of the economic and social life of humans.

Rethinking the implications of Rifkin's book, the TIR can have genuine impacts on ports and hinterland networks:

1 During the economic transition from globalisation to continentalisation, the need for long-distance transport will be reduced amid increasing intra-continental transport.
2 To mitigate the impact of carbon emissions on environment, more and more attention is paid to reduce transport needs, especially irrational and inefficient transport. Transport minimisation and prevention is the goal. Meanwhile, new energy is employed to achieve zero emissions.
3 Much of the energy needed for transport will come from the supply network of distributed energy generated by single buildings or households.

Hence, we can roughly figure out what the basic impacts of the TIR will be on transportation. Zhang et al. (2015) identifies:

1 No longer will technological hardware (the technique) be the determining factor for innovations in ports and port-related activities such as transportation, but factors such as process management, orgware and software.
2 Continentalisation leads to local production, making the production base closer to the customer market to reduce logistical costs and carbon emissions.

3   Cargoes are moving towards both ends of the supply chain. Primary processed products are clustered at the places of origin of raw materials, while finished products for customers are clustered at the places of customer markets, indicating the potential rise of average transport costs of the cargoes in the middle of the supply chain. Hence, we come to a conclusion that as the economy develops, the long-distance transport needs, or traffic turnover, will see slower growth and may even experience a decline.

4   The quality of spatial displacement triggered by cargo transport relies more on cargo quality. The dependency on the quality of transport vehicles will be reduced because of the transport model of multiple small batches, as well as the innovation of carriers. The amount of energy carried will also be reduced with shortened distances and the adoption of new energy.

5   As a kind of major energy for transport vehicles, renewable energy will gradually transform the transport energy pattern dominated by petroleum, which will not only reduce the energy needed for carrying fuel, but may also significantly cut carbon emissions, thus protecting the eco-environment.

The impact of the TIR on transport will definitely be passed to ports, which are the locations where trade, logistics and production converge. As such the TIR will have an important impact on the development of ports in the following respects (Zhang, 2015):

1   To reduce transport volume, more raw materials will be processed at places close to the production area, making primary processed products clusters at the places of origin of the raw materials to reduce the transport needs of waste, which will result in the transport of fewer primary products and less port throughput. But the added value of seaborne cargoes will be improved. At the same time, distributed production based on separate continents and communities will transform the current production model of 'centralised production and global distribution', that is, to manufacture with raw materials sourced worldwide and sent to different places in the world. It will carry out local production and local selling to reduce long-distance transport volumes, which are now extremely high, with cargoes carried by large ships to cut transport costs. When this change happens, port throughput, especially the throughput of cargoes for long-distance transport, will be reduced significantly and fewer berths designed for large ships will be needed. What deserves our attention is that such change is set to bring higher expectations for flexible services and lean operations of ports. Rapid operation and adaptable ships will be highly valued. The roll-on/roll-off ships (drop and pull transport) for short-distance transport will increase significantly.

2   With cargoes moving towards both ends of the supply chain and the adoption of local renewable energy, the ocean trunk transport needs for dry bulk and liquid bulk will be reduced, as it will for finished products. The cargoes in the middle of the supply chain will dominate. Therefore, future ocean trunk transport will have the following characteristics:

    a   Compared with economic growth, the growth of total transport volume will be slower, mainly because of the decreased transport of large bulk and liquid bulk.

b Most cargoes in the middle of the supply chain are cargoes that can be put into containers, so container transport will continue to boom. Despite fewer ocean trunk transport needs of finished products, cargoes that can be put into containers will increase as the products at the place of origin of raw materials are extended to the downstream of the supply chain. Of course, the overall container transport volume of ocean trunk will see slower growth (and even decrease).

3 In view of the aforementioned changes of seaborne cargo structure, terminals will continue to focus on containers. The terminals designated for bulk will be fewer in number, while the percentage of terminals accommodating container ships will continue to rise. Meanwhile, the structure and variety of containers will change a lot in the future. The number of special containers and small containers will increase. The transport units that are more convenient for carriage and dismantling may come into being.

4 In contrast with less ocean trunk transport, the volume of offshore transport and domestic transport will increase significantly. River transport will be revitalised and account for a higher percentage because it is more suitable for the balanced development of the eco-environment. Therefore, the cargo flow structure of ports will change. Ships will become smaller, because large ships are more suitable for ocean transport, but their development is restricted with rising onshore transport and the vigorous development of roll-on/roll-off ships. Meanwhile, the market demand for flexibility and just-in-time supply is increasing, making transport with relatively smaller ships more popular.

5 The current extensive port development will be restricted and the land for port development will be under strict control. With respect to port development, the use of extensive land has been a long-term issue. In view of future developments, available land, a non-renewable resource, is increasingly rare. Consequently port development needs to face transformation. Economical land development is receiving more and more attention.

It is not easy to define the port or logistic models of development during the TIR, because the future development of ports will be affected by many uncertain factors. But in the era of the TIR, it is clear that the port, as a key element of continental and regional communication, will have to develop in harmony with its surroundings and become an eco-friendly port for its adoption of zero-emissions renewable energy; a collection, allocation and distribution centre to improve the connection of materials within its region; a key channel to clear a logistics service supply chain; a flexible port to meet individualised needs; and a smart port with highly integrated data.

20.8 Conclusions and challenges for the future; lessons to be learned

Some scholars are sceptical about this technology optimism. Large infrastructure constructions, for example, underground (freight) transport systems, encounter the limits of what is technologically and financially feasible. And when technologies become available, there occurs the problem of implementation. Moreover, the complexity of decision making and the many parties (functionally) involved increasingly create a barrier for the decisive development of new technologies, and this can mean that promising technological developments never reach the market. Some scholars are even pessimistic when they conclude that technology development also has to be placed in the context of the following four seemingly counterproductive trends:

1   The improvement of environmental performance by technological means is in danger of eroding due to the strong autonomous growth of the transport sector.
2   There is a trend in modal shift from less polluting modes of transport towards more polluting modes, which is partly due to technological innovations.
3   There is no prospect of a short-term technological fix. Consequently, it is expected that the government will have to intervene to assure the collective interest.
4   Finally, we see that many present technological improvements are misdirected. Technological features like the turbocharger, intercooler and so forth are aimed at improving the performance of the car (faster acceleration, heavier and larger cars, etc.) rather than providing a more environmentally friendly vehicle. And we know that the car industry is involved in serious frauds with respect to the fulfillment of emissions standards (see Chapter 18).

We can conclude that there are many obstacles for the successful large-scale introduction of process-integrated technologies, as there are technological, economic, institutional and political barriers. Besides, there are some important paradoxes that have to be faced. Technological solutions can be politically preferable as they involve the least amount of government intervention and restriction. Unfortunately, a first paradox must be faced here. If improvements are to be marketable, they must be seen to be economically beneficial to both equipment suppliers and purchasers. But if improvements to the most environmentally damaging modes also have the effect of reducing operating costs, they may simultaneously attract traffic from less to relatively more damaging modes and increase the total amount of transport. Such effects would be counterproductive. Hence, we emphasise the importance of technology policy and technological development as part of strategic cooperation between government and industry.

Another paradox is in a more subtle respect, in which increased fuel efficiency may actually threaten the environment. Increased fuel efficiency reduces fuel costs, and hence the costs of operation. If a reduction in operating costs generates longer trips, the same amount of fuel expenditure will increase the total of distance travelled by vehicles. Hence, the total of all those environmental impacts are related to the total amount of vessel travel rather than to the total fuel consumption. The two paradoxes indicate that we have to be careful in promoting this technology optimism for the transport sector.

One of the most important areas in which action can realise impact on the rate of useful technological advance relates to setting standards. Standards are seen as important for two different reasons. First, the setting of technical standards is seen as an aid to cost-effective development of new technologies. For example, the establishment of communications protocols in telematics and physical compatibility standards for components in vehicle design have an enabling effect. Second, standards act as constraints on behaviour. Necessary work on performance standards covers a wide range of topics. We see, for instance, that standards have not only been introduced for the performance of terminals and the emissions of deep-sea vessels but also inland shipping and so forth.

To conclude, it can be observed that many books and articles have been published that have attempted to predict the future. Some of the authors, like Jules Verne, frankly acknowledged that they hardly used any serious methods. But since the early 1980s, there is increased attention for the potentials of technological innovation and a better understanding of technology dynamics. Parallel to this, there is increasing concern for the environment. However these two developments took place quite independently from each other; the transportation sector is a sector of continuous technological innovations. At the same time there is a common understanding by all stakeholders (government and industry) that innovation can contribute to address challenges identified. An inspiring example is the use of ICT in logistics, which leads to optimization in transport planning and consequently less transport and lower costs, but also less energy consumption and therefore less CO_2 emissions. These kind of measures underline the opportunities that can be created when we search for new alliances and 'win-win' situations in making new and sometimes unexpected combinations. At present the maritime transport sector (deep sea as well as inland shipping) is lagging behind compared to other sectors in exploiting its innovative capacity. However, there is no 'technological fix' to be expected; the adaptation of technological innovations can become a game changer. The need for a global R&D policy and commitment of all stakeholders involved would be a first good step.

Notes

1 Diffusion entails the acceptance of an innovation into the market of buyers and competitors.
2 The first such cycle resulted from developments in steam power, spinning, weaving and smelting iron. The second wave was based on railroads and the Bessemer steel conversion process; the third on chemicals, electricity and the beginnings of the automobile industry; and the fourth on electronics and aerospace. Each wave was roughly 55 years in length, and the end of the third wave brought us to about 1940. We are now, perhaps, at the end of the fourth and the beginning the fifth wave.

References

Abernathy, W. J., and Utterback, J. M. (1988) *Innovation Over Time and in Historical Context; Patterns of Industrial Innovation*. Washington, DC: National Science Foundation.

Geerlings, H. (1998) *Meeting the Challenge of Sustainable Mobility; the Role of Technological Innovations*. Heidelberg/Berlin, Springer Verlag.

Malthus, T. (1798). *An Essay on the Principle of Population An Essay on the Principle of Population, as it Affects the Future Improvement of Society with Remarks on the Speculations of Mr. Godwin, M. Condorcet, and Other Writers*. London, Printed for J. Johnson, in St. Paul's Church-Yard, 1798. Rendered into HTML format by Ed Stephan, 10 August 1997.

MARIN (Maritime Research Institute) (2014) *Innovation Agenda of the Maritime Cluster*. Marin, Wageningen.

Ricardo, D. (1819, 1978) *On the Principles of Political Economy and Taxation*. John Murray, London.

Rifkin, J. (2012) *The Third Industrial Revolution: How Lateral Power Is Transforming Energy, the Economy, and the World*. Beijing, CITIC Press.

Schumpeter, J. A. (1935). The analysis of economic change. *Review of Economics and Statistics*, *17*(4), 2–10.

Schumpeter, J. A. (1939). *Business Cycles* (Vol. 1, pp. 161–174). New York: McGraw-Hill.

Seiffert, U., and Walzer, P. (1989) *Automobiltechnik der Zukunft*. VDI-Verlag, Düsseldorf.

Smith, A. (1937). *The Wealth of Nations* [1776].

Wiegmans, B., and Konings, J. W. (2015) Intermodal inland waterway transport: Modelling conditions influencing its cost competitiveness. *Asian Journal of Shipping and Logistics*, 31(2), 273–294.

Zhang, M., Van den Driest, M., Wiegmans, B., Tavasszy, L. (2014) The impact of CO_2 pricing or biodiesel on container transport in and passing through the Netherlands, *International Journal of Shipping and Transport Logistics*, 6(5), 531–551.

Suggestions for further reading

Besides the papers provided as references in the text, we recommend the following texts as suggestions for further reading.

European Commission (2014) *Innovation Union Competitiveness Report 2013*. Staff Working Document Directorate-General for Research and Innovation. EUR 25650 EN. Publications Office of the European Union, Luxembourg.

OECD (2014) *OECD Science, Technology and Industry Outlook 2014*. Paris, OECD Publishing.

Rosenberg, N. (1994) *Exploring the Black Box; Technology, Economics and History*. Cambridge, Cambridge University Press.

Tidd, J., Bessant, J., and Pavitt, K. (2001) *Managing Innovation* (2nd ed.). London, Wiley & Co.

Suggestions for further reading – Chinese scholars

Before 2014 there was no literature on the concrete effects the Internet of Things could have for ports and logistics development. Zhen et al. provided the first overview of the potential impacts.

Zhen, H., Linjie, H., Qianwen, L., and Geerlings, H. (2015) Research on the impact of the third industrial revolution on port development. *Review of Economic Research*, 20, 48–55 (in Chines language).

This article forms the basis for Section 20.8. Other articles dealing in particular with the Chinese situation are:

Cullinane, K., & Wang, T. F. (2007). Port governance in China. *Research in Transportation Economics*, 17, 331–356.

Geerlings, H. M., and Hou, J. (2016) Dynamics in sustainable port and hinterland operations: A conceptual framework and simulation of sustainability measures and their effectiveness, based on an application to the Port of Shanghai. *Journal of Cleaner Production*, 135, 449–456.

Qiu, M. (2008). Coastal port reform in China. *Maritime Policy & Management*, 35(2), 175–191.

21

THE LABOUR MARKET OF PORT-CITIES

Jaap de Koning, Kees Zandvliet and Arie Gelderblom

21.1 Introduction

This chapter deals with the labour market of port-cities. Labour markets of port-cities share some common characteristics making them different from other cities and regions. First, ports require specific activities like transshipment. Second, ports attract other economic activities like the chemical industry for which location near or inside the harbour area is advantageous. In turn, the location of the latter industries may enhance other economic activities such as services industries. Hence, ports may have first-order and second-order effects on employment in their neighbouring city and the latter's surrounding region. Also macroeconomic effects of ports on production and employment can be envisaged.

These characteristics of labour markets in port-cities refer to labour demand. However, also labour supply in port-cities has some special features. Traditionally, port-cities have a relatively low-skilled population. In the past, port-related sectors like the shipbuilding industry have attracted large numbers of unskilled migrants, who have stayed although the employment they were initially attracted to has largely disappeared. Recently, ports and port-cities have attracted new migrants from Eastern Europe. Two issues have arisen from these developments. First, does the relatively low skilled labour force in port-cities lead to a qualitative imbalance between labour demand and labour supply leading to problems in filling vacancies? Second, do the new migrants from Eastern Europe displace the migrants from earlier waves, as well as the latter's descendants?

The structure of the chapter is as follows. In Section 21.2 the structure of labour demand in port-cities is analysed. To what extent does this structure differ from that in the rest of the country? What is the employment share of activities directly or indirectly related to the port? Section 21.3 discusses some specific employment aspects of ports, that is, (a) the development of employment in transshipment and its determinants and (b) the regional and macro employment effects of ports. The imbalance between labour demand and labour supply is the subject matter of Section 21.4. This section also discusses the role of migrant labour.

21.2   The structure of labour demand

21.2.1   The impact of ports on the employment structure of port-cities

The labour market of port-cities is as varied as any other regional labour market, as companies and inhabitants in the region have regular needs for goods and services, housing, health care and the like. Table 21.1 illustrates this point. It compares the employment structure of the Rotterdam region, which houses the biggest port of Europe, with the national employment structure of the Dutch economy and Antwerp with Flanders. Table 21.1 shows that the employment structure of the Rotterdam region is not very different from the national structure. The manufacturing industry is somewhat smaller in the Rotterdam region, but this is largely compensated by a larger construction sector. The latter has to do with the harbour in Rotterdam with its relatively large volume of construction and installation activities. If we break down the manufacturing sector into subsectors, it appears that in Rotterdam some subsectors like the chemical industry have a higher representation than in the national economy. The main difference between the national employment structure and that of the Rotterdam region is the larger share of the transport sector, which has obviously to do with the Rotterdam harbour.

TABLE 21.1 Employment structure: Flanders (2012) versus the Antwerp region (2012) and The Netherlands versus the Rotterdam region (2012)

	Flanders	Antwerp region	The Netherlands	Rotterdam region
Agriculture	4.1%	0.8%	1.3%	0.9%
Manufacturing	14.0%	11.1%	10.4%	8.8%
Construction	6.9%	5.5%	4.3%	5.1%
Trade	17.1%	17.8%	16.6%	16.6%
Transport	5.4%	7.0%	4.8%	7.6%
Accommodation and food service	4.8%	4.9%	4.3%	3.8%
Business and financial services	16.8%	21.8%	24.1%	24.7%
Public and personal services	30.8%	31.0%	34.1%	32.6%
Total	100%	100%	100%	100%

Sources: Statistics Netherlands; VDAB (2014).

The same conclusion applies to Antwerp compared to Flanders as a whole, although the difference between the share of the transport sector is less pronounced than in Rotterdam compared to The Netherlands as a whole.

21.2.2   The employment structure in Europe's main port-cities compared to Los Angeles and Melbourne

The employment structures of Europe's main port-cities have much in common. The main differences are that compared to Antwerp and Hamburg, Rotterdam is somewhat more oriented towards business services and financial services, while the employment share of the manufacturing industry is somewhat smaller in Rotterdam. However, if we compare the shares of the manufacturing industry in the national economies, the same pattern emerges. It should also be kept in mind that when comparing employment shares between different countries, different definitions may play a role. We lack recent information about Le Havre, which is another major port region. Recent information about the broader region in which Le Havre lies, Haut Normandy, and older information about the Le Havre region (2003), suggests that manufacturing and construction are more important in Le Havre compared to Antwerp, Hamburg and Rotterdam, while commercial services are less developed.

If we compare the employment structure of the main European employment structure with Los Angeles and Melbourne, it appears that Melbourne has a fairly similar employment structure to Europe's main port-cities. The main differences are the somewhat lower shares of transport activities, business and public services and the high shares of hotels, restaurants and catering, and especially construction (Table 21.2). The differences with Los Angeles are bigger. Particularly, business and financial services play a more important role in Los Angeles compared to the other port-cities.

It is likely that a more detailed classification would reveal considerable differences between port-city regions in employment composition within broad sectors such as the manufacturing sector. The previous section mentions the importance of the chemical industry in Rotterdam. The Rotterdam region does not house a steel industry (which in The Netherlands is located in the port of Amsterdam region) or an automotive industry like some other port regions elsewhere in Europe have. The automotive industry, for example, is relatively important for Le Havre.

21.2.3   The employment share of harbour-related activities

What is the importance of harbour-related activities to their employment share in port-city regions? It is customary to distinguish between activities that are directly harbour related and activities that are indirectly related to the harbour.

Activities that are directly harbour related are:

- All maritime activities, like shipping, inland shipping, fishery, shipbuilding and harbour-related (civil) engineering;

TABLE 21.2 Employment structure regions of Los Angeles, Melbourne, Antwerp, Hamburg and Rotterdam (2012, except Melbourne [2011])

	Los Angeles (a)	Melbourne (b)	Antwerp (c)	Hamburg (d)	Rotterdam (e)
Agriculture	0.1%	0.7%	0.8%	0.1%	0.9%
Manufacturing	9.6%	11.8%	11.1%	13.1%	8.8%
Construction	2.7%	8.5%	5.5%	3.4%	5.1%
Trade	10.1%	15.3%	17.8%	16.3%	16.6%
Transport and communication	4.1%	5.4%	7.0%	9.0%	7.6%
Hotels, restaurants and catering	8.5%	6.3%	4.9%	3.7%	3.8%
Business and financial services	30.4%	22.3%	21.8%	54.3% (a)	24.7%
Public and personal services	34.6%	29.8%	31.0%		32.6%
Total	100%	100%	100%	100%	100%

Sources: (a) Employment Development Department, Los Angeles County; (b) ABS Labour Force Survey, Australia; (c) VDAB, Socio Economische gegevens, Belgium; (d) Bundesagent für Arbeit (2014) for Hamburg; (e) Statistics Netherlands for Rotterdam.

- Harbour activities, like storage and transshipment, expedition, ship brokering, stevedoring, customs and harbour services;
- Harbour-related transport by road, train and pipes.

More indirectly related are activities for which location near to the port is more or less necessary are non-maritime activities, necessarily located at the harbour, like various manufacturing activities (petrochemistry, chemical products, metallurgic industry), energy production, maintenance, various trade activities and various specialised commercial services (judicial, economical, and the like).

The labour market region 'Groot Rijnmond' in which the port of Rotterdam is located, supports about 650,000 jobs (CBS, Statline), of which about 90,000 are regarded as 'harbour related' (Nijdam et al., 2014). This means that 14% of regional employment is connected to the harbour, while this ratio (for all Dutch ports taken together) is 3.3% at the national level. Comparable figures can be found for the port of Antwerp in Belgium (Antwerp Port Authority, 2014) and the port of Hamburg in Germany (Hamburg Port Authority, 2014). In 2006 the share of direct harbour employment in total employment in the Le Havre region was about 10% (Notteboom, 2010).

As Table 21.3 shows, in Rotterdam haulage and manufacturing are most important in terms of employment, followed by transport services and storage and transshipment. Comparable structures of (harbour-related) employment can be

TABLE 21.3 Direct harbour employment, total location-related employment and total employment in the port-city region of Rotterdam (in thousands; 2012)

Direct harbour employment	
Shipping	1.7
Inland shipping	6.9
Haulage (road transport)	27.6
Train transport	1.3
Pipe transport	49
Transport services	14
Storage and transshipment	9
Subtotal	60.5
Total location-related employment	
Manufacturing	17.3
Energy production	2.1
Trade	7.2
Commercial activities	5.1
Subtotal	31.7
Total harbour related employment	92.1
Total employment	650

Source: Havenmonitor (2012); Nijdam et al. (2014).

found in other major seaports in the north-western part of Europe (Le Havre, Antwerp and Hamburg). Differences are related to the specific location of ports within supply chains and production networks. For instance, in Rotterdam petrochemistry and chemical products are relatively important, while automotive industry and metallurgic activities are less important than in other port regions (Notteboom, 2010).

The low share of storage and transshipment illustrates that the traditional port worker no longer dominates the labour market of the port of Rotterdam. Even the percentage of direct harbour labour is relatively low (15%); as a percentage of total employment in the Rotterdam region it is only 1.4%. This is not only the case in Rotterdam. This occupational group is estimated at only 110,000 workers in 22 maritime member states of the European Union taken together (Van Hooydonk, 2013). European ports are increasingly functioning as major economic centres (transport hubs) within global supply chains and global production networks (Notteboom, 2010).

21.2.4 The role of human capital and flexible labour

The main ports in Europe are strongly competing with each other. Competition drives up productivity. Therefore, as in other economic activities, human capital is increasingly the deciding factor in the competition between ports. Skill levels have increased as a result of it. Firms also increase productivity by making their labour force more flexible. In the competition process wage levels are also important, particularly for the remaining low-skilled labour.

In terms of competences required, harbour employment in Rotterdam can be characterised by:

- A relatively high but declining share of unskilled and low-skilled demand, related to transport and logistic activities (drivers, machine operators, logistic employees, etc.). In these activities technical innovation leads to an ongoing shift towards more capital intensive production. This shift is accompanied by a growing demand for skilled personnel (operators).
- The same trend, a rise in the skill level required, occurs with further automation of the handling of goods (use of IT in administrative work).
- A relatively high share of manual skilled demand, related to manufacturing, maintenance, energy production and automated (capital intensive) logistical activities.

Technical innovation has reduced peak demand for labour, but increased the demand for shift labour. Harbours in Belgium, Germany and France still have specific pools for flexible labour demand, with special labour arrangements. In the UK and The Netherlands, flexible labour is hired on the basis of general labour legislation and recruited through private and public employment services. The following four points are also important with respect to flexible labour:

- Shift labour and irregular working hours. Especially petrochemistry and chemical products require not only regular maintenance, but also more incidentally large-scale improvements and adaptations of plants, causing short periods of peak labour demand for skilled labourers.
- As a consequence of outsourcing, the number of workers directly employed by plant owners has decreased considerably. Providers (of maintenance and equipment) can offer their personnel less certainty about a job, as a consequence of strong competition between providers.
- Flexibility is partly achieved by hiring migrant workers, who are better prepared to work under flexible arrangements and for low pay. Section 21.4 discussed the question to what extent migrant workers displace the local population. Displacement is not obvious as migrants partly fill (semi)skilled jobs in occupations suffering from a shortage of (semi)skilled indigenous workers. Furthermore, in those cases where migrant workers are filling low-skilled jobs, factors other than poor working conditions may explain why the indigenous unemployed do not fill these jobs.
- Labour unions in the north-western part of Europe (UK, The Netherlands, Germany, Belgium and France) have joined forces in order to prevent further deterioration of wages and labour conditions (www.fnvhavens.nl/

fnv-havens/150-internationaal/85-havenbonden-sluiten-de-rijen). Employers tend not only to improve labour productivity by means of automation, but also reduce labour costs, by offering less favourable wages and labour conditions, as a response to international competition at the European level. By combining their forces, labour unions from the countries concerned try to withhold this trend of wage reduction.

21.3 Different aspects of employment in port-cities

21.3.1 Factors affecting employment in transshipment

One might associate employment in port-cities particularly with manual workers involved in transshipment. However, this is a picture of the past. We already saw that nowadays transshipment only constitutes a relatively small part of harbour employment. As a percentage of the total employment in port-city regions it is very small. In the Rotterdam region the employment share of storage and transshipment combined together is less than 2%. Notteboom uses the term 'dockworkers' for those workers involved in cargo handling. This is probably even a smaller group. In 2013 in Antwerp the total number of dockworkers was only 8,000, which was less than 10% of the total direct harbour employment. Not only did the number of dockworkers decline, but also cargo handling was largely mechanised.

Although dock work is not dominating harbours in terms of numbers of workers anymore, it still fulfils a vital function. Port performance still highly depends on the efficiency by which cargo is handled. Notteboom gives a description of the technological changes in ships and cargo handling, which led to a change from man load to unit load and bulk cargoes. In this process manual labour was substituted by cranes and containers, which implied:

- A considerable increase in productivity and a fierce reduction in employment;
- A rise in skills required for most of the remaining workforce.

This is not a recent process. Already in the 1960s and '70s much of the manual work disappeared. In 1980 the number of dockworkers in Antwerp was already reduced to 10,000. Since then the decline has not been so large (from 10,000 to 8,000). In Rotterdam the number of dockworkers reached its peak in the 1960s. It then amounted to almost 17,000 workers (Notteboom, 2010). Nowadays it is about half of this figure.

Technological development has severely reduced the number of dockworkers but at the same time considerably improved their working conditions (see Box 21.1 for a description of important technological innovation and automation in the port of Rotterdam). In earlier years, particularly before the First World War dock labour was largely casual labour. Fluctuations in labour demand led to irregular incomes. Furthermore, a structural oversupply of labour had a strong downward effect on hourly wages. About 100 years ago the first institutional arrangements like labour pools were introduced to improve the situation of the workers. Such pools provided the labour

---

BOX 21.1 MAJOR TECHNOLOGICAL INNOVATION AND AUTOMATION IN ROTTERDAM

The construction of the Maasvlakte 2, a new port area, incorporated innovation in construction of marine infrastructure, in combination with more efficient use of energy, both fossil and renewable. For instance the new facilities are connected to existing plants by means of a large system of pipelines, laid out at great depth beneath the new port area, allowing efficient transport of electricity, fluids and gases.

The new harbour facilities, close to the port entrance, together with local expertise in the development of software and automated systems for remote operation, allowed the introduction of the remote-operator ship-to-shore container crane at the terminals of APM Terminals and RWG. This drastically reduces labour needs and handling time. As a consequence even the biggest container ships can be handled very efficiently.

Another important development is the further automation of the administrative process connected to cargo handling. Up-to-date software packages for cargo handling, track and trace, warehouse management and route planning (including modal shifts) become more and more available and implemented by shipping agents and brokers, transport and logistic services. This trend will be accompanied by reduced labour demand and a rise and shift in skills required.

For further reading and details, see www.portofrotterdam.com.

---

demanded by the different employers. As the fluctuations in total labour demand as exercised by all employers taken together are relatively smaller than the fluctuations faced by each individual employer, a pool of people working for different employers makes sense. However, technological developments, particularly the introduction of containers reduced the need for this type of flexibility. The number of workers under contract by individual firms working fixed weekly hours in shifts increased and the number of people working in pools decreased. In the course of time a large number of casual, unskilled and low-paid harbour workers have been replaced by a much smaller number of relatively well-paid skilled workers with much better working conditions. As a result the bargaining power of the workers and the unions representing them has increased. However, competition between port-cities has become stronger, which gives a counterweight to union power. The position of labour has generally weakened and dock labour is no exception to this development.

Increased capital intensity and automation of cargo handling has created a demand for a second category of workers, namely maintenance workers. The latter perform an equally vital role in modern ports as dockworkers. Both types of workers can be categorised as skilled technical workers. With the general decline in technical work, also the number of students in technical vocational education has diminished sharply.

In spite of the decline in the demand for technical work, the supply of technical workers tends to be insufficient to fill the available vacancies. This poses a considerable problem for ports and also for manufacturing and installation firms located in the port area. This problem will be further addressed in the next section.

21.3.2 Employment effects of ports

Ports are part of a broader infrastructure of transporting goods from producers to consumers and to other producers. In a general sense it is fairly clear that transport infrastructure has an added value to the economy. Trade is not possible without transport. And trade gives an important contribution to the wealth of nations. However, this conclusion is not able to be so easily drawn for specific forms of infrastructure. Investment in roads can be taken as an example. Within the framework of the European Regional Development Fund, investment in road infrastructure has been a key activity aimed at stimulating economic development in less developed regions. The reasoning behind this type of investment is that making regions better accessible will more or less automatically enhance regional economic growth. However, if a region is better accessible it is also easier for the inhabitants to go to other regions. Commuting, for example, becomes easier. So, it is possible that the effect on the economy of the less developed regions has a negative rather than positive affect. If the region benefits, it is also questionable whether the economy as a whole is affected positively. More recent evaluations of this type of infrastructural investment do not always point to positive economic effects.

It is not obvious whether ports have positive macroeconomic employment effects. Studies suggesting strong positive effects often use methods like input-output analysis, which usually do not take substitution between products and production factors into account. Obviously, if tomorrow the port of Amsterdam stopped functioning, the Dutch economy would be affected negatively. However, in the long run the economy might adapt to the new situation. Other ports and other types of transport may gradually take over the role of the port of Amsterdam. In the end the macro-economic affects may not be negative. Even for the city of Amsterdam the initial negative effects may disappear as other economic activities could take over the harbour-related activities. However, closing a port in total for economic or sustainability reasoning is not very likely and has not happened in recent times.

A more realistic situation would be if a port wants to grow and investments are planned to realise this growth, for example, when the port of Rotterdam expanded by developing a new port area known as the Maasvlakte 2. Whether this investment constituted a positive macro-economic return is difficult to say. It also depended on the investments taking place in the competing port-cities. A recent study has thrown light on the impact of port-cities on regional employment. It uses state-of-the-art econometric methods to assess the impacts of ports on regional employment. Data from about 560 regions in Western Europe including 116 ports was collected during the period 2000–2006. The study combined cross-section and time series analysis. The study examined the relationship between port throughput and regional employment. It took into account that this is a simultaneous relationship: port throughput may affect regional output and thus regional employment. The study found clear evidence for a causal effect of throughput on regional employment. Similar conclusions were found when separate analyses were carried out for manufacturing employment and service sector employment, providing some evidence that manufacturing employment is more affected than service sector employment. According to the OECD (2014) and Erasmus Smart Port Rotterdam (2011), it is particularly the high-skilled service employees that leak away to other regions. Thus the results support the hypothesis that ports are important for regional employment.

A different matter is whether port-cities benefit from their ports. People working in port-related activities do not necessarily live in the port-city associated with the port. In many port-cities unemployment is much higher than the national average. Port-related activities offer relatively few jobs to the population within the city. One of reasons for this situation is the decline in manual port labour. Furthermore, the city population in many port-cities in Western Europe contains a large proportion of migrants from Morocco, Turkey and former colonies, who do not have much affinity for typical harbour-related activities. This point will be further discussed in the next section. Hence, the benefits of ports for their accompanying port-cities may be limited. Port-cities, on the other hand, invest a lot of money in their ports. So, most likely the region surrounding the port-city and perhaps also the national economy as a whole may benefit more from the port than the port-city. If this is true, it would be an argument in favour of some sort of subsidy for ports.

21.4 The connection between labour demand and labour supply in port-cities

21.4.1 Shortages of technicians

Technicians form a considerable part of employment in port-city regions. First, process operators play a crucial role in transshipment which is, as we discussed earlier, mostly automated. People trained in logistics are also badly needed in these areas. Furthermore, technicians are important for the manufacturing industry located in the port area. Last but not least, maintenance is crucial for both transshipment and the manufacturing industry. An obvious example of an industry needing maintenance is the (petro)chemical industry, which plays an important role in port regions like the Rotterdam region.

Hence, for the economy of port regions it is of vital importance that labour supply is of sufficient quantity and quality. However, already for two decades the manufacturing sector and other sectors complain about shortages of technicians. This phenomenon can be found in many Western countries. It is also prevalent in countries like Germany and the United States. The most important factor causing this shortage is the decline in manufacturing employment. Most port-cities used to have a considerable shipbuilding industry, for example, which has largely moved to low-wage countries. Another industry (mostly located in non-port areas) that disappeared from The Netherlands is the textiles industry. Furthermore, manufacturing industries that have remained active in The Netherlands, like the metalworking and electrical engineering industries, have considerably increased in productivity. Despite continued output growth, employment has declined in this industry. As employment opportunities in manufacturing decline, most of the young have opted for non-technical types of occupations. However, at some point the number of students in technical vocational education became so low that it was not enough anymore to fill the labour demand in technical occupations. The ageing of the working population and the corresponding increase in the number of older workers leaving for retirement also played a role. The latter implied that even industries with declining employment still needed a considerable inflow of young workers. But the image of a declining sector gives a negative signal to young people. Although job opportunities in this sector are better than in most other sectors, many young people have a negative perception of the possibility to find employment in the manufacturing sector. Furthermore, if a firm in the manufacturing sector is closing or is moving some of its activities to other countries, it will catch the interest of the press much easier than a service sector firm going bankrupt. There are also negative perceptions of working conditions and pay in technical work. This is largely an echo of the past. Nowadays, as a result of technical progress, working conditions are quite good and the pay is often better than in non-technical occupations. Until recently, many people, even high government officials, thought that the manufacturing industry would soon disappear altogether from The Netherlands. Gradually, this view is losing ground. An open economy like The Netherlands must export a considerable part of its production and the manufacturing industry is involved in most of the country's export.

Socio-cultural factors also have an effect regarding the negative view of technical workers. Cities like Rotterdam and Antwerp are home to many people of foreign origin. Some migrants have come recently to Western Europe, but others are second-, third- and even fourth-generation migrants. Among these groups the choice to study and work in a technical occupation is less frequent than among people of Dutch origin. This is partly caused by the fact that a considerable number of the migrants and their descendants are still influenced by the culture of their country of origin. In this culture technical work often has a relatively low status compared to white-collar work. Furthermore, first-generation migrants were often employed in the manufacturing sector during the 1960s when the labour conditions in this sector were quite bad. Although these conditions have improved considerably, the negative experiences still play a role as fathers pass them down to their children and the latter to their children.

In a study by De Koning et al. (2014), large-scale surveys have been conducted among students in Preparatory Vocational Education (PVE) and their parents in The Netherlands. PVE prepares children between 12 and 16 years of age for Secondary Vocational Education (SVE). A SVE diploma marks a young person as a skilled worker. During the PVE children decide about which field of education to obtain their first vocational education in. The results confirm that the poor image of technical work is partly based on misperceptions regarding job opportunities and pay in this type of work. There is also evidence confirming the hypothesis that children from migrant families have less interest in technical work than children of Dutch origin. Furthermore, the results confirm that parents particularly from migrant families play an important role in the educational choices made by their children. Parents tend to point their children in the direction of white-collar work. As migrants constitute a large part of the population of port-cities like Rotterdam, this reduces the possibilities to place the local population in harbour-related jobs. The fact that on average third- and fourth-generation migrants have a low educational level add to this problem. This leads to the situation in which on the one hand firms in the harbour area have problems in finding enough skilled workers, while at the same time unemployment rates in the neighbouring port-city are very high. In some quarters of Rotterdam unemployment rates among young people of migrant origin are higher than 30%.

However, the study mentioned in the previous section also provides results that might be used to find ways of solving this problem. First, it appears that the lack of information is an important reason why children and parents opt for non-technical work. If schools have a program stimulating technical education, provide labour market information to children and parents and give clear advice about what is the best field of education for a child, all of this can have a positive and significant effect on the choice of a technical education. However, the study also indicated that prejudice against people of migrant origin is quite common among employers. Although it concerns a minority of employers, the group involved is far from negligible. Hence, also measures aimed at employers are needed.

Traditionally, girls have a very low interest in technical work. For some time, programs have been carried out to try to convince girls as to the opportunities of following a technical vocation. For a long time these programs did not have much success, but recently it has been observed that there are growing numbers of girls and young women opting for a technical education. One of the factors making girls and women hesitant to work in the manufacturing sector is that the latter still have fairly traditional labour relations and are reluctant to offer the type of flexibility in working hours women require.

It is a recognised fact that programs aimed at interesting young people in science and technology should primarily be aimed at primary school–aged children. During this age, preferences are to a large extent formed. However, the previously cited study indicates that even after primary school educational choices can be influenced.

Other measures that could help to solve the problem are:

- To reduce the number of school-leavers with a technical education opting for a non-technical job;
- To reduce the mobility of technical workers to non-technical jobs;
- To prevent unemployment and forced inactivity among older technical workers;

- To develop retraining programs for the unemployed and for technical workers in industries facing restructuring;
- To have programs that influence firms' human resource management and hiring policies regarding women and migrants.

21.4.1.1 Is the port-city's population displaced by new groups of migrants?

To a large extent employers have solved their problems in finding a sufficient number of technicians by attracting new groups of migrants. Most of the people recruited come from Eastern Europe and particularly from Poland. For the port region of Rotterdam the number of these migrants is estimated at 50,000, which is 8% of this region's employment.

Some people claim that these new migrants are displacing the existing population, of which a large proportion consists of 'old' migrants and their descendants. Unemployment among the latter groups is extremely high. One might argue that it would be both socially and economically desirable that a larger share of employment in the port-city region should be filled by its inhabitants (both native Dutch and 'new' Dutch). This would improve the latter's income position, which in turn would stimulate the regional economy. Furthermore, the local government would save on unemployment benefits and on expenses resulting from the problems associated with high unemployment.

However, the fact that it would be desirable that a larger share of the available jobs is filled by the local population does not imply that the new migrants are displacing the former. The questions that should be asked are: What would have happened if the new migrants had not come? It is likely that the local population would have filled the jobs in that case? Cases have been reported where workers from Eastern Europe were hired under labour conditions that were much worse compared to the standards prescribed by Dutch collective agreements and Dutch labour law. This can be seen as unfair competition between ('old' and 'new') Dutch workers and the new foreign workers. In those cases it is fair to speak of displacement. The Dutch government clearly states that unfair competition between Dutch and non-Dutch workers is unacceptable. New legislation has been put in place to fight arrangements by which employers can avoid paying foreign workers according to the standards required by Dutch collective agreements and Dutch law. However, the government cannot avoid firms from moving some of their activities to low-wage countries. Depending on the extent to which this takes place, the new legislation may not be entirely effective. For port-related activities, off-shoring may be less probable, as the location of the port itself is fixed.

Results from studies in other countries have suggested that new migrants do not displace local workers. The same conclusion has been drawn in a study that was carried out by The Netherlands Bureau for Policy Analysis. These studies do not specifically focus on port-cities. However, these results may well apply to local workers in port-cities too. We have seen that the local population in Rotterdam does not have a particular high interest in technical labour. Furthermore, most of the unemployed lack the required skills. While employers do not hesitate to hire migrants from Eastern Europe, the attitude of a considerable minority among them towards migrants from earlier waves, as well as the latter's descendants, is quite negative. This regards not just a matter of skill deficiencies among the older migrants, although many of the latter certainly lack a good vocational education giving access to occupations that are in demand in the labour market. Improving the employment rate among migrants from earlier migration waves requires not only an improvement in their skills, but also a change in attitude, in both of these groups and their employers.

21.5 Conclusions

The structure of employment in port regions is not very different from the national employment structure. The main differences are larger employment shares of the transport industry and the construction and installation industry. Furthermore, ports often house specific manufacturing industries and specialised service industries for which location in a port area is efficient or even necessary. In Rotterdam, for example, the (petro)chemical industry is over-represented. In other port regions this is the case for the steel industry, the automotive industry, or energy production and distribution.

Nowadays, the type of labour traditionally associated with ports, dock labour, only has a small share of harbour employment. In the main European ports it constitutes between 10% and 15% of the direct port employment and not more than 10% of the total harbour employment. This low percentage is the result of a strong productivity increase owing to the introduction of containers and the mechanisation and automation of cargo handling. However, this low employment share of harbour labour does not mean that the importance of dock labour has diminished. It still plays a vital role in ports. Higher capital intensity and automation have created a new group of workers that are equally important for the good functioning of ports, namely, maintenance workers.

Port-related employment forms an important, but not an extremely major part of total employment in port region. In The Netherlands, if you combine the employment rate of all of the ports, it is equivalent to 3.3% of the national employment. This includes the location-related employment, which may not always be totally dependent on the location of a port area. Hence, one might ask the question: are ports important from a macroeconomic point of view? It is difficult to deny that ports are needed in the supply chain. But it is more difficult to assess the macroeconomic effects of a specific port, or the effects of investments increasing a port's capacity. However, there is reliable scientific evidence that ports have a positive employment effect on the region located near it or surrounding it. However, a considerable part of this employment is filled by workers commuting from other regions and by new migrant flows, particularly from Eastern Europe. The population in port-cities contains a large group consisting of low-skilled people that hardly benefit from the neighbouring port. The majority of this group consists of people that arrived with earlier migration flows. They not only have skill deficiencies, but the people and their employers also have an attitude that helps to hamper their integration into the labour market. Just limiting the number of migrants from Eastern Europe, if at all possible, will not solve this problem.

Technicians play an important role in port areas as dock workers, maintenance workers and workers in manufacturing firms located in these areas. Many Western countries face shortages of technicians. It seems logical to find ways of having more of these jobs filled by the port-city's population which suffers from very high unemployment rates. This requires the joint effort of ministries, municipalities, employers, schools, organisations representing migrant communities and of course the affected unemployed people. An awareness for this type of cooperation has grown. On the south side of Rotterdam, where the disadvantaged groups are concentrated, a large program has been launched aimed at revitalising this part of Rotterdam. Getting more people employed in harbour-related activities is one of its aims.

References

Bundesagent für Arbeit (2015) *Regionalreport über Beschäftigte*. Land Hamburg, juni 2015.

De Koning, J., Gelderblom, A., and Gravesteijn, J. (2014) *How Can Children be Stimulated to Choose a Technical Education?* SEOR Working Paper 2014/1, Rotterdam, SEOR.

Dombois, R., and Wohlleben, H. (2011) The negotiated change of work and industrial relations in German seaports; the case of Bremen. In Dombois, R. and Heseler, H. (eds.) *Seaports in the Context of Globalisations and Privatisation*. Bremen, University of Bremen, pp. 45–65.

Erasmus Smart Port Rotterdam (2011) Rotterdam World Port Word City. Hoogwaardige zakelijke dienstverlening voor het Rotterdamse haven – en industriecomplex (Provision of high value added commercial services to the harbour – and industrial activities at the port of Rotterdam), Research Commissioned by the Community of Rotterdam (Department of City Development and Economy), Erasmus University Rotterdam and Utrecht University, October 2011.

Hamburg Port Authority (2014) *The Port of Hamburg Facts & Figures 2014*. Hamburg, Hamburg Port Authority.

Nijdam, M., and Van der Lugt, L. (2014) *Havenmonitor 2012*. De economische betekenis van Nederlandse zeehavens, opgesteld voor het Ministerie van Infrastructuur en Milieu, RHV, Rotterdam, Erasmus Universiteit Rotterdam.

Notteboom, T. (2010) *Dock Labour and Port-Related Employment in the European Seaport System: Key Factors to Port Competitiveness and Reform*. ITMMA, University of Antwerp.

OECD (2014) *The Competitiveness of Global Port-Cities: Synthesis Report*. Paris, OECD.

Van Hooydonk, E. (2013) *Port Labour in the EU: Labour Market Qualifications & Training & Health & Safety*. Antwerp, Ghent.

VDAB (2014) *Socio-Economische Dataset Provincie Antwerpen & de Antwerpse RESOC's*.

Suggestions for further reading

There is not much recent literature on port labour and port labour markets. A relatively recent study about institutional aspects of port labour is Van Hooydonk (2013). A lot of factual information on a broad range of topics is provided by Notteboom (2010). Dombois and Wohlleben (2011) discuss industrial relations in German seaports, especially in Bremen. But their paper mostly deals with developments before 2000.

Dombois, R., and Wohlleben, H. (2011) The negotiated change of work and industrial relations in German seaports; the case of Bremen. In Dombois, R. and Heseler, H. (eds.) *Seaports in the Context of Globalisations and Privatisation*. Bremen, University of Bremen, 45–65.

Harding, A. S. (1990) *Restrictive Labour Practices in Seaports*. Washington, The World Bank.

Notteboom, T. (2010) *Dock Labour and Port-Related Employment in the European Seaport System: Key Factors to Port Competiveness and Reform*. ITMMA, University of Antwerp.

Van Hooydonk, E. (2013) *Port Labour in the EU: Labour Market Qualifications & Training & Health & Safety*. Antwerp, Ghent.

22

PORTS IN TRANSITION

Derk Loorbach and Harry Geerlings

22.1 Introduction

Many of the world's welfare today has been produced or at least facilitated by ports and its related activities: ports are the locations where trade, logistics and production converge. This book illustrates very clearly that ports have developed over the last decades, in line with the emerging global economy, into global hubs for large-scale efficient trade and shipping. In many ports, this development has gone hand in hand with the emergence of large-scale fossil-based industries in port areas, ranging from refineries to chemical industries and energy plants. This has led to ports that are completely based on large-scale fossil fuels and efficient mass logistics. In our current age, however, many aspects of this historically developed basis are starting to erode. Global (economic) developments are putting pressure on the mass-logistics model, the market for fossil resources and fuels is becoming unstable also due to the scale gas discussion, and the location choice and investments in chemical industrial complexes has become an issue of debate. Therefore in many ports, questions arise as to how to reinvent the economic model behind the port and how to move forward towards more sustainable ports based on green chemistry, renewable energy and added value. This requires a more fundamental change than can be achieved only through technological innovation, optimisation or planning: a transition towards sustainable port activities.

The chapter is structured as follows. In Section 22.2 a synthesis is presented of the current theoretical understanding transitions and its context. A brief reflection on transition management and how the insight into the dynamics of transitions can be translated into a framework for governance is presented in Section 22.3. In Section 22.4 we present an empirical case study in which we explore different transition pathways and strategies for moving towards a sustainable inland shipping. In Section 22.5 we zoom out and frame the challenge of sustainable ports: what could be a sustainable port, and what then are the main transition challenges? In Section 22.6 we synthesise our chapter and look into the future.

22.2 Transitions

Transition studies refers to a field of research that focuses on 'transitions', generally defined as non-linear processes of social change in which a societal system is structurally transformed. One of the central premises in transition studies is that persistent problems are symptoms of unsustainable societies, and that dealing with these persistent problems in order to enable more sustainable systems requires transitions and system innovations. Transition studies is an emerging field of research that seeks to integrate insights from areas such as innovation studies, economy, sociology and environmental science to better understand large-scale systemic change in societal systems and explore possibilities for influencing the speed and direction of change in these systems.

It originates from a policy debate in The Netherlands around the fourth National Environmental Policy Plan in 2001. In this debate, the concept of transitions was proposed as an approach to better understand the failure of policy and markets in delivering a fundamental reorientation of the development pathway of modern societies and an opportunity to explore new ways to achieve breakthroughs. Since then, the field of transition studies has become an international research field comprising a multitude of core concepts, approaches and intervention methods. In this chapter we take stock of one of these streams of research, transition management, and discuss experiences of more than fifteen years of action-oriented research in this area.

In essence, transition management studies complex adaptive societal systems (such as societal sectors, regions, cities or ports) that go through fundamental non-linear changes in cultures, structures and practices. Transitions are defined as the result of co-evolving processes in economy, society, ecology and technology that progressively build up towards a revolutionary systemic change during the very long term. Because of this complexity, transitions are impossible to predict, fully comprehend, or steer directly, but they are seen as a pattern of change that can be anticipated. These processes can be adapted in such a way that the inevitable non-linear shifts and associated crises provide massive windows of opportunity for accelerated reorientation towards sustainability.

The multilevel model (see Figure 22.1), originating from innovation and technology studies, is taken as a point of departure. The central level is the meso-level at which the so-called regime is located. The term 'regime' refers to the dominant culture, structure and practice embodied by physical and immaterial infrastructures (not only roads and power grids, but also routines, actor-networks, power-relationships and regulations). These institutionalised structures give a societal system stability and guide decision making and the individual behaviour of actors. At the same time, the regime has a certain level of rigidity that normally prevents innovations from fundamentally altering the structure. The second level is the micro-level

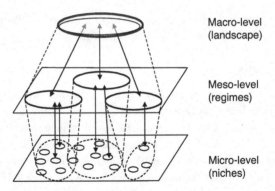

Macro-level
(landscape)

Meso-level
(regimes)

Micro-level
(niches)

FIGURE 22.1 Interaction between different scale levels

Source: Grin et al., 2010.

of innovations; inside so-called niches novelties are created, tested and diffused. Such novelties can be new technologies, new rules and legislation, new organisations or even new projects, concepts or ideas. The third level is the landscape, the overall societal setting in which processes of change occurs. The landscape consists of the social values, political cultures, built environment (factories, etc.) and economic development and trends. The landscape level typically develops autonomously but directly influences the regime level as well as the niches by defining the room and direction of change (see Figure 22.1).

Although transitions are characterised by non-linear behaviour, the process itself is a gradual one. Transitions can be described in terms of 'degradation' and 'breakdown' versus 'build up' and 'innovation', or in terms of 'creative destruction'. The central assumption is that societal structures go through long periods of relative stability and optimisation, followed by relatively short periods of structural change. In this process, existing structures (values, institutions, regulations, markets, etc.) fade away while new ones emerge. Historical analyses of societal transitions suggest that transitions go through different stages. Four phases have been currently distinguished and are represented by an S-shaped curve. The nature and speed of change differ in each of the transitional stages (see Figure 22.2):

- In the predevelopment phase, there is very little visible change on the societal level but there is a lot of experimentation.
- In the take-off phase, the process of change gets under way and the state of the system begins to shift.
- In the acceleration phase, structural changes take place in a visible way through an accumulation of socio-cultural, economic, ecological and institutional changes that react to each other; during this phase, there are collective learning processes, diffusion and embedding processes.
- In the stabilisation phase, the speed of societal change decreases and a new dynamic equilibrium is reached.

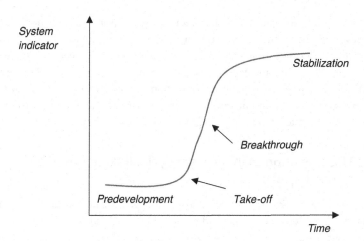

FIGURE 22.2   Four phases of transition

Source: Grin et al., 2010.

The S-shaped curve is a highly simplified model that represents a complex process like a transition. Behind the smooth S-curve, multiple and interrelated innovations also take place at different speeds and levels. In systemic terms, transitions are the result of interacting system innovations (at the level of the subsystems), which in their turn result from (product and process) innovations. For example, the transition in Eastern European countries consists of interrelated innovations in their institutional, economic, socio-cultural and technological systems, while these can only come about as the result of new regulations, organisations, infrastructures and so forth. 'Transition' is thus a collective term referring to a wide range of interconnected innovations at different levels, which is called a 'cascade of innovations'. The S-shaped curve is therefore not so much a model that predicts for example changes from one phase to the other, it is a conceptual (descriptive) model to reflect upon possible dynamics of a societal system and possible future trajectories. The central message of the S-curve is that structural change is not a gradual, linear process and that in the longer term structural change is to be expected under certain circumstances. This message underlines the need for conceptual, cognitive and operational approaches that are able to deal with non-linearity and unpredictability.

Core concepts in transition studies are regimes and persistent problems. The basic idea is that over time dominant configurations (structures, cultures and practices) emerge that are dynamically stable but because of their historical development are also locked-in and path dependent. Our current societal regimes, such as our systems of energy, food, mobility and housing, have emerged out of the era of industrial transformation and developed around central control mechanisms, fossil resources and linear models of innovation and knowledge production. They have brought welfare and growth but also our current sustainability challenges. Many policy and market-based strategies have been developed to stimulate sustainable development and innovation, but the efficiency gains (for example, more efficient industries) have been curbed by growing levels of consumption. The transition premise is that in order to achieve the levels of sustainability needed (e.g., an 80% reduction of CO_2 emissions) requires more than just improving what currently exists; it requires systemic changes or transitions. From this perspective, the challenge to move towards sustainability is thus understood as the need to achieve fundamental systemic change, implying disruptive power shifts.

22.3 The transition management cycle: linking descriptive to prescriptive

The starting point for thinking about transition governance is therefore that we need to develop alternative ideas and approaches to stimulate sustainability transitions, besides policy or market-based approaches. The hypothesis underlying transition management is that (collective) understandings of the origin, nature and dynamics of transitions in particular domains will enable actors to better anticipate and adapt to these dynamics so as to influence their speed and direction. In research practice, this has led to a multitude of practice-based governance experiments, in which structured multi-actor processes have simultaneously informed science in terms of insight in transition dynamics, actor perceptions and strategies, and helped actors to develop visions, strategies and interventions in their respective transition contexts. This governance experimentation based on the understanding of transition dynamics has informed the approach of transition governance and management.

To translate this basic idea into an operational approach, the transition management framework was developed based on empirical studies into the role of a transformative agency in past transitions as well as the conceptual understandings of transitional dynamics. A basic starting point is that a threshold in transitions can only be achieved through *selective participation*, as incumbent actors within a regime context will initially seek to sustain the equilibrium. The framework distinguishes between different clusters of activities that are recognisable throughout any governance processes during long-term societal change. These are the typical phases identified by many policy-process models, but fundamentally different in their focus on societal processes, persistent problems, and normative direction. This process model has been developed by Loorbach and Rotmans, based on iteration between theoretical reflection and practical experiments with new systemic instruments. Such experiments include regional transition arenas, the Dutch national energy transition program, and two transition arenas on resource transition and sustainable housing in Flanders, Belgium.

The systemic instruments are captured in a cyclical process model as a basis for implementing the transition management approach. It thus offers the basis for a normative approach based on an analytical framework, and theoretically offers the perspective of actively influencing the 'natural' self-steering and governance activities present in society. This so-called transition management cycle (see Figure 22.3)

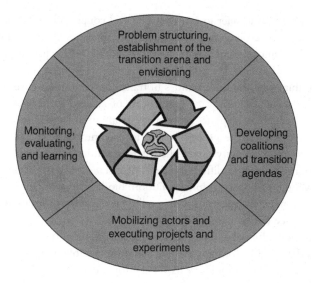

FIGURE 22.3   The transition management cycle

Source: Loorbach (2007).

consists of the following components: (1) structure the problem in question, develop a long-term sustainability vision, and establish and organise the transition arena; (2) develop future images and a transition agenda, and derive the necessary transition paths; (3) establish and carry out transition experiments and mobilise the resulting transition networks; (4) monitor, evaluate and learn lessons from the transition experiments and, based on these, make adjustments in the vision, agenda and coalitions. In reality there is no fixed sequence to the steps in transition management. The cycle only visualises the need to connect activities and presents some possible logical connections, but does not suggest a sequential order of activities.

As societal transitions in large-scale systems are by definition complex, uncertain and contested, they are also unmanageable from a traditional perspective, not only because there is no one point of origin for interventions but every actor influences the dynamics, but also because most of the more powerful actors are part of the system facing structural changes. This means they are by definition adverse of disruptive and swift transitions. A governance approach for transitions thus requires the following new perspective towards governance of planetary problems:

- Action needs to tap into and be in confluence with ongoing dynamics in order to steer the system with incremental steps towards a new sustainable (system) state (co-evolution rather than revolution).
- Small-scale actions need to be directed to domains in which a small intervention can result in tipping towards larger changes or, simply, seek for such changes that can cascade towards broader system's innovation (tipping innovation's cascade).

- Actions also need to refer to processes that will couple with or reroute ongoing processes in a co-evolutionary continuum (feed in and onto co-evolutionary processes) taking into account existing power relations as well as path dependencies.

Given the ambiguities and uncertainties in transitions, transition governance views sustainability as a characteristic of the process of change itself, both in terms of the extent to which it helps to move away from unsustainability and in terms of the extent to which the process is able to accommodate and generate diversity, inclusivity and reflexivity. Given that there are different approaches to dealing with transitions focused on different aspects of sustainability (for example, climate change, recycling, carbon emissions, pollution, biodiversity), the aim is to achieve as many interconnected aspects of sustainability as possible within the scope of the intervention space (Grin et al., 2010).

Based on the multilevel and multiphase understanding of transitions in complex systems and the insights from governance literature, the following tenets for a form of governance based on complexity have been formulated:

- The dynamics of the system should create a feasible and non-feasible means for steering: this implies that content and process are inseparable. Process management on its own is not sufficient; insight into how the system works is an essential precondition for effective management.
- Long-term thinking (at least 25 years) as a framework for shaping short-term policy in the context of persistent societal problems. This means backcasting and forecasting: The setting of short-term goals based on long-term goals and the reflection on future developments through the use of scenarios.
- Objectives should be flexible and adjustable at the system level. The complexity of the system is at odds with the formulation of specific objectives and blueprint plans. While being directed, the structure and order of the system are also changing, and so the objectives set should change too.
- The timing of the intervention is crucial. Immediate and effective intervention is possible in both desirable and undesirable crisis situations.
- Managing a complex, adaptive system means using disequilibria as well as equilibria. Relatively short periods of non-equilibrium therefore offer opportunities to direct the system in a desirable direction (towards a new attractor).
- Creating space for agents to build up alternative regimes is crucial for innovation. Agents at a certain distance from the regime can effectively create a new regime in a protected environment to permit investment of sufficient time, energy and resources.
- Steering from 'outside' a societal system is not effective: structures, actors and practices adapt and anticipate in such a manner that these should also be directed from 'inside'.
- A focus on (social) learning about different actor perspectives and a variety of options (which requires a wide playing field) is a necessary precondition for change.

- Participation from and interaction between stakeholders is a necessary basis for developing support for policies but also to engage actors in reframing problems and solutions through social learning.

22.5 Applying transition management; the case of inland shipping

In this section we illustrate the transition perspective and how it can help to frame, analyse and explore a complex sustainability challenge in the port of Rotterdam, namely the position of inland shipping. The port of Rotterdam is the largest container port in Europe. In 2012, 11.9 million TEU were handled, of which 7.4 million TEU had its origin or destination in the hinterland of the port, with a modal share of 35% for inland waterway transport (2.6 million TEU). In the Port Vision 2030 (Port of Rotterdam, 2011), the Rotterdam Port Authority expects that the throughput of containers will continue to grow up to 33 million TEU in 2035. Therefore, the port authority invested in the development of new port capacity, called the Second Maasvlakte—an extension of the port including large-scale container infrastructure. A major problem related to this port extension is the accessibility of the Second Maasvlakte and the effects on the quality of life in the surrounding urban areas.

A powerful tool to the environmental impact in the port area is the modal shift policy which was introduced with the aim to shift cargo from the truck, the most used transportation mode for container transport in seaports worldwide, to inland waterway transport and railway transport. Modal split policy is reflected at several policy levels. The Rotterdam port authority has the ambition to realise a modal split of 45% for hinterland transport of containers by barge for the total Maasvlakte port area in 2033. In view of this, the barge volume to be handled in the port may be increased up to 8.1 million TEU in 2033, starting from 2.6 million TEU in 2012.

The inland container shipping's market share in the supply and transport of containers to the Maasvlakte is currently remaining stable (around 40%), but the inland container shipping's share in the total modal split in both the inland and international transport is decreasing. And even more alarming is that the favourable environmental performance per unity transported weight was one of the sector's traditional strengths. However, relatively speaking, these performances are decreasing compared to the road transport. Furthermore, we see that in certain market segments of the inland waterway transport, including container transport, there is overcapacity which has an impact on tariffs, and the profitability of skipper-owners is currently under a lot of pressure. This leads to the conclusion that the sector is insufficiently capable of managing the level of the sector as a whole to be able to improve these performances. There are many loose initiatives taking place, but at present none of the actors involved in the sector is capable, or willing, to lift the inland shipping sector as a whole, if necessary in coordination, to a higher plan. However the sector's performance is continually changing, technology, infrastructure, regulation and cultural environment are changing slowly too – in the case of the cultural environment, very slowly. This means that a change in the system cannot be realised in the short term.

If we take a look at all aspects, this situation raises the question if the sector can develop towards an inland container shipping industry that is performing in a clean, safe, efficient and trustworthy way and that can serve the predicted 45% of the hinterland flows from and to the two Maasvlaktes. In spite of the big strategic and economic importance of a well-functioning inland container shipping industry, the large number of stakeholders and the worrying situation with regards to the (relatively lagging) performance in the sector at this moment, there is no common and supported vision about the direction in which the inland waterway transport should develop.

Without system innovation the sector will not be able to play the anticipated dominant role in freight transport in the future. Stabilisation of the sector equals stagnation; it will not increase the share of the inland container shipping in the modal split and it will lead to a marginalisation of the sector, especially when other modes of transport do go through a development stage. The following three paths, which are likely to play a role in the transition, have been identified: (1) a shift to large-scale industrial corridors, (2) radical greening and (3) a fine-grained distribution network. The third path departs most radically from the present system and is thus a long-term option, to be presently only further explored on a small scale.

22.5.1 Transition path 'large-scale industrial corridors in the logistics chain'

This transition path involves an increase in the scale of operations, integration of administrative and logistics processes into 'synchromodal' logistical systems. This path requires not only limited innovations in 'techware' but large innovations in 'orgware', including the underlying culture. This track requires new organisation (and corresponding systems) of the sector that are facilitating 'climbing in the value chain' and that offer the customer (shipper) services at service and functional level. Significant barriers can nevertheless be identified: the current culture and organisational structure of the sector lacks flexibility and hampers the sector to be a competitive partner in inter- and synchromodal arrangements.

22.5.2 Transition path 'radical greening'

In the transition path 'radical greening', inland water transport maintains (or regains) a clear competitive advantage over other modes of transport in environmental performance. As road transport is continuously reducing its relative emissions, this would not entail much more than attempting to catch up with the present state of the art in road transport. In this path, a large role is attributed to technical suppliers, such as the boat engine sector and the willingness and ability of the shipping industry itself to invest in such technology. The role of government policy would be in stimulating the setting of standards (also in the international context) and commit resources from public technical research.

22.5.3 Transition path 'fine-grained distribution network'

The transition path 'distribution network' focuses on complementing the trend of the ever-increasing scale of water shipping, by a reverse trend of reintroducing the inland container shipping for distribution, which is focused on the regional markets, the inland waterway transport on small waterways and/or for the benefit of partial loads (pallets/LCL), outside the main corridors from Rotterdam (to Germany over the Rhine and to the neighbouring port of Antwerp).

The Dutch inland shipping sector is at a critical point. We are likely to observe an 'inland waterway transport paradox': on the one hand, many independent stakeholders are undertaking a range of activities to optimally develop the inland container shipping's potential, whereby their own agenda is the central focus. At the same time, we see that through a lack of a shared vision, 'sense of urgency' and the willingness to cooperate, the sector's 'fundamentals' are weakening. The high expectations have existed for more than 50 years now, but they never come through. Regarding these fundamentals, we should be thinking about the development of the market share, the environmental performance, the economic output and the finance structures.

This requires a transition and integrated approach. The transition agenda for the inland container shipping must presume three levels: strategic, tactical and operational. At the strategic level it is especially important to form a shared vision. In the aforementioned problem analysis it became clear that currently there is a lack of such a vision. The transition approach provides tools to facilitate a similar vision procedure, like a 'transition arena'.

The assessment makes clear that there are means and innovations available to make progress on a tactical level. It is however also clear that these means are especially available for the transition path 'large-scale industrial corridors' and less for the other paths. On the operational level, several initiatives can already be noticed, varying from extended gates and corporations/franchises in the large-scale transition path towards an LNG-vessel in the greening path and experiments with small-scale distribution shipping in the dense distribution network path. But only when this transition is a common attempt based on a shared vision of which there is a sense of urgency now will there be chances for success. Due to the existing path dependencies, in practice, this is often a difficult condition to meet and it is a challenge to create space for exploring more radical innovation pathways. In the next section, we highlight four such difficulties facing typical modern ports and we explore how to start creating such space.

22.6 Applying transition management; the challenges for future port development

Ports are typical examples of modern regimes that have co-evolved with societal changes since trade and shipping started. Current ports are predominantly shaped by the developments over the past decades which is driven by a globalising economy, the emergence of a global chemical industry, a global fossil fuel and energy system, and modern transport systems. They are in many ways engines of economic growth and employment as described in Parts 1 and 2, dealing with strategies and operations. At the same time it becomes increasingly clear that ports face substantial challenges such as the impact on climate change, the port-city relation, security and so forth, as addressed in Part 3 of this book. But this is not a black and white distinction; effects are interrelated, there are spin-offs, trade-offs and so on. Much effort is therefore made for an integrated approach to improve the competitive position of ports, increase efficiency, reboot economic growth or mitigate negative environmental effects.

A transition perspective, however, offers an alternative way to understand the challenges of ports from a sustainability perspective. The current situation can clearly be understood as a lock-in; enormous investments in infrastructures have been made, routines have been built up, regulations and institutions have been developed to facilitate the current regime and many stakes are involved. Understandably, most actors involved will seek to stimulate innovation that will help to improve the existing systems. However, given the shifting external landscape context (think of climate change, volatility of the resource and energy markets, information technology and sustainability concerns in public and policy) on the one hand, and maturing radical alternatives on the other (circular design, biobased chemistry, clean mobility technologies, renewable energy and resources, new production methods like 3D printing), it is likely from a transition perspective that over time tensions in the current regimes will lead to abrupt and shock-wise systemic change.

Such change could threaten the current regimes and lead to breakdown, divestments and loss of jobs, but can also be seen as a necessary precondition to create space for more radical sustainability transitions. The challenge transition management if put central is how, besides improving the existing alternative pathways, can be experimentally developed which will open up the possibility for sustainability transitions. Starting from developments that emerged over the last decades, we can sketch four possible transitions ports could explore: transport, chemical industry, fossil fuels and area development. Each of these transitions has its own characteristics, time horizons and challenges, but when successful they could lead to the port of the twenty-first century: one that is clean, diverse, resilient, socially connected and value-producing. Ports are facing clear challenges. Four of these challenges are described next.

The first challenge, connected to broader global shifts, is *the transition from fossil-based to renewable energy systems*. Besides being a hub for fossil fuel trade, logistics and refinery, ports also frequently facilitate the large-scale production of fossil-based energy. Large investments and built-up infrastructures have led to a fossil lock-in, which is now increasingly challenged by the growing uncertainties in the fossil fuel energy market and the increasing competition of alternative renewable options. A recent example in the port of Rotterdam is the construction of two new coal-fired power plants which were economically unviable from the moment of opening. Companies like Eon suddenly decide to abandon coal-based energy production and follow the shift towards renewable markets. Overall we see in Europe and the

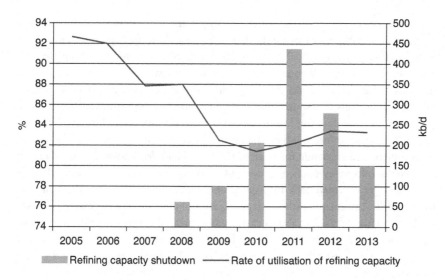

FIGURE 22.4   Utilisation rate of the European Union refining capacity (%) and capacity shutdown

United States a reduction in refinery capacity and the closure of small refineries. By definition, the newly emerging energy market is much more distributed (decentralised, off-shore) and diverse (biomass, wind, solar, tidal and other). For ports, this creates a challenge to their business model and practice, as for a number of the alternatives ports are no longer necessary; they do not require bulk input of resources and a centralised location. This is illustrated by Figure 22.4.

A closely related challenge is that of the *position of the chemical industrial complex*. Based on fossil fuels, very large-scale plants produce base and fine chemicals for all the products we use. Enormous infrastructures have evolved around these complexes, and payback times are often over 30 years. In many locations, generous taxing regimes are in place so that very low taxes are paid (for example, for energy consumed) and disposing of waste heat and emissions is very cheap. There have been substantial developments in biobased chemistry, where the initial first generation biomass use (for example, corn) is now being replaced by second generation biomass (organic by-products and waste). There have been notable successes in pilot plants, but these now enter the phase of upscaling, creating investment challenges besides technological and institutional ones. In practice it has proven difficult to scale-up biobased experiments in a context of fossil chemical complexes: they threaten the existing business models and do not correspond well with the dominant culture and practices.

A third transitional challenge concerns *sustainable transport and logistics*. In the previous section we briefly sketched the inland shipping transition, but the same goes for road transport and sea shipping. Not only do the emissions and congestion issues lead to a push to more sustainability, but also logistical efficiency could benefit from innovation. As with the other challenges there are existing infrastructures, routines and institutional contexts that primarily allow for optimisation around ideas of efficiency and modal splitting. However, from a transition perspective it is not unlikely that we will enter periods of relatively low economic growth and the demand for large-scale (bulk) transport will decrease (for example, by decentralised production or shifts in resource flows). The possibilities to shift towards zero-emission transport that does not lead to negative societal impacts need to be positioned against this background in which the dominant culture and practices are locked in.

A fourth core challenge for ports is *the issue of area development and its relation to the hinterland.* In many ports, as they grow or move outward, older industrial zones need to be transformed back into sustainable areas. While old industrial buildings have a certain appeal, often these areas are also polluted and not necessarily considered as attractive areas. But also in existing port areas, the environmental and spatial quality of the area needs to be safeguarded and often improved. It requires a strong commitment and vison from port authorities to develop green port areas, but also new ways to reconnect to the city. In many cases, ports have more or less disconnected from their original urban context because of the enormous expansion and industrialisation. This has also led to disconnection in terms of actor networks and orientation, often requiring quite intensive governance processes to reconnect port and city and to redevelop areas towards sustainable ones.

These four transitional challenges are in many ways beyond the direct control of ports and port authorities. The transitions in the energy system and chemical industry are semi-autonomous and especially in the energy system quite disruptive and uncertain. The core governance challenge that comes forward from this transition perspective is therefore how to develop new strategies to play into these transitions and if possible seize the new opportunities that come along with them. As we have argued, the historical successes of ports co-evolved with the historical developments such as the continuing growth and expansion of fossil-based systems. As the structure of the economy changes and new technological alternatives mature, the question is whether current port structures, cultures and practices are resilient and adaptive enough to play into these changes.

22.7 Outlook

A core recommendation coming from transition management is the need to create space for strategic experimentation and exploration of transition pathways, facilitating participatory innovation and learning processes in which change-inclined incumbent and niche-level firms and organisations are brought together. This is not only a challenge for industries and port authorities, but also for governments, non-governmental organisations and other stakeholders. Needed are the development of new technologies, new routines, new institutions, new infrastructure and a new culture. As international regulations and leading firms will continue to demand more sustainability (as for instance in the food chain, where such a development is already taking place), the need to reconnect to the hinterland grows, and more and more markets will transform fundamentally. At the end of the day a license to operate will be required, as we see already happening with the introduction of ISO standards. To not only safeguard the existing status quo, but also to proactively develop a transformation strategy, will lead to a truly sustainable port (Geerlings et al., 2015).

By making this statement at the end of this book, we try to open up the discussion towards the longer-term future of ports. We have argued that the currently there is a dominant focus on existing port systems and their optimisation neglects that broader societal changes and challenges we see from a perspective of transition. Our argument is that the levels of (environmental) sustainability that can be achieved by improving upon what already exists, are limited by definition and much higher potential benefits could be seized by moving towards more inherently sustainable systems (renewable energy production, circular resource systems and biobased industries). We have also tried to highlight the complexities involved in opening up the pathways to such alternative futures: the existing path-dependent development allows very little space for radical innovations. But simultaneously we argued that broader socio-technical changes will over time inevitably threaten the now dominant options. A sustainability strategy for modern ports, we argued, therefore needs to combine a strategy of improving the existing with exploring the new.

To conclude this chapter, we see that transition management can be considered as a specific policy discourse and a field of academic research that contributes to a new governance model for sustainable and innovative port developments. Using a business or governance approach helps to make the future manifest from current decisions, by adopting longer time frames and exploring alternative trajectories, and it opens avenues for system innovation (as well as system improvement). This concept has been already successfully applied in various sectors, such as the energy sector, agriculture, water management and even nowadays the transport sector (see the recent developments in e-mobility, for example, where Tesla considers it and presents its software as open source). Although every transition management process is unique in terms of context, actors, problems and solutions, the next challenge is to apply this approach to come to a sustainable port development that aspires to the highest levels of environmental and social quality and combines this with economic vitality and resilience. In the longer term this is inevitable and it will lead to new value propositions: a proposition that is not solely based on the currently dominant paradigm of economic growth but also on growth in terms of added value to the economy, society and environment. This clearly requires a process of exploration and innovation that will take decades and requires all knowledge and talent available. We hope this study book can contribute to this 'grand transition'.

References

Geerlings, H., and Avelino, F. (2015) The value of transition management for sustainable transport: An illustration from the port extension in Rotterdam. In: Hickman, R., Givoni, M., Bonilla, D., and Banister, D. (eds.) *An International Handbook on Transport and Development*. Cheltenham, Edward Elgar.

Grin, J., Rotmans, J., and Schot, J. (2010) *Transitions to Sustainable Development; New Directions in the Long Term Transformative Change.* New York, Routledge.

Loorbach, D. (2007) *Transition Management. New Mode of Governance for Sustainable Development.* PhD thesis. Utrecht, International Books.

Suggestions for further reading

Besides the papers provided as references in the text, we recommend the following texts as suggestions for further reading.

Geerlings, H., Stead, D., and Shiftan, Y. (Eds.). (2012) *Transition Towards Sustainable Mobility: The Role of Instruments, Individuals and Institutions.* London, Ashgate.

Loorbach, D., and Rotmans, J. (2010) The practice of transition management: Examples and lessons from four distinct cases. *Futures*, 42(3), 237–246.

Rotmans, J., Kemp, R., and Van Asselt, M. (2001) More evolution than revolution: transition management in public policy. *Foresight*, 3(1), 15–31.

INDEX

Page numbers in *italics* indicate a figure, table or box on the corresponding page.

Printed in the United States
by Baker & Taylor Publisher Services